· · · · ·

PRACTICAL
PROTEIN
CRYSTALLOGRAPHY

· · · · ·

◆ ◆ ◆ ◆ ◆

PRACTICAL
PROTEIN
CRYSTALLOGRAPHY

◆ ◆ ◆ ◆ ◆

DUNCAN E. McREE

Department of Molecular Biology
The Scripps Research Institute
La Jolla, California

ACADEMIC PRESS, INC.

A Division of Harcourt Brace & Company

San Diego   New York   Boston   London   Sydney   Tokyo   Toronto

Cover ilustration: Electron density map (2Fo-Fc, 2.0 Å) of endonuclease III superimposed on the model of [4Fe-4S] cluster and surrounding protein. The map has been colored according to the gradient of the density in increasing order, (green, red, yellow, blue) in order to color carbon density green, oxygens red, sulfurs yellow, and irons blue. This picture was made using XtalView and AVS (Advanced Visual Systems) software by Mike Pique, Maria Thayer, and John Tainer of The Scripps Research Institute.

This book is printed on acid-free paper. ∞

Academic Press, Inc.
1250 Sixth Avenue, San Diego, California 92101-4311

United Kingdom Edition published by
Academic Press Limited
24–28 Oval Road, London NW1 7DX

Library of Congress Cataloging-in-Publication Data

McRee, Duncan Everett, Date.
    Practical protein crystallography / Duncan E. McRee
       p. cm.
    Includes bibliographical references and index.
    ISBN 0-12-486050-8
    1. Proteins--Structure. 2. Crystallography.   I. Title.
QP551.M366   1993
574. 19' 245--dc20   93-13272
                                          CIP

PRINTED IN THE UNITED STATES OF AMERICA
93   94   95   96   97   98   BB   9   8   7   6   5   4   3   2   1

CONTENTS

····· 3 ·····
COMPUTATIONAL TECHNIQUES

◆ ◆ ◆ ◆ ◆ ◆ ◆ ◆ ◆ ◆ ◆ ◆ ◆ ◆

PREFACE

◆ ◆ ◆ ◆ ◆ ◆ ◆ ◆ ◆ ◆ ◆ ◆ ◆ ◆

This book is a practical handbook intended to be used by anyone who wants to solve a structure by protein crystallography. It should prove useful both to new protein crystallographers and to old hands. While users of this handbook should be familiar with at least one of the texts in the Suggested Reading list that follows, *Practical Protein Crystallography* examines several subjects that are not covered in these other texts, but which are widely used today, for example, area detectors and synchrotron radiation sources. The topics covered in this book are well-tested, robust methods commonly used in our laboratory and in others. The informed crystallographer will note, however, that many techniques and methods are not mentioned or are mentioned very briefly. These have been omitted due to space, less common use, difficulty of application, the need for special expertise, or perhaps oversight on the author's part. The exclusion of a method should not be taken in any way to be a disapproval; it is simply not possible for one book to include everything.

The book is divided into four chapters: (1) Laboratory Techniques, (2) Data Collection Techniques, (3) Computational Techniques, and (4) Protein Crystallography Cookbook. Techniques and methods to be practiced at the lab bench are in Chapter 1, methods used for collection of diffraction data are in Chapter 2, and those done at the computer keyboard are in Chapter 3. Chapter 4 contains examples of projects to provide "recipes" for solving a project. It presents a means for tying together several different methods and includes real numbers to give a better understanding of results. There is no need to read the book in any order; chapters that use information from other chapters will reference those chapters when necessary. In fact, the best use of the book may be to read the Cookbook first and then refer to other chapters as needed.

In Chapter 1, only cursory information on crystal growth is presented because several excellent texts on crystal growth already exist. A fair amount of attention is given to protein sample handling, because proper attention to the sample can often mean the difference between a successful project and a failure. Proteins are delicate materials that demand special handling and are very difficult to purify in large quantity. Protein crystals are also very delicate and require special handling techniques different from those of small molecule crystals.

Chapter 2 bridges the gap between the laboratory and the computer. Special emphasis is placed on the newer techniques using area detectors and synchrotron sources that, except for special cases, will soon replace the older film and single counter diffractometer methods, which are covered only briefly. Because of the high cost of these systems, the user will probably have access to only one. Experience has shown that the user will then come to regard this one as best and will defend it vehemently against all others. Emphasis is placed less on a specific system and more on general techniques relevant to all area detectors.

Chapter 3 openly acknowledges the integral role that computers play in modern crystallography. Not so many years ago, protein crystallographic software was stored on punched cards and data was stored on paper tape. Now, most of us can't find a working card reader. While the hardware used by protein crystallography has surged ahead, much of the software being used was originally intended for punch cards. Hopefully, the reader will use the information in this book as a guide into the modern age of computing. The variety of software used by different groups is enormous and no book could hope to cover even a small portion of it. General information that should be applicable to most techniques is given. Although this book can never substitute for the individual manuals of each program, it does give guidelines that will allow the reader to make intelligent choices among the program options. In order to allow discussion of specifics, the XtalView system is used. This system employs a visual interface that is easy to grasp and illustrates well the possible options at a given step. XtalView is available from the Computational Center for Macromolecular Structure, San Diego Supercomputer Center, P.O. Box 85608, San Diego, CA 92186-9784, phone (619) 534-5100. Send e-mail to CCMS-HELP@SDSC.EDU requesting more information on XtalView. If you do not have access to e-mail then write to the above address.

Chapter 4 contains examples drawn from the experience of the author and his colleagues, which provides some examples of protein structures solved by various methods. These examples can be used as guides for the user's own projects and to give a feel for how to apply the varied methods. Real numbers are given as a basis for interpreting the user's own. By following the examples in multiple isomorphous replacement users can, with luck and perseverance, solve their own structure.

Appendix A contains formulae commonly used in protein crystallography, but with a twist: the formulae are coded in both FORTRAN77 and C. These two languages easily account for 99% of all protein crystallographic software. This will be of great aid to users in writing their own software and in understanding other software. Also for those of us who understand a computer language better than we do math, this appendix explains the formulae. One goal of this book is to provide enough information so that the computer neophyte can write a simple program to reformat or filter data. Unfortunately, because of the incredible variety of software available and the consequent large variety of file formats, this is a necessary skill. Appendix A provides information to help write programs that will continue to be useful on different operating systems and for other projects. It is surprising how many programs that were originally written to be used once for a single problem are now commonly used by a dozen laboratories — all wishing the original writer had made the program more general.

Appendix B contains the manual of the XtalView system. Much practical information is contained here that will answer specific questions about file formats and options available. Again, this is only one of many systems available, but it will serve to give a practical handle on the nature of crystallographic computing.

Rules of thumb are provided throughout to serve as a guide. Like all rules, these are made to be broken and should not be taken too literally. Every protein has its vagaries and its own special difficulties — otherwise several thousand protein structures would have been solved to date instead of a few hundred. The low ratio of proteins with published coordinates to the number of proteins reported as crystallized makes it plain that many protein structures take years to be solved and that many will never be solved. Perseverance and a willingness to try new methods have been keys to solving many structures. Protein crystallography is not a turn-key science. But that is part of its fascination and challenge.

Suggested Reading

Crystallography

Stout, G. H., and Jensen, L. H. (1989). *X-Ray Structure Determination: A Practical Guide*, 2nd Ed. Wiley, New York.

Ladd, M. S. B., and Palmer, R. A. (1985). *Structure Determination by X-Ray Crystallography*, 2nd Ed. Plenum, New York.

McKie, D., and McKie, C. (1986). *Essentials of Crystallography*. Blackwell Scientific, Oxford.

Protein Crystallography

Blundell, T. L., and Johnson, L. N. (1976). *Protein Crystallography*. Academic Press, San Diego.

Wyckoff, H., ed. (1985). In *Diffraction Methods for Biological Macromolecules, Methods in Enzymology*, Vols. 114 and 115. Academic Press, San Diego.

Helliwell, J. R. (1992). *Macromolecular Crystallography with Synchrotron Radiation*. Cambridge Univ. Press, Cambridge.

Crystallization

McPherson, A., Jr. (1982). *The Preparation and Analysis of Protein Crystals*. Wiley, New York.

Protein Structure

Dickerson, R. E., and Geis, I. (1969). *The Structure and Action of Proteins*, Benjamin/ Cummings Publishing Co., Menlo Park, California.

Fersht, A. (1985). *Enzyme Structure and Mechanism*. W. H. Freeman, New York.

Oxender, D. L., and Fox, C. F., eds. (1987). *Protein Engineering*. Alan Liss, New York.

◆ ◆ ◆ ◆ ◆ ◆ ◆ ◆ ◆ ◆ ◆ ◆ ◆ ◆

ACKNOWLEDGMENTS

◆ ◆ ◆ ◆ ◆ ◆ ◆ ◆ ◆ ◆ ◆ ◆ ◆ ◆

My thanks go to a number of people for their assistance in preparing the manuscript of this book. In particular I thank David Stout, Yolaine Stout, and Michele McTigue for critically reading the manuscript and making many helpful suggestions. Any errors in the book are, of course, my responsibility. I thank John Tainer, Elizabeth Getzoff, Hans Parge, Susan Redford, Gloria Borgstahl, Brian Crane, David Goodin, Ch-fu Kuo, Zhong Ren, Andrew Arvai, Peter Lauble, Art Robbins, and David Stout for allowing me to use their work as examples in this book. I also thank the large number of patient people in the structural biology group at The Scripps Research Institute who beta-tested XtalView for me. I hope that this book answers a few of their questions about what to do with all those buttons and sliders. The Scripps Research Institute has provided me with an outstanding environment and the freedom for learning and practicing protein crystallography. I especially thank Michael Pique who patiently taught me computer graphics over the years and who, along with Lynn Ten Eyck and Andrew Arvai, provided me with routines for XtalView.

I have worked with many people over the years who have provided the intellectual stimulus and help that made this book possible. David Richardson was my Ph.D. advisor and started me out on the right foot. Jane Richardson has been a major inspiration over the years. Wayne Hendrickson took me under his wing and taught me much about anomolous scattering, phasing, refinement, scaling, and critical thinking. Bi-Cheng Wang and Bill Furey taught me all about phase modification in a couple of visits to the University of Pittsburgh. Fred Brooks made me think a lot about the virtues of a good user interface and the real power of computer graphics. John Tainer and Elizabeth Getzoff have been invaluable as partners in solving an ever-increasing number of interesting protein structures.

XtalView distribution through the Center for Macromolecular Structure at the San Diego Supercomputer Center is funded by Grant DIR 8822385 from the National Science Foundation.

Finally, I thank my wife, Janice Yuwiler, and her father, Art Yuwiler, for always believing that the book would one day be finished. And last, but not least, I want to thank my sons Alex and Kevin for putting up with me while I spent so many evenings late at work.

···· 1 ····
LABORATORY
TECHNIQUES

◆ ◆ ◆ ◆ ◆ ◆ ◆ ◆ ◆ ◆

····· 1.1 ·····
PREPARING PROTEIN SAMPLES

A protein sample must be properly prepared before it can be used in a crystal-growth experiment. There are many ways that this can be done to accomplish the same goal: to put the protein at a high concentration in a defined buffer solution. Methods that have worked well in our laboratory are outlined here, but if you know of a quicker, easier method then by all means use it. The steps needed to prepare a protein sample are detailed below.

History and Purification

Since it is not uncommon for one batch of protein to crystallize while the next will not, it is vital to keep a history of each sample and to track each batch separately. Your records may provide the only clue as to the differences between samples that produce good crystals and samples that are unusable. For example, there have been several cases where the presence of a trace metal is needed for crystallization. The most famous case is insulin. It seemed that the only insulin that would crystallize was that purified from material collected in a galvanized bucket. It was eventually discovered that zinc was required, and later it was added directly to the crystallization mix. In our lab any sample received is logged into a notebook with a copy of any letters or material sent with the sample.

Samples should be shipped to you on dry ice and kept frozen at $-70°$ C until they are ready for use. Ask the person sending the sample to aliquot the protein into several tubes and to quick-freeze each one. This way you can thaw one aliquot at a time without having to repeatedly freeze–thaw the entire sample, which can damage many proteins. Keep a small portion of the sample apart and save it for future comparison with samples that do not crystallize or crystallize differently.

Always keep protein samples at $4°$ C in an ice bucket to prevent denaturation and to retard bacterial growth. Perform all sample manipulations at $4°$ C either in a cold room or on ice. When the samples are finally set up they can be brought to room temperature. Proteins are usually stabilized by the presence of the precipitants used in crystallization and agents are added to retard microbial and fungal growth. A common anti-microbial agent is 0.02% sodium azide. Other broad-spectrum cocktails are sold for use with tissue culture that are quite effective.

Exchanging Buffers

If the desired buffer of the sample is not already known, the sample should be placed in a weak buffer near neutrality. A good choice is 50 mM Tris-HCl at pH 7.5, 0.02% sodium azide. Some proteins will not be stable at low ionic strength, so a small portion should be tried first to see if a precipitate forms. Be sure to wait several hours before deciding if the sample is stable. Observe the sample in a clear glass vial and hold it near a bright light to detect any cloudiness in the sample.

There are two methods for exchanging the buffer solution of the protein sample: dialysis and use of a desalting column. The desalting column is the fastest method, and if disposable desalting columns such as a PD-10 column from Pharmacia-LKB is used, it is very convenient. A single pass through the column will remove 85–90% of the original salt in the sample and, if this is not enough, two passes can be used. Every time the sample is passed over the column it will be diluted about 50%. Unless the sample is very concentrated to begin with, it may be necessary to concentrate the sample after desalting.

Dialysis on small volumes is best carried out in finger–shaped dialysis membrane such as a colloidicon (Fig. 1.1). Always soak the membrane first in the buffer to remove the storage solution. A minimum of two changes of dialysis buffer 12 h apart is recommended. The dialysis buffer should be stirred and should be at least 100 times the sample volume. Use membranes with a molecular weight cutoff less than half of the sample molecular weight,

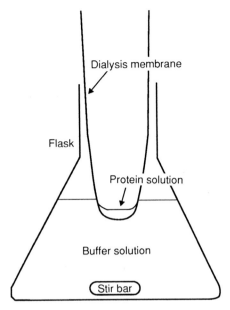

Dialysis membrane

Flask

Protein solution

Buffer solution

Stir bar

FIG. 1.1 Dialyzing samples to exchange buffers.

otherwise you risk losing a substantial portion of the sample. When removing the sample, wash the inside of the membrane with a small amount of buffer to recover the sample completely.

Concentrating Samples

First measure the starting concentration of the sample. The simplest way is to measure the absorbance of a 50-times dilution of the sample at 280-nm wavelength and assume the concentration is simply the absorbance times 50. While this method is not very accurate, it is reproducible, and the sample should be pure enough to warrant the assumption that all absorption is due to the protein.[1]

Always use buffer as a blank and check the buffer versus distilled water to be sure it does not have significant absorption. Some buffers have a significant amount of absorption at 280 nm, which can greatly reduce the accuracy of different absorbance measurements. The diluted absorbance must be below 1.0 or the measurement will be inaccurate. Concentrate the sample to

[1] For further information on determining protein concentration see Scopes, R. K. (1982). "Protein Purification, Principles and Practice." Springer-Verlag, New York.

10–20 mg/ml. If you have enough sample, it is better to concentrate to 30 mg/ml, wash the concentrator with one-half the sample volume, and then add the wash to the sample to make a final concentration of 20 mg/ml. A Centricon is one of the best ways to concentrate the sample. An Amicon will also work well. Another method is to dialyse against polyethylene glycol (PEG) 20,000 using a finger-shaped dialysis membrane. The advantages of this method are that it can be combined with dialysis and that the same membrane used to dialyze the sample can be transferred directly to the PEG-20K for concentrating. The dialysis tube can be put directly onto solid PEG-20K. The water in the sample will be quickly removed, so check the sample often. However, be aware that PEG is often contaminated with salts and/or metals and this may or may not be desirable. In many cases, though, it has actually contributed to crystallization. You may want to keep a sample of the particular batch of PEG you use. If, in the future, a new batch causes problems, you can analyse the differences.

Another method often used is precipitation with ammonium sulfate. If an ammonium sulfate step has already been used in the purification procedure, this may be an easy way to achieve a high concentration. You will want to use a high level of ammonium sulfate to insure that the entire sample is precipitated. The ammonium sulfate should be added slowly to the solution while kept cold. Let the solution sit for at least 30 min after all ammonium sulfate has dissolved. Spin down the precipitate and remove the supernatant. The pellet can be redissolved in a small amount of buffer. However, since the pellet will contain some salt, a dialysis step will be needed before the protein is ready.

It is not uncommon for a protein to precipitate at high concentrations. If this happens while you are concentrating, add buffer back slowly until all the sample dissolves. Raising the level of salt by using a more concentrated buffer or by adding sodium chloride can often help stabilize protein solutions. If a precipitate forms, examine it carefully to make sure it is not crystalline. Amorphous precipitates are cloudy and have a matte appearance. Crystalline precipitants are often shiny and if from a colored protein are brightly colored with little cloudy appearance. Two proteins have been crystallized in our lab accidentally during concentration. One was found in an ammonium sulfate precipitation step and the other during concentration on an Amicon to lower the salt concentration.

Storage of Samples

The entire sample will not be used all at once and the remaining protein solution should be aliquoted and stored frozen at $-70°$ C. Divide the sample into 100- to 200-μl aliquots in freezer-proof tubes (not glass, which becomes

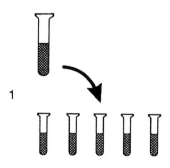

1

2

FIG. 1.2 Storage of samples. (1) The procedure involves aliquoting samples into several tubes, then (2) quick-freezing each sample in an acetone–dry ice bath and store at −70° C.

brittle and shatters at low temperatures) and quick freeze the tubes in an acetone–dry ice or liquid nitrogen bath (Fig. 1.2). Label each tube with the date, a code to identify the sample and the particular batch of the sample, and your initials. Cover the label with transparent tape to prevent the ink from rubbing off when you handle the frozen tubes later. Place the tube in a cardboard box and store in the freezer. It will harm protein samples to be freeze–thawed; although often they may withstand several cycles of freeze–thawing, it is best not to find out the hard way. Thaw the samples in an ice bucket or the cold room when they are to be used. If some sample is left over and it will be used the next day, it can usually be stored at 4° C overnight.

Ultrapurification

While it is beyond the scope of this handbook to cover purification techniques, the crystallographer has one special technique that is usually not tried by others to further improve the sample: recrystallization (Fig. 1.3). We will assume that you have succeeded in finding conditions that will grow small crystals but are having trouble growing larger ones. It may be worthwhile to recrystallize the sample to improve the purification. A large sample

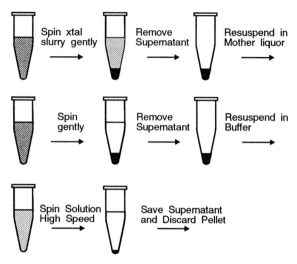

FIG. 1.3   Redissolving crystals.

of the protein can be set up in the crystallization mixture and seeded with a crushed crystal. After crystals have grown, you may wish to add slightly more salt to push more protein into the crystalline state. Gently centrifuge down the crystals, or allow them to settle by gravity, and remove the supernatant. Resuspend the crystals in a mother liquor higher in precipitate by about 10% to avoid redissolving and to wash them, and then remove the supernatant. Resuspend the crystals in distilled water to dissolve them. If you have a large amount of precipitate present with the crystals, this method will not remove the precipitate unless it settles the crystals. In these cases the author has resuspended the crystals in 2 ml of artificial mother liquor in a petri dish, then picked up individual crystals with a capillary and manually separated them from the precipitate.

For the crystals to redissolve well, they should be freshly grown. Old crystals that still diffract well often will not redissolve even in distilled water because the surface of the crystal has become cross-linked. This is especially true of crystals grown from polyethylene glycol.

..... 1.2 .....
PROTEIN CRYSTAL GROWTH

Several excellent texts have been published on methods for growing protein crystals (see Suggested Reading in the preface) and I will not repeat this material here except briefly, to add some of our own experience. Like fine wine, protein crystals are best grown in a temperature-controlled environment. Most cold rooms have a defrost cycle that makes them especially poor places to grow crystals. Investing in an airconditioner for a small room to keep it a few degrees colder than the rest of the laboratory is the best way to keep a large area at a constant temperature for crystal growth. To ensure that room is tightly regulated, get a unit with a capacity larger than needed. Another alternative is to use a temperature-controlled incubator. However, a room is best because you will need to examine your setups periodically at a microscope. In a room everything can be kept at the same temperature.

Invest in a good, dissecting stereomicroscope and remove the light bulb in the base. Substitute a fiber-optic light source so that the base does not heat up and dissolve your crystals as you observe them. Even with the fiber-optic source be careful not to put the setup down near the fiber-optic light source, which gets hot during operation. Have a Plexiglas base built over the dissecting scope base (Fig. 1.4) to provide a large surface on which to place setups so that they do not fall off the edges during examination. This will also provide a base to steady your hands during delicate mounting procedures.

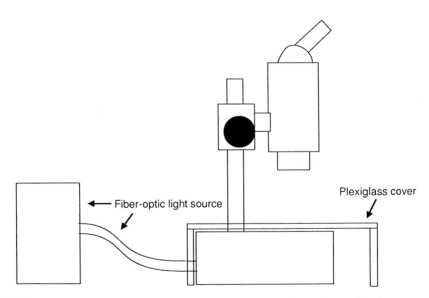

FIG. 1.4 Modified dissecting scope. A Plexiglas base is put over the scope to make a larger area, providing a place to rest your hands during mounting operations and prevents tipping hanging-drop plates over the edge. A fiber-optic light source is used instead of the built-in light to prevent the base from heating and damaging crystals.

Initial Trials

In all other aspects of protein crystallography except initial crystal trials, the more past experience you have the better. Beginner's luck is definitely a factor in finding conditions for crystallizing a protein the first time. This is partly because beginners are more willing to try new conditions and will often do naive things to the sample, thus finding novel conditions for crystal growth. This is also because no one can predict the proper conditions for crystallizing a new protein. There are conditions that are more successful than others, but to use these exclusively means that you will never grow crystals of proteins that are not amenable to these conditions. So fiddle away to your heart's content. What is needed is to observe carefully what does happen to your sample under different conditions and to note carefully the results. The least experienced part-time student can outperform the most expensive crystallization robot because he or she has far more powerful sensing faculties and reasoning abilities. Leave the "shotgun" setups to the robots. Having said all this, I present in Table 1.1 a recipe to use for initial trials.

The most commonly used methods for initial crystal trials are the hanging-drop and sitting-drop (Fig. 1.5) vapor-diffusion methods. The batch method can actually save much protein if done properly. In the hanging-drop method many different drops are set up. Most of these will never crystallize. It is hoped that just the right conditions will be hit upon in a few of the drops. A method that I have used successfully for many years is to place a small

TABLE 1.1
Conditions for Initial Trials

Precipitant	Concentration range	Additives
Polyethylene glycol 4000	10–40% w/v	0.1 M Tris, pH 7.5
Polyethylene glycol 8000	10–30% w/v	0.2 M ammonium acetate
Ammonium sulfate, pH 7.0	50–80% saturation	
Ammonium sulfate, pH 5.5	50–80% saturation	
Potassium phosphate, pH 7.5	$0.5–2.5\ M$	
2-Methyl-2,4-pentanediol	15–60%	$50–200\ \mathrm{m}M$ potassium phosphate, pH 7.8
Low ionic strength	Dialysis	
Sodium citrate	$0.5–2.5\ M$	

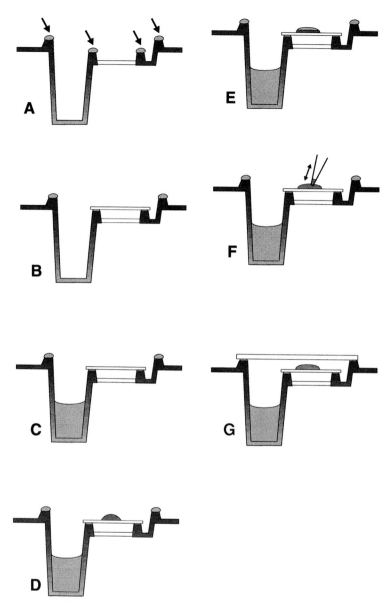

FIG. 1.5 Crystals setup using ACA plates. A cross section through a single well is shown. (A) The lips of the wells are greased where the coverslips will later be placed. High-vacuum grease can be used alone or mixed with about 20% silicone oil. The addition of oil makes the grease less viscous so that it flows more easily. (B) The lower coverslip is pressed into place. Make sure there are no gaps in the grease for air to leak through. (C) Place the reservoir solution in the well. (D) Put the protein solution onto the lower coverslip. (E) Carefully layer the precipitant (often some of the reservoir solution) onto the protein. (F) Mix the two layers together quickly by drawing up and down with an Eppendorf pipette. (G) Put on the upper coverslip to seal the well completely. Again check that there are no gaps in the grease. Wait several hours to several weeks for crystals to appear.

amount of protein in a 1/4-dram shell vial with a tightly fitting lid (caplug). Small aliquots of precipitant are added slowly. After each addition, the shell vial is tapped to mix the samples, then held up to a bright light. When the protein reaches its precipitation point, it will start to scatter light as the proteins form large aggregates that cause a faint "opalescence." Slowly make additions to the sample, waiting several minutes before each addition to avoid overshooting the correct conditions. If two vials are used, they can be leapfrogged so that when one reaches saturation, the other will be just below. For example, set up the first vial with 20 μl of protein plus 2 μl of precipitant and the second with 20 μl of protein plus 4 μl of precipitant. Then add 4 μl to the first and then 4 μl to the second and so forth, so that one of the vials is always 2 μl ahead of the other. When opalescence is achieved in one vial, put both away overnight to be observed the next day. If the precipitation point is overshot, a small amount of water may be added to clear the precipitate and the precipitation point can be approached again more slowly. The reason this method is less wasteful is that it allows a finer searching of conditions in just two shell vials substituting for the large number of hanging drops needed to do as fine a scan. It also encourages more careful observation of the samples. Finally some proteins do not fare well during the evaporation that occurs in hanging drops.

It is impossible and impractical to systematically scan every possible precipitant that has been used for growing protein crystals. Therefore, another approach is an incomplete factorial experiment.[2] A small subset of all possible conditions are scanned in a limited number of experiments by combining a subset of solutions. These drops are scanned for crystals or promising precipitates. If anything is found, then a finer scan can be done to find better growth conditions. A particularly successful version of this method was developed by Jancarik and Kim[3] and has been optimized to 50 conditions combining a large number of precipitants and conditions. A kit is available from Hampton Research[4] that contains all 50 solutions premixed, so all one has to do is set up 3–5 μl of protein sample with each of the solutions. This method recommends that you first dialyze the protein against distilled water to allow better control over pH and other conditions. Try this on a small sample first. Many proteins will not tolerate distilled water and will precipitate (or sometimes crystallize). Use as low a concentration of buffer as you can. Phosphate buffers will give phosphate crystals in several of the drops

[2] Carter, C. W., Jr., and Carter, C. W. (1979). Protein crystallization using incomplete factorial experiments. *J. Biol. Chem.* **254,** 12,219–12,223.

[3] Jancarik, J., and Kim, S.-H. (1991). Sparse matrix sampling: A screening method for crystallization of proteins, *J. Appl. Crystalogr.* **24,** 409–411.

[4] Hampton Research, 5225 Canyon Crest Drive, Suite 71-336, Riverside, CA 92507, Tel. (909)789-8932.

that contain divalent cations. We have used this method with a fair amount of success. While you may not get usable crystals on the first trial, you may get some good leads. Some of the drops may stay clear for a couple of weeks. You can raise the precipitant concentration in the drop by adding saturated ammonium sulfate to the reservoir (but not to the drop). This will cause the drop to dry up somewhat. More ammonium sulfate can be added until the drop either precipitates or crystallizes.

Another crystallization method not often tried approaches the crystallization point from the other end by first precipitating the protein and then slowly adding water until the critical point is reached. Often when the protein is precipitated microcrystals are formed. As the precipitant is lowered, protein is redissolved and crystals large enough to see may grow out of the precipitate using these microcrystals as growth centers. Also, if the excess solution is removed from the precipitant, the result is a high concentration of the protein, which may force crystals. This can be done on a micro basis using a variation of the hanging drop. For example, to use this method with ammonium sulfate, mix the drop with 10 µl of protein sample and 3 µl of saturated ammonium sulfate and set this over a reservoir of saturated ammonium sulfate. The drop should dry slowly and the protein precipitate, which will give a final concentration about three times higher than at the start. Every few hours add some water to the well to lower the ammonium sulfate concentration. Keep careful track of the amount added. If the drop starts to clear, slow the addition down to once a day and add water very slowly. I have grown a large number of crystals using this method. Although they are rarely suitable for diffraction, they can be used as seeds to grow better crystals. This method allows searching a large number of conditions with a small amount of sample.

Never give up on a setup unless it is completely dried; it may take several months for crystals to appear. Proteins that are not stable in buffer are often stabilized by high precipitant concentrations. Also, the presence of precipitant in a setup does not preclude crystallization. Often a crystal will grow from the precipitant. Nucleation is a rare event and may require a very long time to occur if you are near, but not right on, the correct crystallization conditions.

Growth of X-Ray Quality Crystals

The elation that you experience following the first crystals of a new protein can be short-lived. It is often discovered that the first crystals grown are of insufficient quality to use for data collection. A long series of experiments may be needed before large, single crystals can be obtained.

The first step to try is a very fine scan of conditions nearest those used initially to find the optimal conditions for growing only a few large crystals in a single setup. Vary the precipitant concentration, the protein concentration, the pH, and the temperature. You may also want to try varying the buffer used and its concentration. Using different types of setups will vary the equilibration rate, which can often lead to improved growth. What are needed are conditions where nucleation is rare and crystal growth is not too rapid. Do not look at your setups too frequently—once a day is enough—since disturbing them can result in the formation of extra nuclei. Leave fine-scan plates alone for a week before disturbing. Since nucleation is a stochastic process, preparing a large number of identical setups will often yield a few drops that produce nice crystals by chance. This is most useful only if you have sample to waste.

If nucleation is unreliable, then seeding is often the answer. Two methods of seeding are used: microseeding and macroseeding. In microseeding small seeds obtained by crushing or those usually present in a large number in old setups are introduced into a fresh drop of preequilibrated protein. Seeds will usually grow in conditions where nucleation will not occur. An extreme example of this is photoactive yellow protein, where seeds will grow in ammonium sulfate solutions at 71% of saturation but nucleation will not occur at concentrations less than about 100%. The microseeds are diluted until only a few will be introduced. This usually requires serial dilutions and can be very difficult to control. Another method is to place a very small amount of solution from a drop in which a crystal has been crushed at one point of a fresh drop without any mixing. A mass of crystals will grow at this point but often a few seeds will diffuse to another part of the drop where a large crystal may develop. The author has found that 30–100 μl sitting drops are good for this technique.

The steps involved in microseeding are illustrated in Fig. 1.6A. The first step in microseeding is to establish the proper growth conditions. Drops with precipatating agent of increasing concentration are set up and preequilibrated overnight. Then a crystal is crushed with a needle so that the entire drop will fill with microscopic seeds. A whisker or eyelash glued to a rod is then dragged through the solution to pick up a small amount of the liquid containing microcrystals. The whisker is then streaked or dipped into the preequilibrated drops. After several hours or days, crystals should grow in the drops with sufficiently high precipitant concentration. To prevent unwanted nucleation, it is desirable to use the lowest concentration that will sustain crystal growth. When proper growth conditions are established, several drops are then preequilibrated to this concentration. A crystal is then crushed as before and a few microliters of the mother liquor in the drop is then pipetted into the first of a series of test tubes with stabilizing solution and mixed well. These are then serially diluted about 10–20 fold so that each

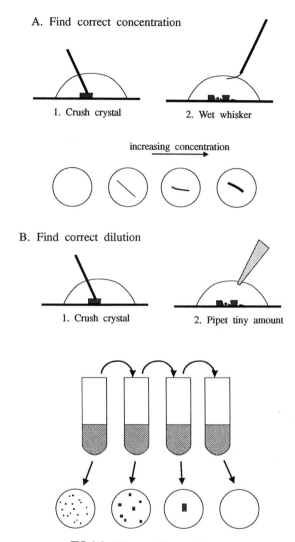

A. Find correct concentration

1. Crush crystal   2. Wet whisker

increasing concentration

B. Find correct dilution

1. Crush crystal   2. Pipet tiny amount

FIG. 1.6   Microseeding techniques.

successive tube contains fewer microcrystals. A few microliters of each of
tube is then put into the preequilibrated growth drops and after several days
examined for growth. Each drop should contain progressively fewer crystals.
The goal is to find a dilution that will provide just a few crystals per drop. If
the microcrystals are stable enough, it may be possible to seed many drops
from this same tube to grow many large crystals.

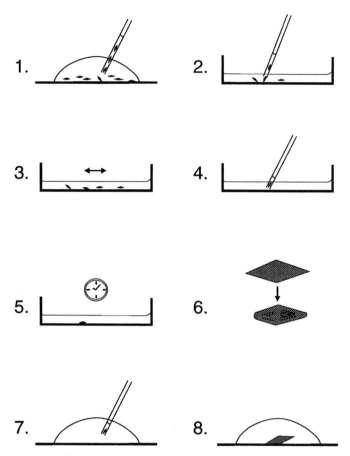

FIG. 1.7 Macroseeding method of growing larger crystals. First, two solutions are prepared in small petri dishes, a storage solution (usually a few pecent higher than the growth concentration) in which the crystals are stable for a long time and an etching solution in which the crystals will slowly dissolve over several minutes (usually a few percent lower than the growth concentration). About 2 ml of each is needed and they should be kept covered to prevent evaporation. (1) Using a thin capillary draw up several small but well-formed seed crystals. (2) Transfer these crystals into the petri dish with the storage solution. (3) Gently rock and swirl the storage solution petri dish to disperse the seeds throughout the dish. This dilutes microseeds and separates the crystals from each other and from any precipitate that might have been transferred with them. (4) Pick up a single seed from the storage solution and transfer it to the etching solution, bringing with it as little of the storage solution as possible. (5) While observing the crystal through a microscope, let it sit and occasionally rock the dish gently. (6) The corners of the crystal should start to round and the faces may etch, leaving scars and pits. (7) Pick up the crystal with as little solution as possible and gently transfer it to a fresh drop of protein preequilibrated to the growth conditions overnight. Often the crystal will fall out of the transfer capillary of its own accord so that no solution need be added to the drop. (8) Over several hours or days the crystal should grow larger.

In macroseeding a single seed is washed and placed in a fresh, preequilibrated drop (Fig. 1.7). The seeds need to be well washed in 2 ml of artificial mother liquor in a plastic petri dish. The dish is gently swirled to dilute any microseeds. The seed is then transferred with a minimum amount of solution to another dish with a precipitant concentration (found by experiment) in which crystals slowly dissolve. This produces a fresh growth surface on the seed and dissolves any microcrystals. The crystal is then transferred with a small amount of solution and placed in a fresh setup. Microseeds can be broken off by mechanical disturbances. Because protein crystals are soft and fragile, a gentle technique is necessary for this method to work. Let the crystal fall in to the fresh drop and settle of its own accord. Do not disturb the crystal after placing it in the growth drop. Often the less-dense dissolving solution will layer on top of the drop and mix only slowly, allowing any microseeds to dissolve. Several problems can be found with this technique and it will not work in every case. Sometimes the crystal may be so fragile that a trail of microcrystals is left in its wake. Often the seed will not grow uniformly; instead, spikes form on the seed surface. Other times the new growth will not align perfectly with the seed, causing a split diffraction pattern. In this case try using a smaller seed that will not contribute substantially to the overall scattering. Often the dislocation between the old and new crystal can be seen. It may be possible to expose only a region of the crystal away from the old seed. The seed should be freshly grown and well formed. If imperfect seeds are used, then you will only grow larger imperfect crystals. Often several generations of seeding will be needed to produce single crystals. Multiply twinned crystals can be crushed and fragments macroseeded until single crystals are obtained.

Better crystals can sometimes be obtained by further purifying the protein sample to make it more homogenous. Isoelectric focusing is especially useful. If you have a large amount of sample (>100 mg), then you can use preparative isoelectric focusing. Smaller amounts of sample (<20 mg) can be chromatofocused on Pharmacia-LKB XX media. This method has proved successful in several cases. Be aware, though, that both these methods introduce amphylytes into the solution that can be difficult to remove.

The last method for improving crystal quality, and often the best, is simply to look for more conducive conditions. For example, *Chromatium vinosum* cytochrome *c'* can be crystallized easily from ammonium sulfate, but these crystals are always so highly twinned that they are unusable, even for preliminary characterizations. By searching for new conditions, it was found that PEG-4K at pH 7.5 produces usable crystals that can be grown to large dimensions and will diffract to high resolution. This is very common for proteins. If they crystallize under one condition, chances are they will crystallize under another condition in a different space group and in a different habit. If there were only one condition out of all possible ones, I doubt that very many proteins would ever crystallize.

..... 1.3 .....

CRYSTAL STORAGE AND HANDLING

Protein crystals can be stored for a few years and still diffract. Some precautions will help increase lifetime. Some proteins can be simply left in the drops in which they grew. Others, though, will grow small unaligned projections on their surfaces if kept in the original drop. These need to be transferred to an artificial mother liquor for storage. The artificial mother liquor must be found by experimentation. Usually, raising the precipitant a few percent is all that is needed. Do not use mother liquor with precipitant at the growth level. The protein in the crystal is in equilibrium with the protein in solution, and if mother liquor at the growth conditions without protein is substituted, the crystal will partially or wholly dissolve to reestablish equilibrium. Higher precipitant concentrations drive the equilibrium toward the crystal. For the same reason do not store the crystal in a volume larger than necessary. Too high precipitant concentration will result in cracked crystals because the change in osmotic pressure will cause them to shrink. Change the reservoirs in vapor-equilibrium setups to prevent drying. Observe the crystal in the artificial mother liquor for several days before committing more crystals to it. Ideally, a crystal in a new artificial mother liquor should be examined by X-ray to confirm that no damage to the diffraction pattern has occurred.

Keep the crystals in the dark. Light causes free radical chain reactions in the solution which will cross-link and eventually destroy the crystals. This is especially true of polyethylene glycol. Commercial PEG contains an antioxidant to retard polymerization caused by light that will slow, but not completely prevent, oxidation of PEG solutions. Solid PEG and PEG solutions must be stored in the dark at all times.

..... 1.4 .....

CRYSTAL SOAKING

To solve the phase problem, the most common method used is multiple isomorphous replacement. In this method a heavy atom(s) is introduced into the structure with as minimal a change to the original structure as possible. This gives phasing information by the pattern of intensity changes. A heavy atom must be used to produce changes large enough to be reliably measured. Only minimal changes, or isomorphism, are necessary because the primary assumption of the phasing equations is that the soaked crystal's diffraction pattern is equal to the unsoaked crystal's diffraction pattern plus the heavy atoms alone. For more details, see later chapters and the suggested readings.

In order to introduce heavy atoms or substrates into protein crystals, they are usually soaked in an artificial mother liquor containing the reagent of choice. The compound is prepared in an artificial mother liquor solution at about 10 times the desired final concentration and then one tenth of the total volume is layered onto the drop containing the target protein crystal. Diffusion occurs within several hours to saturate the crystal completely. With some heavy atoms, secondary reactions often occur that can take several days. Heavy atom compounds are usually introduced at $0.1-1.0$-mM concentration. A typical protein in a 10-μl drop requires roughly micromolar concentrations for equimolar ratios. Many compounds will not dissolve well in the crystillization solution. In these cases it may be beneficial to place small crystals of the compound directly in the drop. Also, many heavy atom compounds will take several hours to hydrate. If they do not completely dissolve at first, be patient. Gentle heating may speed dissolution.

Soak time is more difficult to determine. If the soaking drop has several crystals, they can be mounted at different time intervals. Some heavy-atom reagents that are highly reactive may destroy the crystals and yet be useful if soaked for a short time. Other heavy-atom compounds undergo slow reactions that may produce a new compound that will bind. For instance, a platinum compound in an ammonium sulfate solution will eventually replace all of its ligands with ammonia. As a crystallographer you are not as concerned with exactly what binds to your protein so long as something heavy binds at a few sites in an isomorphous manner. A good way to check that something is binding is to place the crystal in a capillary so that when you invert the capillary the crystal will slowly settle. When heavy atoms bind to the protein they will increase its density and cause it to settle faster. Similarly, if an artificial mother liquor can be sufficiently concentrated so that it has a slightly higher density than the protein crystals, the crystals will float. When a sufficient number of heavy atoms bind, the crystal will sink. This can be used as a way to screen a large number of solutions quickly and has the advantage that small crystals can be observed underneath a microscope. The change in osmotic pressure of the increased density mother liquor may cause the unit cell of the crystal to shrink and, if so, any changes found in the diffraction pattern may be due to this effect rather than heavy-atom binding. In any case, it is a simple matter to resoak a fresh crystal in the usual mother liquor.

What heavy atoms should you try? Table 1.2 presents a partial list of heavy-atom compounds in the order that I usually try them. (Whenever you

TABLE 1.2

Useful Heavy-Atom Reagents and Conditions

Reagent	Conditions
Platinum tetrachloride	1 mM, 24 h
Mercuric acetate	1 mM, 2–3 days
Ethyl mercury thiosalicylate	1 mM, 2–3 days
Iridium hexachloride	1 mM, 2–3 days
Gadolinium sulfate	100 mM, 2–3 days
Samarium acetate	100 mM, 2–3 days
Gold chloride	0.1 mM, 1–2 days
Uranyl acetate	1 mM, 2–3 days
Mercury chloride	1 mM, 2–3 days
Ethyl mercury chloride	1 mM, 2–3 days

meet another crystallographer, first you swap crystal growing tales and then you always ask what heavy-atom compounds he or she has had particular success with.) Because most of the heavy-atom compounds are extremely toxic, extreme caution is in order. Some people experience respiratory distress and allergic reactions when exposed to these compounds. Therefore, always wear gloves and work in a well-ventilated area.

For reproducibility it is best to use fresh solutions; old solutions may oxidize and/or dismutate with time. Solutions must be kept in the dark and preferably under argon. The bottles of reagents themselves should be stored in a well-ventilated area with the caps sealed with parafilm.

····· 1.5 ·····
ANAEROBIC CRYSTALS

Many proteins lose activity if exposed to air and so they must be grown anaerobically. In other cases in order to reduce the protein to its active confirmation it must be kept anaerobic. The easiest method is to use an anaerobic hood. Solutions are passed in and out an airlock and crystals can be set up and handled with conventional techniques. For large-scale work this is by far the best method but not all of us have access to an anaerobic hood. Another method uses a glove bag. This is a plastic bag with gloves that you can put your hands in to manipulate samples. The bag must have everything inside that you are going to use before you seal it. This can present some logistical problems.

A very simple anaerobic apparatus invented by Art Robbins consists simply of a capillary filled with degassed solution into which you float a crystal (Fig. 1.8). One end of the capillary is sealed by melting and the other is sealed with a layer of diffusion pump oil. Dithionite crystals can be dropped into the oil layer through which they will float into the lower liquid. Any residual oxygen will be destroyed by the dithionite and the oil layer prevents the entry of new oxygen. In our laboratory Cu,Zn superoxide dismutase crystals have been reduced in this manner and they have stayed reduced for over a year. The data are collected by mounting the capillary directly on a goniometer head. The size of the capillary in this case is chosen so that the crystal will wedge part way down. The extra solvent decreases the diffraction due to absorption but it was still possible to get a 2.0-Å data set using a large crystal.

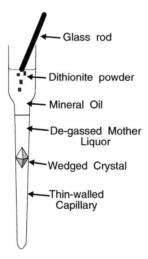

Glass rod

Dithionite powder

Mineral Oil

De-gassed Mother Liquor

Wedged Crystal

Thin-walled Capillary

FIG. 1.8 Simple anaerobic apparatus. Degassed mother liquor is placed in capillary and then a crystal is introduced. The crystal should be large enough so that it will wedge itself in the tapered portion of the capillary as it sinks. Mineral oil is then layered over the mother liquor to form a seal. The top few millimeters of the mother liquor, which have been exposed to air, can be drawn off with a capillary inserted through the oil layer. Solid reductant, such as dithionite, is then placed on top of the oil and allowed to sink through the oil into the mother liquor. Overnight the dithionite will diffuse to the crystal and reduce it. Excess oxygen is destroyed by the dithionite. The data can then be collected by mounting the capillary on a goniometer head as is normally done. Crystals reduced in this manner have remained oxygen-free for over a year.

2

DATA COLLECTION TECHNIQUES

····· 2.1 ·····
PREPARING CRYSTALS FOR DATA COLLECTION

Protein crystals must be kept wet or they will disorder. Since solvent forms a large portion of the crystal lattice, a large change from the crystallization conditions will cause the crystals either to dehydrate and crack or to melt. Crystals are usually mounted in thin-walled glass capillaries.[1] The thin glass wall minimizes absorption of the scattered X-rays and also minimizes background from the glass. For protein work use the glass capillaries. Quartz capillaries are stronger, but the quartz scatters strongly around 3-Å resolution in a sharper band where the glass scattering is diffuse. Solvent contributes to background, which is always bad, and so as much solvent as possible must be removed without letting the crystal dry up.

Crystal-Mounting Supplies

Before mounting a crystal, make sure you have all the supplies you need at hand:

• **Capillaries.** Thin-walled capillaries are needed in a variety of sizes. You will want to have a large supply of suckers previously made to pick from.

[1] Available from Charles Supper Company.

21

- **Tweezers.** Two pairs of tweezers are needed: a pair with straight ends and a curved pair for prying up cover slips.
- **Scissors.** A sharp pair of surgical scissors is needed for cutting capillaries and another pair for cutting filter paper in thin strips small enough to fit inside capillaries. Do not cut paper with the pair meant for cutting glass or they will quickly dull, and once dull they shatter rather than cut the glass capillaries.
- **Capillary sealant.** Dental wax and other types of low-temperature wax are the traditional means of sealing capillaries. Recently, 5-min epoxy has become popular. The epoxy requires no heat, sets quickly, and even before it sets it forms an immediate vapor barrier. The handiest kind is clear, comes in a dual-barreled syringe, and is quite fluid before it hardens. Avoid types that are thicker and more like clay before hardening.
- **Plasticine.** This is also known as non-hardening modeling clay and is available in toy stores. It is very useful for sticking capillaries to goniometers and for holding them in position while mounting crystals. When warmed by rolling with the fingers, Plasticine can be wrapped around thin capillaries without breaking them. An alternative is to use pins that are sold for use with goniometers by Supper and Huber. The glass capillary is inserted into a hole in the pin and held in place with wax or epoxy.
- **Filter paper strips.** Cut Whatman #1 filter paper into strips thin enough to fit into capillaries for drying. The ideal strip is about 50 mm long, tapering from about 1.5 mm at one end to a fine point. The strips tend to curl when cut and can be straightened by gently curving in the opposite direction with fingers. Thin paper points originally meant for dental work are available from Hampton Research. As these come, they are too short to reach into capillaries. Mount the fine size into the end of a #18 syringe needle and then they will reach the crystal in most capillaries.

Mounting Crystals

There are several ways to mount a crystal in a capillary but they all accomplish the same goal. The method used almost exclusively in our lab is as follows. A capillary at least twice the width of a crystal is used. It is shortened by breaking with a pair of sharp tweezers so that it is about 4 cm long. If it is not shortened it will be too long to fit onto most X-ray cameras. The broken end is sealed with either melted dental wax or 5-min epoxy. The large funnel-shaped end is left open; the crystal will be placed in the capillary through this end. A ring of wax or epoxy is placed where the funnel end narrows to make a place where the capillary can be cut later. Without the ring, the capillary may shatter completely. A small ball of Plasticine is warmed in the fingers and gently wrapped around the capillary to serve as a mounting base. The capillary is put aside, using the Plasticine to stick it in a handy spot where it can be reached later.

The crystal to be mounted is selected and another capillary that will fit inside the first into which the crystal can be sucked is readied (Fig. 2.1). We use a piece of rubber tubing that fits over the capillary at one end with the

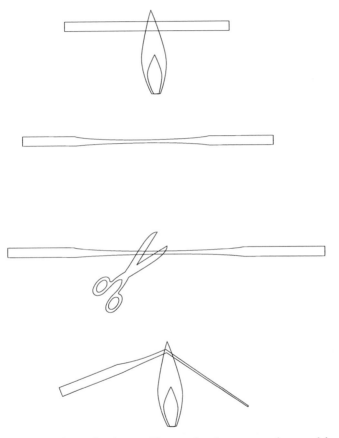

FIG. 2.1   Making crystal-transfer pipettes. The transfer pipettes or suckers used for mounting crystals are made from 200-µl capillaries or any thin-walled piece of glass about 1/8 inch in diameter. The capillaries are useful because they come with tubing and mouthpieces. The glass is easily softened by holding over a Bunsen burner flame while turning slowly. When the glass softens, it is removed from the flame and the ends pulled apart. Hold it still for a moment to cool and then put it aside. Then draw out several more. Each drawn-out capillary is then cut in the middle to form two pieces. The cut capillary is then bent by holding briefly over a flame until the end droops. With this technique you should be able to produce a number of different sizes. The ends are usually tapered and the proper bore can be obtained by cutting them off at the appropriate length.

other end going into a mouthpiece. It takes a little practice to get the knack of the sucking operation. The liquid will tend to stick at first because of surface tension, and then it comes all in a rush, requiring a little back pressure. Instead of mouth pressure, a syringe can be used to suck up the liquid. It is harder to control the syringe, however, and it takes one of your too-few hands. For toxic solutions, such as heavy atom soaks, it is never advisable to use a mouthpiece. Before the crystal is sucked out of the drop, a small amount of reservoir solution is sucked up and placed in the bottom of the previously prepared capillary (Fig. 2.2). Often a thin piece of filter paper is then pushed down the capillary to hold the reservoir liquid and to prevent it from moving. The crystal is then sucked up into the transfer capillary (Fig. 2.3). This frequently requires blowing liquid gently back and forth over the crystal to free it from the surface it grew on. More stubborn cases can be removed by very gently inserting the sharp point of a surgical blade between the crystal and the surface to pry it loose. Once it is freely floating, it can then be sucked up. The sucker is removed from solution and then a little air space is drawn in. This helps prevent the liquid from being drawn out by capillary action at the at the wrong time, wedging your crystal between the sucker and the mounting capillary. The sucker is then guided into the capillary, with you observing this through a dissecting microscope. The crystal is then gently blown into the capillary and the sucker quickly removed. Some mother liquors are easier to handle than others and some will insist on sweeping the crystal between the sucker and the capillary wall, catastrophically crushing the crystal. This can be avoided by first placing a band of reservoir liquid in the capillary into which the end of the sucker is inserted and then the crystal can be gently blown out. This leaves a large amount of solution to be removed later. In general, you want to blow out the crystal with as little solution as possible. The next step is to remove the solution around the crystal (Fig. 2.4). A very thin, fine capillary about 0.1–0.2 mm in diameter works the best for removing large amounts of liquid. Use one small enough that the crystal will not fit inside. Start removing the liquid at the edges first. Many liquids with a high surface tension cannot be fully removed this way and require further removal with a strip of thin filter paper. This can be worked up next to the crystal where it will slowly absorb all the free liquid. Leave a small amount of liquid between the crystal and the capillary to hold it in place. The capillary is then sealed with either dental wax or epoxy. Epoxy has the advantage that it is cool; there is some danger with dental wax that heating will hurt the crystal. This can be minimized by laying a strip of wet tissue on the capillary over the crystal before the melted wax is applied.

Another common mounting technique is to fill the capillary with liquid and float the crystal down into it. The crystal can be picked up in a mini

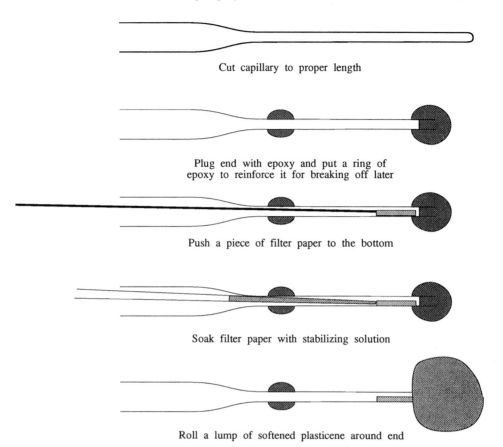

Cut capillary to proper length

Plug end with epoxy and put a ring of
epoxy to reinforce it for breaking off later

Push a piece of filter paper to the bottom

Soak filter paper with stabilizing solution

Roll a lump of softened plasticene around end

FIG. 2.2 Preparing X-ray capillary for mounting. If at any time during mounting you want to temporarily seal the capillary you can simply plug it with a softened piece of Plasticene. This gives you an opportunity to find something you forgot or to take a break. Sometimes it is necessary to allow viscous liquids time to bead up again before you can fully remove them.

pipette and placed into the tube held vertically. Or the crystal can be sucked directly up into the capillary along with the mother liquor. Both methods require a large amount of liquid to be removed before the crystal is ready. However, some liquids are very difficult to dry completely as they tend to stick to the glass and an excessive amount of time and effort may be required to dry the capillary. These methods are easier and may also be gentler on the crystal. A disadvantage of the method is the necessity of making an artifiical mother liquor with which to fill the capillary. This artificial mother liquor

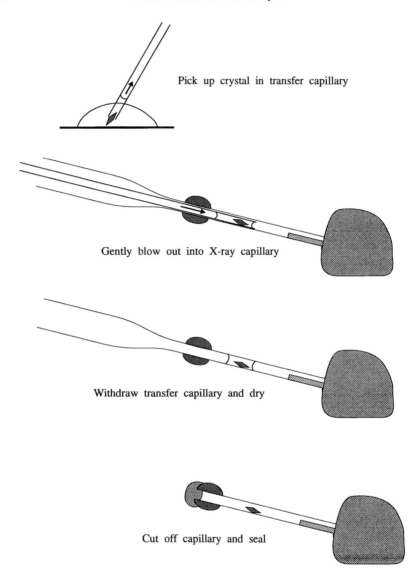

Pick up crystal in transfer capillary

Gently blow out into X-ray capillary

Withdraw transfer capillary and dry

Cut off capillary and seal

FIG. 2.3   Crystal mounting illustrated.

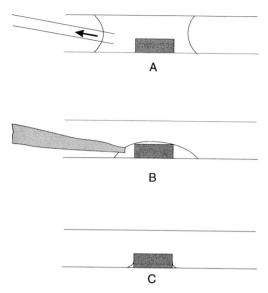

FIG. 2.4 Drying crystals. (A) Remove large amounts of liquid by drawing up into a drawn-out pipette by capillary action or by sucking. (B) Final drying is done with a thin piece of filter paper. (C) The crystal should have a small amount of liquid to keep it wet and to help it adhere to the capillary walls.

sometimes damages the crystal when it is transfered. The gentlest technique of all is to grow the crystal in the capillary and then remove the growth solution for data collection. This has been necessary for some protein crystals with very high solvent contents.

Drying Crystals

How dry does the crystal need to be? This depends upon the particular protein, and therefore requires experimentation with. If some crystals are left too wet, they will dissolve slowly. If too dry, some crystals may crack. If a low-temperature apparatus is to be used, then temperature gradients may cause the liquid to distill around the capillary, either cracking or dissolving the crystal. To prevent this, use a short capillary with as little free liquid as possible. A piece of filter paper may be used to wick solution around the capillary and reequilibrate it.

Polyethylene glycol solutions are amazingly tenacious, and a layer that slowly beads up around the crystal will remain bound to the glass, destroying your careful drying work. One remedy for this is to place the unsealed capillary in a sandwich box with a reservoir of crystallization liquid to keep the crystal wet while you wait about half an hour for the liquid residue in the capillary to draw up around the crystal and rewet it. This time when you remove the liquid the crystal will stay dry and you can seal it. The sandwich box is also handy to have to give yourself a break if you think that the crystal is getting too dry. It can be reequilibrated in the box before you continue.

Preventing Crystal Slippage

Crystals are held in place by the surface tension of the thin film of liquid between the crystal and the capillary wall. In most cases this is adequate, but sometimes the crystals will slip slowly or suddenly. Some methods of data collection are gentler than others and the crystal is less likely to slip. If the crystal does slip during data collection, it is always a problem. The crystal can leave the center of the X-ray beam or it can rotate, changing the pattern of the diffraction.

There are several ways to avoid this situation. First, dry the crystal thoroughly; excess liquid around the crystal encourages slipping. (See the preceding note on viscous liquids and drying crystals.) The slippage may be due to excess liquid that builds up around the crystal after data collection begins. For instance, if you use a low-temperature device there may be a temperature gradient along the capillary that causes water to distill from one end of the capillary to the other. This changes the vapor equilibration point at your crystal and can cause it to get wetter. In the worst cases a bead of liquid may form above the crystal and slip down onto it and dissolve it. To avoid this keep the capillary as short as possible and put a wicking material such as filter paper in the capillary to encourage reequilibration of the liquid.

Second, mechanically holding the crystal in place with fibers may be used. This should be a last resort, as material used to hold the crystal in place will add to the background scattering. Pipe-cleaner fibers have been found to be useful for this purpose.

A third method is to glue the crystal in place with a glue that dries in a thin film over the surface of the crystal and cements it into place. The glue and the method used is described by Rayment.[2] I have no experience with the method but it seems promising. The thin film of glue would not add substantially to the background.

Also consider the shape of the capillary relative to the surface of the crystal you are mounting. If the crystal has a flat face, then mounting inside

[2] Rayment, I. (1985). In "Methods in Enzymology," (Wyckoff, H., *et al.*, eds.), Vol. 114, pp. 136–140, Academic Press, San Diego.

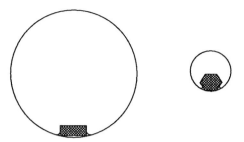

FIG. 2.5   Choose the capillary size to fit the shape of the crystal.

of a large-diameter capillary will provide a better contact between the crystal and the glass (Fig. 2.5). Conversely, a small capillary may be better suited to a crystal with many facets that presents a more curved surface. In fact, by floating crystals down capillaries filled with liquid, it is possible in extreme cases of crystal slippage to wedge them into the capillary where the glass tapers.

····· **2.2** ·····
OPTICAL ALIGNMENT

The next steps will be made easier if the crystal is first aligned optically. This is accomplished using a special goniometer stand called an optical analyzer and the dissecting scope. The crystal in the capillary is placed on a goniometer head that is, in turn, mounted on the optical analyzer. The first operation is to find the center of rotation of the analyzer with respect to the microscope reticules (Fig. 2.6). Rotate the analyzer to 0° and note the position of the crystal, then rotate it to 180° and note the position. The center is the midpoint between these two positions. The crystal can be translated using the slide on the goniometer head to move it to the midpoint. Another check of 0° and 180° is usually needed to fine-tune the centering. Repeat these steps for 90° and 270°. Note that the center is defined by the range of motion as the axis is rotated and not by any particular point in the microscope. The most common source of fustration in alignment is to assume that the cross hairs on a piece of equipment correspond to the center of rotation. Never make this assumption; the center is that position at which the crystal does not move when rotated. If the crystal has a definable axis, you may want to align this with the rotation axis. This is done by comparing the views at 0° and 180° and adjusting the arcs on the goniometer until the crystal axis in

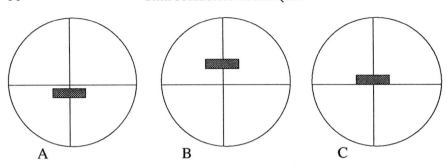

FIG. 2.6 Centering a crystal. This figure illustrates the steps needed to center a crystal on a camera using the crosshairs as guidelines. The view through the microscope is shown in three steps A, B, and C. The rotation axis is horizontal and the direct beam passes through the crystal vertically. The microscope cross hairs in this example are not perfectly aligned with the rotation axis of the camera. (A) View at 0°; (B) view at 180°. The translation on the goniometer is moved so that the crystal is halfway between the two positions observed at 0° and 180°. The final position after correction is shown in (C), and the crystal will now be in an identical position at both 0° and 180°. Note that the cross hairs do not go through the center of the crystal. The center is not defined by the cross hairs but by the center of rotation. If the cross hairs and the center do not coincide, the cross hairs should be adjusted to facilitate future alignments. Never assume the cross hairs are centered unless it has been done recently because high-power microscopes can become misaligned easily, especially if they are frequently moved.

both views is in the same position. This is then repeated for 90° and 270°. The third alignment that needs to be made is the position of the crystal faces relative to 0°. In general, crystal axes are either perpendicular to a face or they pass through an edge (Fig. 2.7). If the goniometer head provides a z-rotation, you may want to rotate the crystal with optical analyzer held at 0° until either a face or an edge is directly facing you. Draw a sketch of your crystal including its dimensions at both 0° and 90° on the optical analyzer. Include two lines for the walls of the capillary. This drawing will be very useful later when you want to correlate diffraction information with crystal morphology (Fig. 2.8).

Other optical properties of the crystal may be noted using polarized light. Protein crystals are birefringent and will appear brightly colored in polarized light. The polarizers should be oriented so that one is below the sample and the other above. The top polarizer is then rotated until the field becomes dark. The crystal should then appear bright since it further rotates the polarized light. As the crystal is rotated there will be four positions where the crystals will appear dark, and it will have maximum brightness in between. The exception is cubic crystals that are so symmetric they will appear dark in all directons. If one is looking exactly down a symmetry axis of the crystal that is centrosymmetric in projection (a centric zone), then the crystal will not be birefrigent and will always appear dark. By noting these direc-

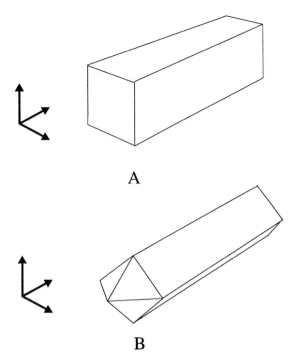

A

B

FIG. 2.7 Crystal axes relative to faces. (A) Faces are parallel to the crystal axes shown on the left. (B) Axes pass through the edges of the crystal and the faces are diagonals of the unit cell. Often crystals are combinations of both.

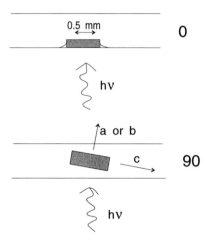

FIG. 2.8 Example of a sketch of crystal on X-ray camera. This rough drawing of the crystal is as it appears 90° apart on the camera, with the probable positions of the axes indicated. Later this sketch can be compared with X-ray photographs to determine the equivalence between morphology and crystal axes.

tions and comparing them with the external morphology and X-ray photographs, it may be possible to identify the directions of the crystal axes. This can be very useful in mounting the next crystal—especially if data need to be collected in a specific direction.

The quality of the crystal may be judged to some degree by the brightness in the dark field. For intance, if soaking in an inhibitor or heavy atom destroys the crystalline order, the crystal will become dark. In other cases it may be possible to see a dividing line between twinned sections of the crystal, although one must be cautious because a change in thickness will have the same effect. In the best cases, twins may be cut apart by applying a sharp scalpel to the line that joins the crystals. A more complete description of optical properties of protein crystals may be found in Blundell and Johnson.[3]

····· 2.3 ·····
X-RAY SOURCES

X-rays for protein crystallography are produced by two methods. The first method is to accelerate electrons at high voltage against a metal target, and the second is to use synchrotron radiation emitted by electrons and positrons in high-energy storage rings. Laboratory sources are limited to the former, whereas the latter is available at several international facilities. A brief comparison of sources is in Table 2.1.

Laboratory sources fall into two types: sealed tube sources and rotating-anode (Fig. 2.9). Both produce X-rays by accelerating electrons to a high voltage of 40–50 kV at a metal target. The limiting factor in the power at which the source can operate is the rate at which heat can be removed from

TABLE 2.1

Some X-Ray Sources

X-ray source	Wheel diameter (in.)	Brilliance	Cost ($)
Sealed tube		1	15K
GX-20	4	4	100K
GX-13	18	12	150K
Storage ring		$10^2 - 10^4$	

[3] Blundell and Johnson, pp. 98–104. See Suggested Reading list in preface for complete reference.

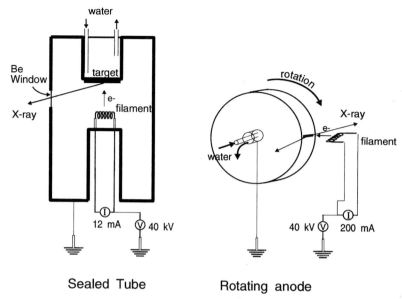

Sealed Tube   Rotating anode

FIG. 2.9 Sealed tube and rotating anode X-ray sources. Both sealed tubes and rotating anodes generate X-rays by accelerating electrons against a metal target. The electrons are boiled off a filament that is heated by a filament current. The filament is 40 kV above the target that produces the electron acceleration. A Be window is used to let the X-rays out. The power of the source is limited by the amount of heat the source can dissipate without melting the target. In the sealed-tube case this is 12 mA, which produces 480 W. By rotating the anode at high speed, more heat can be dissipated so that the rotating anode's current is 200 mA, which produces 8000 W.

the target. A typical sealed tube cannot operate at more than 20 mA of current or 0.8 kW at 40 kV. A rotating anode can reach 8 kW. A brighter source is better because radiation damage to protein crystals is not linear with dose but is a combination of dose and time. Once the crystal is exposed, a series of chemical reactions start that eventually damage the crystal. These are triggered by the ionization resulting from the radiation. Higher powers do not linearly increase the rate at which this damage occurs, and, therefore, more useful data can be collected before the crystal is irreversibly damaged. Another factor in favor of the rotating anode is that many data-collection installations are swamped with projects and the faster speed is needed to satisfy demand.

The choice of the metal target determines the characteristic wavelength at which the X-rays are emitted. The most common choice is a copper target, which emits at 1.5418 Å wavelength (CuK_α). This wavelength is a good compromise between maximum-achievable resolution and absorption. The other factor in favor of copper is its superior heat-conducting properties that allow it to be used at higher powers. In fact, most other metals used in targets are actually overlayed onto a copper base. Copper X-ray radiation is a soft X-ray: it will not penetrate very far through most materials; is absorbed quickly by air; and is scattered efficently by air, water, and glass. The path length through air must be minimized; if the distance from the crystal to the detector is over 150 mm, a helium box should be used. At distances shorter than this the windows of the helium path absorb about as much or more as is recovered by the helium path. Because copper X-rays are scattered by air, the free air path length of the direct beam should be minimized to prevent excessive background scatter. This means placing the collimator and the beam stop as close to the crystal as possible. The amount of glass and water surrounding the sample must be kept to a minimum, this will greatly increase the signal from the sample by reducing both background scatter and absorption of the diffracted rays. Phillips[4] has an excellent discussion of X-ray sources and their optimization for protein work.

In addition to laboratory sources, synchrotron radiation sources are also used. Since they may be extremely bright, they make ideal sources for characterization of small crystals. More information is given later on syncrotron sources and a full discussion of synchrotron sources including a listing of available sources and the equipment at each can be found in Helliwell.[5] Optics at synchrotrons are also generally superior to those found in the laboratory. This is partly due to cost but also the brilliance of the synchrotron beam means that more can be thrown away and still have a very bright beam. This should be remembered when characterizing a new crystal as even very small crystals (around 50 μm) can produce diffraction patterns with the very bright beam available at a synchrotron source.

Nickel-foil Filtering

The X-ray radiation as it comes from the tube cannot be used for data collection without some filtering. The spectrum of a copper source consists of two main peaks at CuK_α and CuK_β and also has white radiation at both higher and lower energies than the characteristic radiation. The simplest filter to remove the CuK_β radiation is a piece of nickel foil (Fig. 2.10). This will remove most but not all of the K_β without attenuating the K_α overly much. A

[4] Phillips, W. C., in "Methods in Enzymology," Vol. 114, pp. 300–316.
[5] Helliwell, J. R., pp. 94–135, and see list on p. 534 for addresses of contacts. See Suggested Reading list in preface for complete reference.

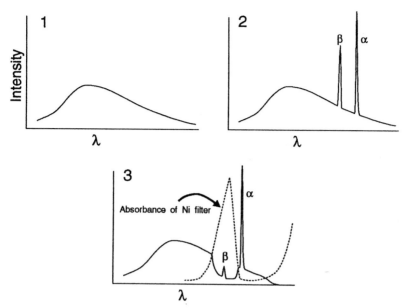

FIG. 2.10   Production of X-rays. Electrons are accelerated in a vacuum to strike a copper target. In one the electron acceleration voltage is below the threshold for characteristic radiation, and the X-rays produced have wavelengths corresponding to the energy of the electrons: $\lambda = hc/eV = 12,398/V$. After the electrons exceed a certain voltage they will have enough energy to displace a K-shell electron. X-rays are then produced as electrons fall from the L-shell (K_α) of M-shell (K_β), producing sharp peaks of X-rays superimposed on the white radiation. In 3, nickel foil is used to filter out most of the K_β radiation. The nickel filter has the absorbance curve indicated by the dashed line. It also filters out most of the higher-energy white radiation.

piece of foil 0.0005 inches thick is used as a filter[6] and is mounted in a holder that makes it easily removable.

Filtering by Monochromators

A single-crystal monochromator can be used to filter the X-rays. It produces a cleaner output than nickel foil, removing more high-energy radiation than the nickel foil lets pass through. Any radiation that is absorbed by the sample and causes damage but does not contribute to the diffraction pattern is wasteful. Experience has shown that monochromators can extend the useful life of a protein crystal's diffraction by removing harmful radiation that is absorbed by the crystal but does not contribute to the diffraction pattern.

[6] Stout and Jensen, pg. 12. See Suggested Reading list in preface for complete reference.

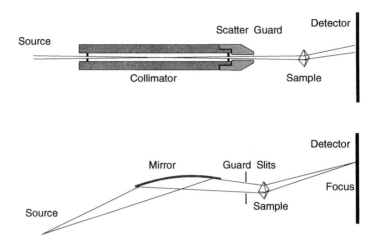

FIG. 2.11 Two types of X-ray optics. Top. With a collimator the beam is divergent. Bottom. With mirrors, the beam is convergent and is better suited for use with small crystals and large unit cells. Two mirrors are used: one in the horizontal plane and the other in the vertical.

A good setup with a monochromator, collimator, and beam stop can produce excellent signal-to-noise ratios. Dramatic reductions in the amount of scattered radiation in the background can be made over the nickel-foil filter by using a monochromator. However, the optics are still divergent. For large unit-cell crystals where the diffracted spots are very close together, it may be better to use mirrors as explained in the following.

Focusing with Mirrors

X-rays can be focused by deflecting at low angle with curved mirrors (Fig. 2.11). Two mirrors are used, one in the horizontal plane and one in the vertical plane.[7] The biggest advantage of a mirror is that the X-rays can be focused, increasing brilliance and allowing resolution of very large unit cells with closely spaced diffraction spots. The mirrors are not 100% efficient, and a substantial portion of the direct beam is lost, although the overall brilliance is increased. For smaller cells the advantage is lessened; indeed, for larger crystals with small cells, mirrors can so focus the diffraction spot as to saturate the detector locally—which can be alleviated by defocusing. If you have a large unit cell (greater than 150 Å), or very small crystals (<0.3 mm in largest dimension), then mirrors are definitely indicated. Mirrors will not re-

[7] Phillips, W. C., and Rayment, I., in "Methods In Enzymology," Vol. 114, pp. 316–330.

flect the higher energy X-rays but do reflect the lower energy X-rays that, because they are efficiently absorbed by the sample, are thought to cause much of the radiation damage to protein crystals.

..... 2.4 .....
PRELIMINARY CHARACTERIZATION

There is nothing more exciting than the first diffraction photos from a new protein crystal. There is nothing more disappointing than to discover that the crystal is really salt! The goals of a preliminary characterization are to discover if the crystal is protein, to what resolution it diffracts, the space group of the crystal, the quality of the crystal, and the relationship of the visible morphologies to the unit-cell axes.

The first photo is usually taken on a precession camera. While there are many tales of a crystal being fully chacterized by collecting all the data on an area detector and then later characterizing the data, for most of us a good precession picture holds more information; and, usually, the area detector is very busy anyway. I will, however, mention that a lot of protein crystals are reported with a low-symmetry space group and later when the the full three-dimensional structure is reported, the space group is of a higher symmetry. This usually happens when an incomplete characterization is performed.

The film orientation in the cassette should be standardized to facilitate alignment. We cut the upper right corner as the film is placed in the cassette and always clip the film holder to the top when it is processed. This allows the top, bottom, left, and right sides of the film to be distinguished. The film can be labeled with a #2 pencil to uniquely identify it. The label will be visible after the film is developed. It is important to keep a careful record of each film. Note the camera and goniometer angles, the exposure time, the exposure type, and the amount of rotation. Make a rough sketch of the crystal that shows how the crystal morphology relates to the camera axes and the direct beam. After each still exposure, mark the direct beam position on the film by closing the shutter, rotating the beam stop out of the way, and opening the shutter for the briefest time possible. This defines the center and allows the alignment angles to be calculated accurately.

The goniometer carrying the crystal is placed on the precession camera and aligned using the microscope mounted on the precession camera. Never assume the goniometer will mate exactly the same with the camera as it did with the optical analyzer. Rotate the spindle until it is at 0 and set the precession angle to 0°. The first shot is a "still" photo taken without the precession motion. If the crystal's smallest dimension is larger than 0.1 mm, usually a 30-min exposure on a rotating anode is suffcent. Small crystals require longer exposures. The exposure time is also dependent upon the unit-cell size. Smaller cells have fewer spots, with correspondingly more energy in each. For a first photo you will not know the unit-cell size, so start with a large collimator and take a 30-min exposure.

Examine this film for diffraction spots. A small-molecule film is shown in Fig. 2.12 for comparison with a protein film. The protein will have more diffraction spots that fall into families of rings. There will be spots at lower resolutions (nearer the center), and since each spot is usually weaker, it will not be as streaked as in the salt case, which has a white radiation streak. If there are no spots on the film, check the crystal alignment and make sure that X-rays are coming out of the collimator and that the camera is aligned. If all these things check out, try a long exposure of several hours. If there is still nothing, you may have damaged the crystal when mounting and it may be worth trying again. In the meantime you may wish to verify that the camera is working properly, using a crystal that is well characterized. Many laboratories keep a lysozyme crystal around for this purpose.

The spots on the photo should suggest a family of concentric circles, commonly referred to as a "zone." The center of the circles may correspond to a principal axis of the crystal but may also be a diagonal of the lattice. The circles may not be well populated and may not be obvious at first glance. If there are no obvious zones then rotate ϕ 30°–45° and take another photo. If there still are no zones, try 20°. If there still are no zones, you are not looking closely enough. A piece of acetate with concentric circles scribed onto it with a compass can be overlaid onto the film to make the rings more obvious and to find the center. Mark the center and measure the distance from the direct-beam spot in the vertical and horizontal directions (Fig. 2.13). The correction needed is $\tan^{-1}(\Delta/F)$ where Δ is the vertical or horizontal distance and F is the crystal-to-film distance (Fig. 2.14). The vertical correction is applied to spindle and the horizontal correction to the horizontal arc of the goniometer. Move the crystal so that an imaginary line from the crystal to the center of the zone aligns with the direct beam. Always remember to check the centering of the crystal after you adjust an arc and, if necessary, correct the goniometer translations.

Precession Photography

After the crystal is aligned using zones, then a small-angle screenless precession, of 2° or 3°, is taken to find the center more accurately. This film should show a circle in the center with a plane of the reciprocal lattice. The center of the camera is at the center of this pattern. If the pattern is perfectly centered, then the crystal is aligned. If not, the error is measured in the horizontal and perpendicular directions as defined in Fig. 2.15, and the proper correction is applied by looking up the error on a chart to find the corresponding

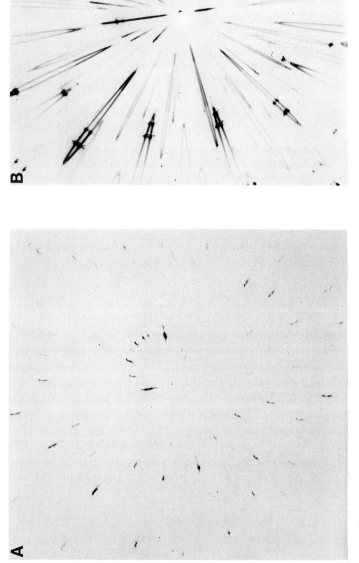

FIG. 2.12 Small molecule X-ray photos compared. X-ray photos of a small-molecule crystal, sucrose, are shown as an example of a buffer or salt crystal. (A) X-ray still photo of sucrose crystal taken with a copper rotating anode source at a crystal-to-film distance of 100 mm without any filters for 5 min. The spots are bright and spaced far apart and there are no pieces of lattice visible. (B) Precession photo (3°) of the same crystal, again without a filter. There is still no lattice visible as there would be for most protein crystals. Note the streaky character of the spots. Similar streaks appear with protein crystals, but they are rarely bright enough to see.

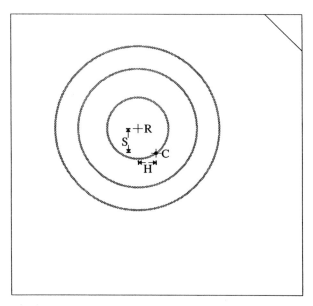

FIG. 2.13 Crystal Alignment Film. This diagram illustrates the method of aligning a crystal by means of a still photo. The spindle axis of the camera is horizontal and the crystal-to-film distance has been set to 100 mm. C, Center of the camera, which has been marked by a brief exposure with the direct beam. The distance from the center of the camera to the center of concentric rings of diffraction spots, R, is measured in the horizontal and vertical directions. The vertical correction that is applied to the spindle is $\tan^{-1}(S/100)$, and the horizontal correction to be applied to the horizontal arc on the goniometer head is $\tan^{-1}(H/100)$.

angular correction or by using the following equation, which is accurate for small errors and small precession angles. Given Δ, the error in millimeters, the angular missetting in minutes is given by $\epsilon \cong \Delta \times 8.5' \times 100/F$. Be careful to measure the errors from where the principal axes meet and not the center of the beam stop. A check photograph at the same precession angle is taken to confirm the corrections. Depending upon how practiced you are at making the corrections, this photo will either show no need for corrections or that small ones are necessary. Repeat the correction process as required. This procedure is illustrated in Fig. 2.16.

The camera is then set up to take a screened precession photo of 12° to 15°. A chart that comes with the precession camera can be used to find the correct screen and screen distance for the precession angle, or you may use the formula

$$CS = r/\tan u,$$

where CS is the crystal to screen distance, r is the radius of the screen, and u is the precession angle. A simple shortcut is to set the crystal-to-screen dis-

Film Error vs. Angular Correction at Small Angles

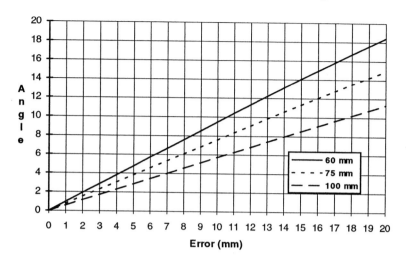

Film Error vs. Angular Correction at Large Angles

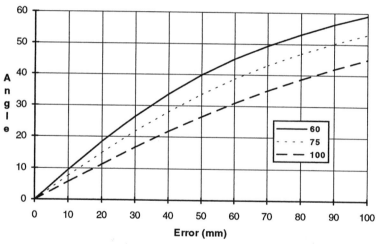

FIG. 2.14 Film error versus angular correction for $F = 100, 75, 60$ mm. If the spindle is near one of the cardinal points, 0, 90, 180, or 270, then the corrections to the arcs are straightforward. The most horizontal arc is corrected and the other arc is left alone. If, however, the spindle is off by more than about 10°, then corrections need to be made to both arcs. This is known as "crossed arcs." Subtract the spindle angle from the nearest cardinal value and take the sine and the cosine. The most horizontal arc is moved by the correction times the cosine, and the other arc is moved by the correction times the sine. For example, if the spindle is at 75° (after correcting the vertical error) and an angular correction of 12° on the horizontal plane is called for, then the arc closest to horizontal is corrected by $\cos(90 - 75) \times 12$, or 11.6, and the other arc by $\sin(90 - 75) \times 12$, or 3.1. The directions are chosen such that an imaginary line from the crystal to the zone on the film is moved to the center.

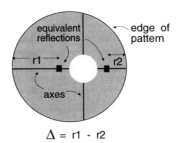

$$\Delta = r1 - r2$$

FIG. 2.15   Measuring Δ on a Precession Photo.

tance to 57 mm, at which distance *r* in millimeters is approximately equal to the precession angle in degrees. Thus, for an 8° precession picture, the radius to use is 8 mm with a screen distance of 57 mm. These values can be confirmed with the equation. The camera is then set to these values and the exposure taken. The exposure time needed for a screened precession is very long because most of the diffraction is blocked by the screen and only the small area of the annulus is exposed at any given time. The exposure time can be estimated from the time used for the small-angle photo by comparing the areas to be exposed. To a first approximation the area at small angles is proportional to the square of the angle. So if a 3° precession required 1 h for a good exposure, then a 15° will require 1 h × (15 × 15)/(3 × 3) or 25 h. The precession angle is equal to the θ of the highest angle reflection that will be recorded. A 15° photo will correspond to a 2θ of 30° or a 3-Å resolution photo for CuKα radiation (Fig. 2.17). An example of a 15° precession photo is shown in Fig. 2.18.[8]

Rotation Photography

Rotation photography, also called oscillation photography, is done by rotating the crystal about a single axis while the photo is taken. A 1° rotation photo is a picture taken while the crystal is rotated through 1° on the spindle axis. Rotation photography is more efficient than precession photography since there is no screen and all the diffracted photons are recorded. Rotation photography has been used extensively for data collection and we will not dwell on the details here as there is much material already written on the subject.

Figure 2.19 shows a rotation photo taken at a synchrotron source. Note that the pattern is of concentric, nearly cirular regions called lunes. The

[8] For more information on precession geometry, see Blundell and Johnson, pp. 260–269.

bottom edge of each lune corresponds to the start of the rotation, and the top edge of the lune is the end of the rotation. The spots in between are "swept out" as the crystal rotates (Fig. 2.20). Each lune arises from a single plane of reciprocal space. Compare this to a precession photo where the entire photo is of a single plane in reciprocal space. In the rotation photo several planes are each partially developed. In order to collect an entire data set, more rotation photos, each adjacent to the other in rotation angle, are taken until the crystal has been rotated through enough reciprocal space to collect all unique data. The spots at the edge of the lunes are only partially developed, and when integrated, either are added to the corresponding partial on the adjacent film or are ignored. In order to maximize the amount of whole, integrable spots, a large rotation angle is desired. At some point, however, the spots will start to overlap on adjacent lunes. This maximum range is resolution dependent. The farther out from the center of the film, the shorter the angle that can be rotated before spots start to overlap. In Fig. 2.19 the lunes at the top and bottom of the film are just starting to overlap. The other limit to the rotation is the amount of background that can be tolerated. The background, given a constant rotational speed, is proportional to the amount of rotation. Thus, a 2° film has twice the background of a 1° film. Each individual data spot, on the other hand, is the same intensity, as long as it is a whole spot, regardless of the rotation angle. In order to maximize the signal (spot intensity)-to-noise (background) ratio, more films of shorter angle need to be taken. Balanced against this is the fact that short rotations require more films; more film handling, which is very time consuming; and more partial spots. These conflicting needs must be reconciled for each individual data collection.

The maximum rotation angle depends upon three factors:

1. Unit-cell size. The larger the unit cell the smaller the rotation angle. Only the axes prependicular to the rotation angle are limiting. The axis that the crystal is rotated about is not limiting.

2. Spot size or mosaic spread. The width of the spot in the rotational direction—typically 0.1° to 0.5°, depending upon crystal quality and X-ray optics.

3. Resolution. The higher the resolution desired, the shorter the permissible rotation angle. The maximum overlap will occur in the area of the film perpendicular to the rotation axis and at the maximum diffraction angle.

The maximum rotation angle can be estimated with the formula

$$\Delta \text{rotation}_{\text{max}} = \tan^{-1}(d_{\text{min}}/\text{cell edge}) - \text{spotwidth}$$

For example, suppose that we want to collect a 2-Å data set on an orthorhombic crystal with unit cell dimensions $a = 100$, $b = 75$, $c = 50$ Å

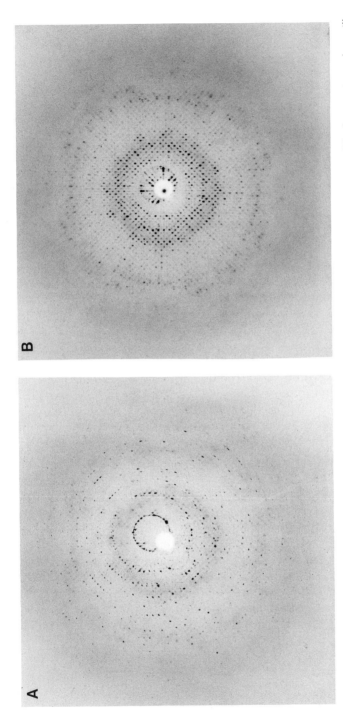

FIG. 2.16 Aligning a precession photograph. All of the following films are taken at a crystal-to-film distance of 100 mm. (A) The first step is to take a still photograph. This is a photograph for iron-binding protein that has a 110 × 110 × 200 A unit cell. Note the concentric rings centered slightly up and to the right. This film is corrected as described in the text and in Fig. 2.13. (B) After applying these corrections, a small-angle screenless precession photo is taken to fine-tune the alignment. The direct-beam was marked on this photo by moving the beam stop out of the way with the X-rays off, and then the shutter was opened as briefly as possible. This marks the center and makes determining the correction easier. Note that the beam stop is not centered. In general, the beam stop is never a reliable indicator of center of the film. The error is measured as described in the text and in Fig. 2.15. This precession photo was taken at 1.5° precession angle because of the large cell spacing of 200 A in the direction perpendicular to the film. Smaller cells are usually taken

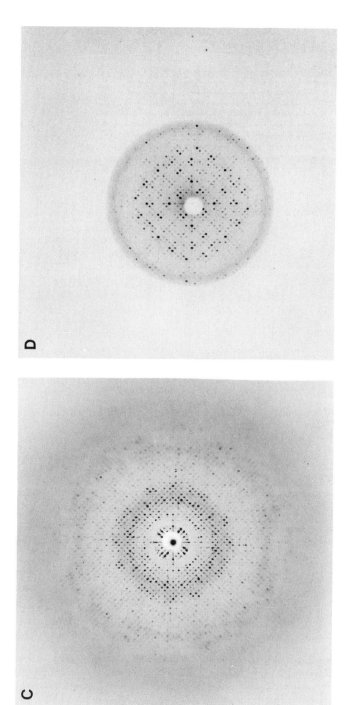

at 2.5°–3° precession angle (C) Before taking a higher-angle photo, a check precession photo is taken. The alignment can be confirmed by noting the position of related spots on opposite sides of the beam stop. This photo still shows a very small misalignment, but it is near the limit of accuracy that the camera and goniometer can be set to. (D) Finally, a high-angle screened photograph is taken. The precession angle was set to 8°, the crystal-to-screen distance was 58 mm, and an 80-mm Δ2 screen was used. In the photo the major axes are tilted 45°, and the pattern shows 4-fold symmetry. A precession photograph taken at 90° to this shows every second and third spot missing along the c axis, indicating that the space group is either $P4_1$ or its enantiamorph $P4_3$. Photographs thanks to Dr. Michele McTigue, The Scripps Research Institute.

45

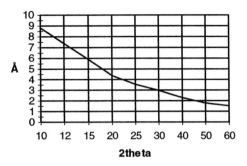

FIG. 2.17   Resolution versus 2θ for CuK$_\alpha$ (1.54-A) source.

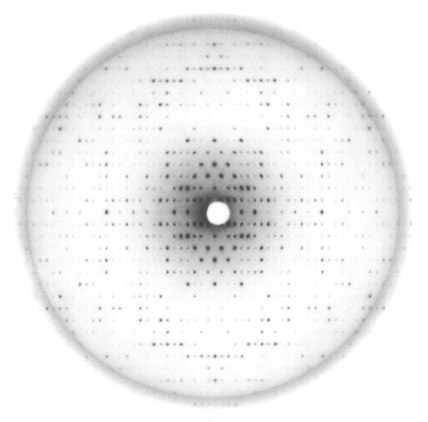

FIG. 2.18   Precession photograph (15°). This precession photograph is of *E. coli* endonuclease III h0l zone. Note the *mm* symmetry with a mirror on each axis. Also note that every odd spot is missing along the two major axes. The other two zones both show the same symmetry and systematic absence, identifying the space group as $P2_12_12_1$. Photograph courtesy of Drs. Chefu Kuo and John Tainer, The Scripps Research Institute.

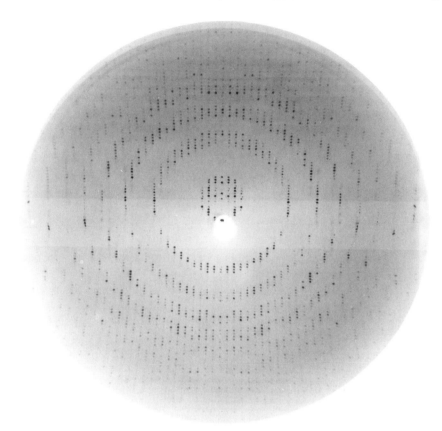

FIG. 2.19 Rotation photograph. A 2° rotation photo of aconitase taken at Stanford Synchrotron Radiation Laboratory on beam-line 7-1. The rotation axis is about the horizontal. The horizontal band across the center of the film is due to a piece of mylar that supports the beam stop. Photo thanks to Dr. C. David Stout, The Scripps Research Institute.

and, given the optics, the spotwidth is 0.3°. The crystal is mounted so that a, the longest axis is along the rotation axis. When we are rotating with c in the plane of the film, the maximum permissible rotation is limited by b, so that $\tan^{-1}(2/75) - 0.3 = 1.2°$, and when c is limiting, the maximum rotation angle is 2.0°. The maximum angle is thus 1.2°. In order to collect a full data set we need $90 + 2\theta_{max}$. With CuK_α radiation (1.54 Å) 2θ is 45° at 2 Å. So, to collect a full data set without overlaps, we need 135/1.2 or 113 films. Some saving can be made by using 2° for the part of the data set where c is limiting and switching to 1.2 when b is limiting.

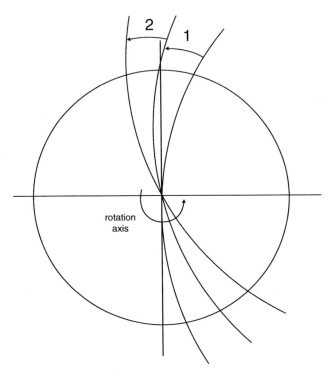

FIG. 2.20 Rotation geometry projected down rotation axis. The volumes swept out by two successive rotation photographs are marked 1 and 2.

Blind Region

Near the rotation axis there is a region of reciprocal space that cannot be collected because the Lorentz correction[9] is very large. The Lorentz correction accounts for the amount of time that a reflection spends in diffracting conditions while being rotated through the Ewald sphere. Near the rotation axis this time gets to be very large, and at the limit a reflection directly on the rotation axis is always diffracting. In order to fill in this data, it is necessary to rotate about another axis in order to sweep out the data near the rotation axis. If you have the crystal mounted on a goniostat, this may mean simply moving chi or phi, otherwise the crystal will have to be remounted. In the absence of symmetry the crystal will have to be rotated by at least $2\theta_{max}$, and with mirror symmetry the angle will be θ_{max}. Another strategy is to rotate with the crystal mounted such that the nearest axis is 20°–30° off the rotation

[9]Blundell and Johnson. pp. 319–320.

axis. If you are using auto-indexing this is the ideal solution as it maximizes the amount of unique data.

Offsetting the axes of the crystals from the rotation axis is always desired for maximizing unique data, but in the past crystals had to be nearly perfectly aligned or the software being used could not index, and therefore could not integrate, the film. For instance, consider the film in Fig. 2.19. It is set so that the mirror symmetry in the vertical direction is along the vertical. This makes for a pretty film but it also means that a partial on the right half has a corresponding mate on the left. If the crystal was rotated so that the mirror was off the vertical by a few degrees, the corresponding mirror-related reflections would be such that when one was partial the other would likely be whole. It may be desirable to have mirror symmetry when collecting Bijvoet pairs in order to use an anamolous scattering signal. In this case, the mirror-related reflections may be Bijvoets, and thus collecting them both at the same time with the same geometry may maximize the accuracy of the small difference between pairs.

An excellent set of three articles on rotation photography can be found in "Methods in Enzymology," including use with large cells, synchrotron radiation, and integration of intensities.[10]

White-Radiation Laue Photography

In Laue photography, the crystal is held still during the exposure and multiple wavelengths, or white radiation, are used rather than monochromatic radiation. This gives many more diffraction spots at once than with monochromatic radiation, and recently, with the availability of synchrotron radiation light sources that give off a continuous spectrum of useful X-ray wavelengths, several groups have successfully used this technique to study reaction intermediates. It promises to give very short exposure times. A single Laue photograph of 100 ms exposure can have nearly an entire data set for high-symmetry crystals. Laue photography is very sensitive to the mosaicity of the crystal. Crystals that are too mosaic to use with Laue photography can still be used with monochromatic radiation.

A Laue photograph is shown in Fig. 2.21. Note that the pattern is that of many rings. Where these rings intersect is called a nodal. The positions of these nodals are used by the Laue software to orient the crystal by searching through all possible positions and finding the one that best predicts that nodal pattern.

---

[10]Harrison, S. C., *et al.*, p. 211; Rossman, M. G., p. 237; Fourme, R., and Kahn, R., p. 281. All in "Methods In Enzymology," Vol. 114.

A

FIG. 2.21 White-radiation Laue photographs. (A) Laue photograph of photoactive yellow protein taken at beam line X26C, National Synchrotron Light Source, Brookhaven National Laboratories. Crystal size: $60 \times 60 \times 30$ μm. Exposure: 30 msec. The image plate used was scanned on a Fuji BAS2000 scanner. The image plate was placed in a cassette with a rubber front that was then mounted on the X-ray camera using a mount similar to those used for film cassettes. (B) Close-up of Laue pattern. The horizontal line across the middle of the image is displayed in cross-section at the bottom of the screen. Because of the low background noise of image plates and their high sensitivity, weak diffaction spots are well imaged. In the center just below the horizontal line is a nodal where several circles cross. Laue patterns are indexed by noting the

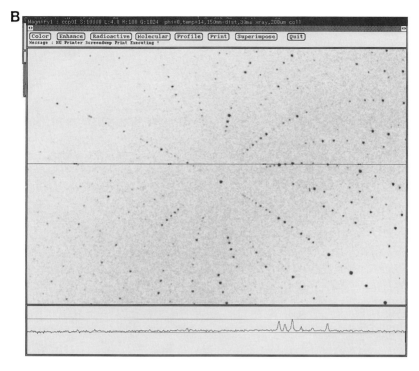

FIG. 2.21—*Continued.*

positions of several nodals. The software then searches through all possible orientations for one that matches the list of nodals. Nodals are low-indices reflections and are usually energy multiples making them especially bright. Photos courtesy of Dr. Keith Moffatt and Zhong Ren, the Univeristy of Chicago.

Space-Group Determination

Space groups are determined by examining the diffraction pattern and noting any symmetry and systematic absences. It is necessary to have a copy of the *International Tables for Crystallography* to look up the symmetry patterns to find the identity and number of the space group. A list of the seven crystal systems and their main features is given in Table 2.2. Since proteins are asymmetric objects and occur only in the L-form they cannot be involved in symmetry elements requiring inversion centers, mirrors, or glide planes. This limits the possible space groups to 65 out of the 230 mathematically possible space groups and leaves 2-, 3-, 4-, and 6-fold axes along with the corresponding screw axes, and centering, as the possible symmetries. We will consider each in turn.

TABLE 2.2
Protein Space Groups

Crystal system	Class	Space groups	Lattice restrictions	Angle restrictions
Triclinic	1	P1	—	—
Monoclinic (2-fold parallel to b)	2	P2, P2$_1$ C2	—	$\alpha=\beta=90$
Orthorhombic	222	P222, P222$_1$, P2$_1$2$_1$2, P2$_1$2$_1$2$_1$ C222, C222$_1$ F222, I222, I2$_1$2$_1$2$_1$	—	$\alpha=\beta=\gamma=90$
Tetragonal (4-fold parallel to c)	4	P4, P4$_1$, P4$_3$ P42, I4 I4$_1$	a=b	$\alpha=\beta=\gamma=90$
	422	P422, P42,2, P4,22, P4$_3$22, P4$_1$2$_1$2, P4$_3$2$_1$2, P4$_2$22, P4$_2$2$_1$2, I422, I4$_1$22		
Trigonal (3-fold parallel to c)	3	P3P3$_1$, P3$_2$, R3	a=b	$\alpha=\beta=90$, $\gamma=120$
	32	P312, P321, P3$_1$21, P3$_2$21, P3$_1$12, P3$_2$12, R32	a=b=c (R)	$\alpha=\beta=\gamma<120$
Hexagonal (6-fold parallel to c)	6	P6, P6$_1$ P6$_2$ P6$_3$ P6$_4$ P6$_5$	a=b	$\alpha=\beta=90$, $\gamma=120$
	622	P622, P6$_1$22, P6$_5$22, P6$_2$22, P6$_4$22, P6$_3$22		
Cubic	23	P23, F23, I23, P2$_1$3, I2$_1$3	a=b=c	$\alpha=\beta=\gamma=90$
	432	P432, P4$_1$32, P4$_3$32, P4$_2$32, F432, F4$_1$32, I432, I4$_1$32		

2-folds. A 2-fold causes the presence of a mirror plane perpendicular to the 2-fold axis in the reciprocal space pattern after the addition of the inversion center (see Fig. 2.22). A 2_1 can be distinguished by the absence of every other spot on the axis that lies in the plane of the mirror. The presence of even one exception to the screw axis absences means the axis is a 2-fold.

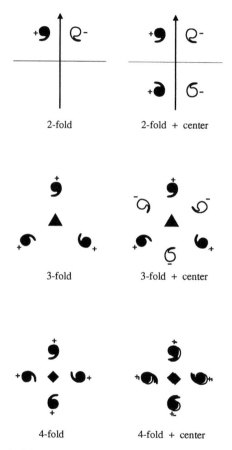

FIG. 2.22 The effect of adding an inversion center to a 2-, 3-, and 4-fold. On the left is the symmetry in real space with a comma used to indicate an asymmetric object. A "+" and black coloring indicate an object above the plane of the page and a "−" and white coloring indicate an object below. Note in the case of the 2-fold that an inversion center forms a mirror perpendicular to the 2-fold. In projection there are two mirrors and, therefore, the 0-level will also show two mirrors, while an upper level will show only one mirror. An inversion center plus a 3-fold still has the same symmetry as before, although in projection it will appear to be a 6-fold. A 4-fold plus an inversion center adds a mirror in the plane of the paper. A 6-fold is similar to a 4-fold but is not shown.

The 2-fold constrains the angles between the 2-fold and the other two axes to be 90°.

3-folds. A 3-fold causes 3-fold symmetry in the reciprocal axis except on the 0-layer, where the inversion center raises the symmetry to a 6-fold. Thus it is usually necessary to take an upper level precession photo to differentiate between a 3-fold and a 6-fold. There are two 3-fold screw axes, the 3_1 and the 3_2. They both give identical absences along the 3-fold axis: only every third spot is present. They cannot be distinguished from each other at this point and must be left to distinguish later. In addition, the 3-fold symmetry gives rise to a hexagonal lattice that constrains the a and b axes to be identical and the angles between axes to be 90° between the 3-fold and the other two axes, and to be 120° (60° in reciprocal space) between the two non-3-fold axes.

4-folds. A 4-fold gives rise to 4-fold symmetry to the diffraction pattern in the plane perpendicular to the 4-fold axis. It also constrains two axes to be identical to each other and all axes to be at 90°. A mirror will be found at the plane passing through the origin and perpendicular to the 4-fold ($hk0$). There are two possible types of screw axes: 4_1 and 4_3, with only every fourth spot present; and 4_2, with every other spot present on the 4-fold axis.

6-folds. A 6-fold gives rise to 6-fold symmetry both on the zero level and on upper levels. In addition, a mirror is found at the plane passing through the origin and perpendicular to the 6-fold axis ($hk0$). Both trigonal (non-rhombic) and hexagonal space groups have a hexagonal lattice; the symmetry of the intensities must be used to tell the two apart. The 6-fold axis itself can have three different screws: 6_1 and 6_5, with every 6th spot only; 6_2 and 6_4, with every second spot only; and 6_3, with every third spot only on the 6-fold axis. Two pairs of the screw axes, 6_1 and 6_5, and 6_2 and 6_4, can only be told apart at a later stage.

Rhombic. Rhombic is a special case of trigonal and is characterized by having all three axes equal and all three angles equal. It is the hardest system to diagnose because of the difficulty in finding the zones and determining their relationships when the axes are not near 90°. In certain cases, C-centered monoclinic may really be R32, so this may be worth checking out.

Centering. Centering can be detected by systematic absences throughout the diffraction pattern. Centering can cause confusion about the direction of the principal axes. Always use the symmetry to determine the lattice. For instance, in a C-centered lattice, spots with $h + k =$ odd are missing. At first inspection, the lattice appears to be running on the diagonals of the cell. Symmetry on the zero level will show the presence of two mirrors that give the correct direction of the two axes. The five types of centering possible are A, B, C, F and I. A, B, and C centering are identical except for the naming of the axes. The convention is to name the axes such that the cell is C centered, so that $h + k = 2n + 1$ reflections are missing, that is, the *ab* face of the crystal is centered. (Thus, A centering has the *bc* face centered and B centering has the *ac* face centered.) F is face-centered so that all three faces *ab*, *ac*, and *bc* are missing the odd reflections. I is body-centered so that an extra lattice point is found at the body center of the lattice. This would be easy to miss by precession photography of 0-levels alone, although a picture of a diagonal zone might reveal it.

Always assume that you may have higher symmetry than you do until proven otherwise. Never trust 90° angles unless held up by symmetry. Never use low resolution photographs to decide the space group. Always keep an open mind about the space group until the structure is solved and refined.

To determine the correct space group it is necessary to take enough precession photos to determine the symmetry elements present, any systematic absences along the axes, and any centering. These are then compared with the diagrams in the International Tables. The tables are grouped according to the highest symmetry present (i.e., if you have a 6-fold then the molecule is found in the hexagonal section). You then search for alternative space groups and ask if the precession photos you have are necessary and sufficient to eliminate all other possible space groups. This may mean taking an upper-level photo to determine the difference between some possibilities such as trigonal versus hexagonal (Fig. 2.23). Determine the size of the unit cell by Bragg's law and compare the volume of the cell in ångstroms cubed with the size of the best estimate of the protein's molecular weight, MW, in daltons. Listed in the International Tables for each space group is the Z number, or the number of asymmetric units in the unit cell. Use this formula to calculate the ångstroms-cubed per Dalton of the asymmetric unit: Volume/(Z $*$ MW).[11] The expected value for this number for protein molecules ranges from 1.7 to 3.0 with the average being about 2.3. If you have a number substantially smaller than 1.7, then it is likely that something is wrong, or it may mean that there is internal symmetry in the protein molecule that corresponds with a crystallographic axis. For instance, spot hemoglobin is a tetramer and has a dimer axis that coincides with a crystallographic axis so that the asymmetric unit contains one-half of a tetramer instead of a full molecule. If the number is substantially larger than 3.0, then there are two possibilities. One is that there is more than one molecule in the asymmetric unit, which is

[11] Matthews, B. W. (1968). *J. Mol. Biol.* **33**, 491.

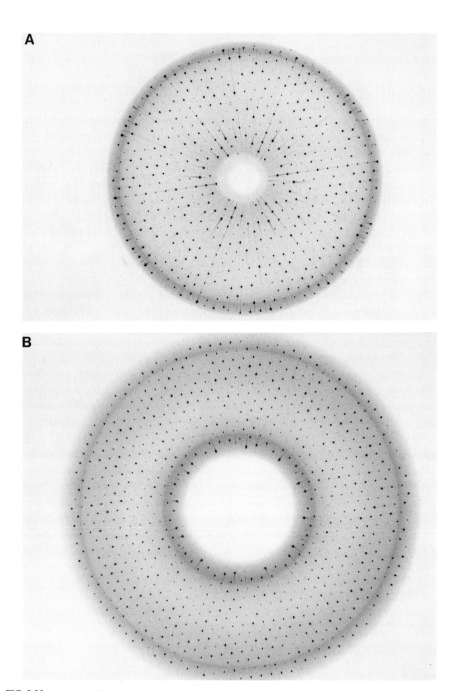

FIG. 2.23 Distinguishing 3- and 6-folds. (A) The 0-level photograph of photoactive yellow protein taken down the *c* axis shows a hexagonal net with 6-fold symmetry. Both a 3-fold and a 6-fold will show 6-fold symmetry in the 0-level. Thus an upper-level precession photograph hkl (B) was taken to distinguish the two possibilities. This photo also shows 6-fold symmetry, confirming the presence of a 6-fold axis. Another precession photograph (not shown) was taken of the 6-fold axis and showed every odd spot systematically missing. There are no mirrors, eliminating the possibility of the class 622. This is enough information to assign the space group as $P6_3$.

very common. The other possibility is that the space group has higher symmetry than you have determined. Try looking for higher-symmetry space groups that contain the symmetry you have already determined to see if there is a precession photograph that you could take that would prove or disprove this possibility. For instance, R32 can be reindexed as monoclinic C2 and it can be difficult to spot the difference. If you are using XENGEN to reduce the data, it is also possible to try reindexing your data in different space groups and to look at the R-merge values. There are also programs that are commonly used by small-molecule crystallographers to search for additional symmetries in a three-dimensional data set.

Finally, a common error is to mistake psuedosymmetry for true symmetry. An example of a crystal with psuedosymmetry is given in Fig. 2.24. Psuedosymmetry appears correct at lower resolutions but breaks down at higher resolutions. Most protein crystals show some pseudosymmetry in the range of infinity to 6 Å. It is unusual not to have low-order reflections on the axis that are virtually extinct, leading one to the conclusion that there is a screw axis. Always confirm screw axes to at least 3 Å or better resolution. If you cannot confirm the screw axis to high resolution, then always keep the possibility in mind that the axis is not a screw axis and try both possibilities. The presence of even a single reflection that breaks the symmetry rules out the presence of a screw axis. If that reflection is weak, however, it may be worthwhile considering that it is an artifact (in particular, K_β radiation can cause artifacts) and do not exclude the 2-fold screw until better evidence is found. One of the best confirmations of the space group is a good heavy-atom Patterson. For example, consider the case of determining whether a 2-fold or a 2_1 screw is present on the a axis. The Patterson map for both possibilities is calculated the same way. Even if you enter the incorrect possibility in the symmetry operators, both a 2-fold and a 2_1 degenerate to a mirror in Patterson space. In the case of the 2-fold, the Harker vectors will be at the plane $x = 0.0$ and for the 2_1 the Harker vectors will be on $x = 0.5$. It is important to plot out the entire Patterson map and look at all possible Harker sections. (For more information on Patterson maps, see the following.)

Unit-Cell Determination

A single still photograph taken along one axis can be used to determine roughly the three directions of the unit cell. You should have a pattern of concentric circles around the beam center. Portions of the lattice will be visible and these can be used along with Bragg's law to determine two directions.

$$d = \frac{\lambda}{2 \, \sin(\tan^{-1}(\Delta/F)2)},$$

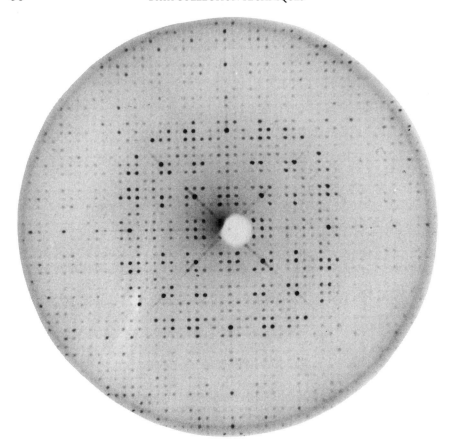

FIG. 2.24 Psuedosymmetry. This precession photograph of iron-binding protein shows true 4-fold symmetry. There are also psuedo-mirrors along the main axes and the main diagonals. Close examination shows these mirrors to be inexact. Photograph courtesy of Andy Arvai, The Scripps Research Institute.

where Δ is the spacing between spots, F is the crystal to film distance, and λ is the the wavelength of X-rays (1.5418 for copper targets). It is best to measure a long row and divide the length by the number of spaces to get a more accurate determination. Also, the closer in to the center the row is the more accurate the approximation of using Bragg's law will be, because the spacing gets stretched the farther out from the center you are.

The third direction can be determined from the spacing of the concentric rings using the equation:

$$d = \frac{n\lambda}{1 - \cos((\tan^{-1}(r/F))},$$

where r is the radius of the nth circle. The circles need to be close to concentric about the beam stop (i.e., an axis aligned along the direct beam) for this equation to be accurate.

The direction determined is correct if you have an orthogonal cell. Otherwise the lengths need to be corrected for the fact that the photographs show d^* instead of d directly. To do this simply divide the distance by the sine of the appropiate angle. For example the correct value of b is $b/(\sin(\beta))$.

More accurate distances can be determined from the undistorted lattices found on precession photographs. For this you will need photographs of at least two zones. Measure a number of spots in a row and divide by the number of spaces as above and use the same equation derived from Bragg's law. Again, for non-orthogonal cells these distances will need to be corrected.

Evaluation of Crystal Quality

The number one piece of information requested about protein crystals is the limit of observable diffraction. The desire to have this number be as high a resolution as possible (i.e., to have the smallest numerical value) has led to some rather creative definitions where the single highest resolution spot is used to report this value. It is better to report the resolution where at least one-third, or so, of the possible reflections are still visible above background. Even this is pushing it, but resolution inflation is common and pervasive.

Another factor to consider is the mosaicity of the crystal. Mosaicity is a measure of the order within a crystal. If a crystal has low mosaicity, the crystal is highly ordered and diffraction spots will be sharp. High-mosaicity crystals will have broader peaks because of lower crystalline order (Fig. 2.25). Since increased mosaicity means that a spot is in diffracting conditions for a larger range of angles, mosaicity may be recognized as broadened lunes on diffraction patterns. Or, if you are using a diffractometer or area detector, the profile of a peak may be directly measured. While increased mosaicity in itself may not be a problem, it may indicate other problems. For instance, if when mounting the crystal you let it dry too much, the mosaicity will be increased. If the crystal is suffering from radiation damage (heating, drying, etc.) it is quite likely that the mosaicity will increase. If the crystals have high mosaicity, it may be worth trying to see if a crystal can be mounted to give a diffraction pattern with less mosaicity. Mosaicity may also indicate twinning, where the crystal is actually made up of several crystals joined together. Unless the twinning can be accurately accounted for, it is not possible to use the amplitudes of a twinned crystal to determine the X-ray structure. Finally, increased mosaicity is usually accompanied by lower-resolution diffraction overall and lower signal-to-noise since the counts are spread over a larger diffraction angle.

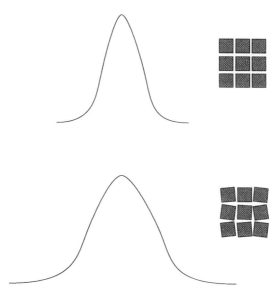

FIG. 2.25 Mosaicity of crystals. Each block is composed of from one to many unit cells. Top. This crystal has high order (right) and thus the diffraction of a single spot is sharper (left). Bottom. This crystal has lower order and broader diffraction spots.

In looking for twinning it is important to look for extra families of circles that cannot be accounted for by a single crystal in the beam (Fig. 2.26). A twinned crystal is one or more crystalline units joined together. Sometimes the joining is apparent in the morphology, but often the only way to tell is from the diffraction pattern. Still photographs or small-angle photos are best for this purpose. Be cautious in assigning twinning solely due to split spots (Fig. 2.27). If the crystal is slightly misaligned about the center of the camera or the camera is misaligned, this can cause split spots because the Ewald sphere is not centered on the camera center. It is possible then for a reflection to occur twice in near proximity and to produce split spots. It may be more fruitful to align the camera carefully rather than to throw the crystal away.

On the area detector or diffractometer, twinning can be recognized from looking at the profile of several spots in different areas of reciprocal space. For area detectors, 10–20 frames of 0.05°–0.10° in oscillation angle are taken on the profile of spots as a function of oscillation angle is plotted. This can be done using the frameview program from XtalView (see following). Diffractometers allow a continuous scan in one of several angles. The presence of split spots indicates a twinned or cracked crystal, which may not be used for data collection.

Are the crystals big enough for data collection? This is a question with so many parameters that it is not possible to give a good answer. It is always

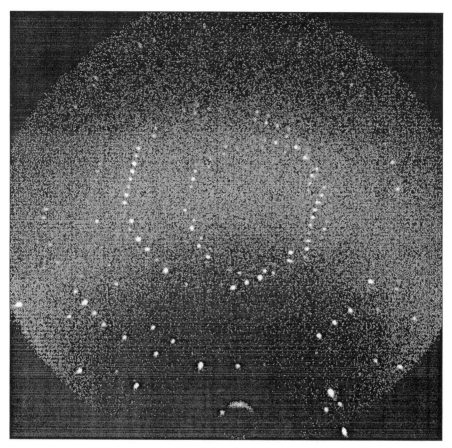

FIG. 2.26 Image of twinned crystal. An image of a crystal with a twinning defect. Two separate, unrelated sets of concentric circles are evident. This crystal cannot be used for data collection.

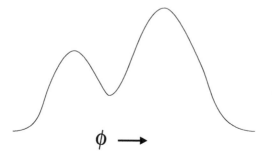

FIG. 2.27 Split-spot profile. A split-spot profile such as this may indicate a cracked crystal or twinning.

possible to grow a larger crystal, although it may take considerable experimentation and hard work. The size needed is determined by the quality of the diffraction pattern, not its physical dimensions. Crystals with large unit cells have weaker diffraction patterns than do similar crystals with smaller cells so that a larger crystal is needed. (Actually they diffract the same amount of photons—in the larger cell these photons are spread out over more reflections.) In the end you have to decide which questions you wish to answer with your experiment. If you want an atomic-resolution structure of a mutant protein to look at small changes from the wild-type, then clearly a 4-Å diffraction pattern is not enough. It may be worthwhile collecting data on a small crystal for now so that you can start working on structure solution at low resolution while you wait for larger crystals to grow. Avoid collecting data just because you can do it. If you collect a 2-Å data set but all the spots beyond 3 Å are below the noise level, then it is really a 3-Å data set and will not give you any information beyond this.

····· 2.5 ·····
HEAVY-ATOM DERIVATIVE SCANNING WITH FILM

The traditional method of scanning for heavy-atom derivatives is to use screened precession photos with a precession angle of 5°–10° or higher. The method is inefficient in that it takes a longer exposure to collect the same number of photons by precession photography than it does to by other methods because most of the diffracted X-rays are blocked by the screen. Shorter exposure times can be used if several degree rotation photos or low-angle screenless precession photos are used. In any case the object is to compare the intensities of the heavy atom film with an equivalent "native" film and to look for intensity changes. The unit cell can be quickly checked by overlaying equivalent rows on the native film. If the unit cell of the putative derivative changed significantly, more than 0.5–1.0%, then the derivative may not be usable. Deciding if there are intensity changes can be difficult for the beginner because it is necessary to differentiate between different exposure times and differences in the rate of fall off for the entire pattern. The best way to convince yourself that the changes are real is to look for reversals where the intensity is greater in one photo and another pair where the intensity differences are reversed (Figs. 2.28 and 2.29). A good heavy-atom derivative has obvious differences. Most photos will not have large differences but may show one or two differences. Remember, "One difference does not a derivative make." The differences should occur at all resolutions. Differences will be found in the lowest-resolution reflections between infinity and 10 Å from differences in solvent contrast because of the presence of the heavy atom in

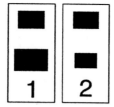

FIG. 2.28 An intensity reversal. Intensity reversal between otherwise identical spots on two films. Note that in film 2 the upper spot is larger than the lower, whereas it is the opposite in film 1.

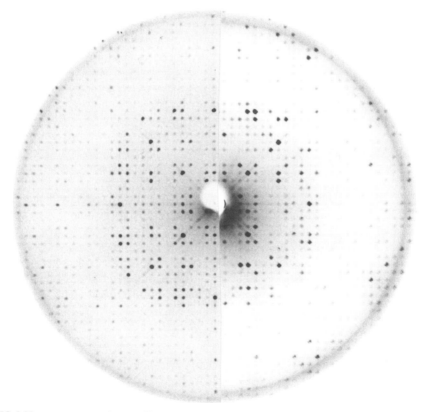

FIG. 2.29 Derivative and native films compared. On the right is native iron-binding protein. On the left is iron-binding protein soaked in iridium hexachloride. There are clear intensity changes, and many examples of reversals can be found. Also note that the pseudo-mirror symmetry between top and bottom has been clearly broken in the derivative.

the solvent, even if there is no binding to the protein. The differences of an isomorphous derivative will fall off slightly with resolution and will increase with resolution if the derivative is not isomorphous. However, unless the pattern of differences is obvious, it is probably better to decide these questions by collecting some data on the derivative and determining the size of the differences with resolution statistically on a large number of reflections.

If rotation photos are used, be careful that you do not compare spots that could be partial in one photograph but not in the other. Examine only reflections roughly perpendicular to the rotation axis and at least one row from the edge of the lunes. Reflections near the rotation axis are probably partial, and very small differences in crystal orientation can cause large intensity changes. Knowing this, it is possible to use rotation photos of about 5° to scan for derivatives for most unit cells. Choose as large an angle as you can without getting overlap below 3 Å. Align the crystal carefully with still photos 90° apart on the spindle. It is not necessary to align the crystal as precisely as for a precession photo, but be aware that a different pattern of partials can be confused with true intensity changes. In any case the worst that can happen is that you falsely identify a derivative and collect an extra data set. This is far better than missing a derivative altogether.

The other common method of finding derivatives is to scan using the area detector. As people are becoming more familiar with area detectors, this is becoming more common. In most laboratories with both detectors and cameras it is easier to get time on the camera, and you can usefully fill time scanning for derivatives with film while waiting for the detector to become available. In using the area detector collect enough frames to index and integrate a small amount of data. This is then merged with the native data and the resulting statistics (see following) can be used to determine if the crystal is derivatized. If it is not, it can be removed and another crystal tried with a minimal waste of time. This is known as the "take-it-off" strategy. Using this method several crystals can be scanned in a single day. It takes overnight to make a precession photo for the same purpose and then one is usually comparing fewer total unique spots and the method is not quantitative.

····· 2.6 ·····

OVERALL DATA-COLLECTION STRATEGY

Unique Data

The essence of data collection strategy is to collect every unique reflection at least once. First you need to determine the unique volume of data for your space group. This is done by considering the symmetry of your space group and including an additional center of symmetry. Thus the space groups

P222, P2$_1$22, P2$_1$2$_1$2, P2$_1$2$_1$2$_1$, C222, and C222$_1$ all have *mmm* symmetry in reciprocal space because both a 2-fold and 2$_1$ screw degenerate to a mirror plane when a center of symmetry is added. To determine the unique data you can look up your space group in the International Tables and determine the reciprocal space symmetry (also called the Patterson symmetry because the Patterson function also adds a center of symmetry). In Table 2.3 the volume needed for each space group is listed. For instance, for orthorhombic space groups we need to collect $0-h_{max}$, $0-k_{max}$, and $0-l_{max}$. If you are using a diffractometer setting, this is simple: just enter the bounds and the software takes care of the rest.

Film and image plates cover a large enough area that you can set up to rotate around any convenient axis by 90°. However, the geometry of diffraction is such that the data occur on the surface of the Ewald sphere. If data collection is started so that the h axis is aligned and we are rotating about the k axis, when we have gone 90° to the l-axis we will be missing a piece of data because of the curvature of the Ewald sphere (Fig. 2.30). To collect all the data we need to rotate another 2θ°, where 2θ is the highest resolution

Table 2.3
Unique Data for the Various Point Groups

Crystal system	Class	Data Symmetry	Unique data
Triclinic	1	$\bar{1}$	$h, -h$; $-k, k$; $0, l$ or $h, -h$; $0, k$; $-l, l$ or $0, h$; $-k, k$; $l, -l$
Monoclinic (2-fold parallel to b)	2	$2/m$	$h, -h$; $0, k$; $0, l$ or $0, h$; $0, k$; $-l, l$
Orthorhombic	222	*mmm*	$0, h$; $0, k$; $0, l$
Tetragonal (4-fold parallel to c)	4	$4/m$	$0, h$; $0, k$; $0, l$ or $0, 1$ and any 90° about c
	422	$4/mmm$	$0, h$; $k \geq h, k$; $0, l$ or $h \geq k, h$; $0, k$; $0, l$
Trigonal (3-fold parallel to c)	3	$\bar{3}$	$0, h$; $-k, 0$; $0, l$ or $0, 1$ and any 120° about c
Rhombohedral	3	$\bar{3}$	$0, h$; $l \geq -h, l \leq h$
	32	$\bar{3}m1$	$0, h$; $0, k$; $0, l$
	32	$\bar{3}1m$	$0, h$; $k \geq -h/2, k \leq h$; $0, l$
Hexagonal (6-fold parallel to c)	6	$6/m$	$0, h$; $0, k$; $0, l$ or $0, 1$ and any 60° about c
Cubic	23	$m\bar{3}$	$0, h$; $0, k \leq h$; $0, l \leq h$
	432	$m\bar{3}m$	$0, h$; $0, k \leq h$; $0, l \leq k$

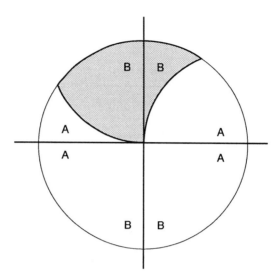

FIG. 2.30 Diagram of reciprocal space volume swept out by 90° of data collection. In this diagram the rotation axis is vertical to the page and revolves around the center sweeping out the shaded area in 90°. The crystal is orthorhombic with *mmm* symmetry as indicated by the letters A and B. The asymmetric unit, the minimal volume that needs to be collected, is any one of the four quadrants. Note that the point A is not included in the sweep. To collect this region the sweep must be greater than 90°.

data to be collected. Additionally, we will have missed a blind zone of data near the rotation axis that can be filled in by rotating around another axis; in this case if *h* is chosen then the crystal should be aligned so that *k* is in the plane of the film and *l* is along the direct beam. Rotation cameras provide no means of rotating the crystal to this new position, but by using a 90° offset on the crystal it can be replaced on the goniometer head with the capillary vertical instead of parallel to the rotation axis. This is hardly very convenient because if the camera spindle is rotated too far the capillary will hit the collimator or the beam stop and break. In polar space groups, such as P6, any 60° around the unique *l* axis will collect all the data except the blind zone. If you are using sophisticated software, the blind zone can be collected by setting up the rotation axis about 25° off the crystallographic axis. Between the four quarters of the film all the blind zone will be collected. However, some data-collection software requires that the crystal be almost perfectly aligned.

    The area detector presents a greater challenge because the area of the detector is smaller than film or image plates. The detector can be swung out to collect higher-resolution data, but its height is not enough to cover all the data needed. It is necessary to rotate the crystal about another axis besides the rotation axis. On a 4-circle machine this is χ or ϕ and on 3-circle this is ϕ. The crystal is "ratcheted" around to collect the data completely. This method is discussed more completely by Xuong and co-workers.[12] In addition, a program, RSPACE[13] is available in which the orientation of the crystal can be entered and the data that can be collected at different angles of the camera can be viewed. A similiar program, ASTRO, is available for Siemens area detectors. Another option is to use XRSPACE (Fig. 2.31), which can be used to show completeness of data but does not display the goniometer parameters. It is not necessary to align the crystal accurately before data collection, although it is more efficient to start data collection with one axis roughly aligned with the direct beam at the start of data collection. In fact, it is better not to be perfectly aligned as this will cause symmetry-related reflections to fall in equivalent positions so that both will be missing if they fall in a blind region. Also, redundancy increases the accuracy of the data. The sigma of n identical measurements of a reflection is[14]

$$\sigma_n = \frac{\sigma}{n}$$

Bijvoet Data

When an anomolous scatterer is present in the crystal, it may be worthwhile collecting Bijvoet pairs. It is then necessary to know which reflections are Bijvoets in your space group. The Friedel mate of a reflection, $-h, -k, -l$ is always a Bijvoet. Other reflections may or may not be. A reflection is a Bijvoet pair of another reflection if it takes an odd number of sign changes to transform its indices and if it is related by a symmetry element. Thus, the reflection h, k, l and $-h, k, l$ in an orthorhombic space group form a Bijvoet pair. The reflections h, k, l and $-h, -k, l$ are not a Bijvoet pair. In a monoclinic space group (b unique) h, k, l and $-h, k, l$ are not related by any space group symmetry, so they are not Bijvoets in this space group. A general rule of thumb is that if the reflections are related by a mirror or inversion center, then the reflections are Bijvoets. If they are related by a 2-, 3-, 4-, or 6-fold they are not Bijvoets. In order to collect complete Bijvoet pairs, approximately twice as many data need to be collected. Centric reflections never have a Bijvoet mate, so they need to be collected only once.

[12]Xuong, N. H., Nielson, C., Hamlin, R., and Andersen, D. (1985). *J. Appl. Chrystallogr.* **18**, 342–350.

[13]RSPACE was written by Mark Harris, Computer Graphics Laboratory, University of North Carolina, Chapel Hill.

[14]The sigma of an averaged reflection is calculated as follows: $\sigma_{avg} = 1/(1/\sigma_1 + 1/\sigma_2 + 1/\sigma_3 \cdots)$, so that if the sigmas are equal this reduces to $1/(n/\sigma)$.

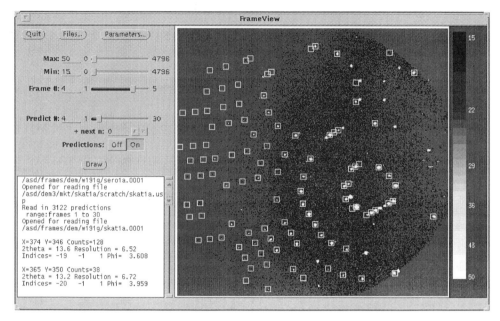

FIG. 2.31   Using XRSPACE to check for data completeness.

Indexing of Data

Indexing of data is normally done in two stages. In the first stage a primitive lattice is used to give each reflection an index that is used to identify each individual point in reciprocal space. In the second stage reflections that are related by the point-group symmetry of the space group are reduced and collected together.

Modern programs can auto-index the data (Fig. 2.32). These fall into two categories. In the first type, the unit cell is entered into the program and all possible orientations of this cell are searched until a match is found. In the second type, vectors are looked at in reciprocal space, three non-coplanar groups are searched for, and the shortest reciprocal space distance (the largest in real space) is taken to be the cell edges. This must be checked against the known cell to find the correct lattice. In either case the cell should be known already—usually from precession photographs. The correct cell is the reduced cell that is consistent with the highest symmetry and the longest axes (in real space).

In most space groups there are ambiguities when it comes to deciding upon a unique volume. In some cases, such as orthorhombic, all possible choices are equivalent and lead to the same answer. In others it makes a

A

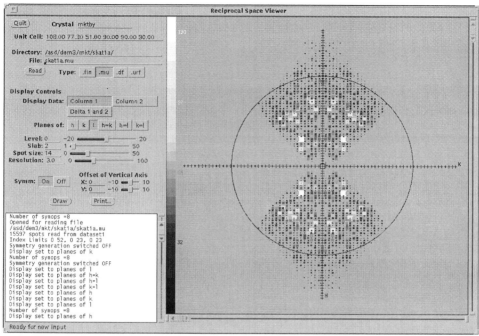

B

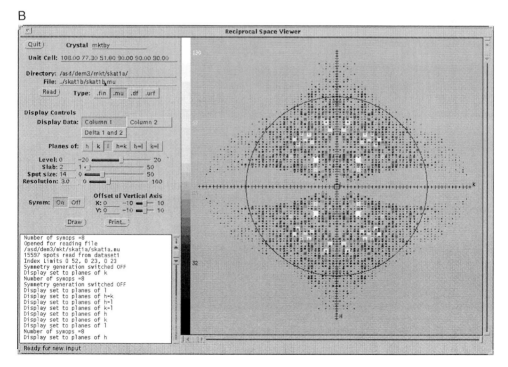

FIG. 2.32 Frames of data with prediction boxes superimposed. Two 0.25° frames from a Siemens area detector are shown with the predicted positions of spots superimposed as boxes. These can be used as a check.

difference which volume is chosen. For instance, in hexagonal space groups there is no way to tell the a axis from the b axis on the basis of the lattice itself. The decision must be made at a later stage when some intensities are known. The first time you choose it makes no difference, but a second run of data must be indexed consistent with the first. One sign of inconsistent indexing is if the R-merge of two data sets is very high, around 30–50%, while the individual R-symm's are in the normal range, say below 12%. Since there is no way to decide on the correct orientation before you integrate the data, the correction must be made at the reduction step. XENGEN[15] provides a handy means of doing this with the $-i$ option of reduce. A 3×4 matrix is entered that is used to alter the indices of the primitive lattice before the data are reduced to the unique indices that will be used to group the data for scaling and merging purposes. For example, to reindex data in point group 32 on a hexagonal lattice (e.g., P3$_1$21) such that a and b are incorrectly chosen, use the matrix

$$\begin{pmatrix} -1 & 0 & 0 \\ 0 & -1 & 0 \\ 0 & 0 & 1 \end{pmatrix} + \begin{pmatrix} 0 \\ 0 \\ 0 \end{pmatrix}$$

This will change h to $-k$ and k to $-k$ and leave l as is. The fourth column is added to the indices and in this case is all 0s. When the data are reduced, the effect of this will be to switch h and k (the same as reassigning a and b) while preserving the handedness of the Friedel pairs. This matrix will also switch a and b but makes the data left-handed and thus switches the Friedel pairs so that they are incorrect:

$$\begin{pmatrix} 1 & 0 & 0 \\ 0 & 1 & 0 \\ 0 & 0 & -1 \end{pmatrix} + \begin{pmatrix} 0 \\ 0 \\ 0 \end{pmatrix}$$

As a final example, here is a matrix for switching a hexagonal space group in point group 6 (e.g., P6$_3$). These space groups are polar, and the essential problem is that it is not possible to tell whether the c axis is pointing one way or the other. To switch l it also necessary to reassign h and k thus:

$$\begin{pmatrix} 0 & 1 & 0 \\ 1 & 0 & 0 \\ 0 & 0 & -1 \end{pmatrix} + \begin{pmatrix} 0 \\ 0 \\ 0 \end{pmatrix}$$

To find the correct matrix the easiest thing to do is to compare equivalent sections of data from both data sets. It is best not to use the 0-level as

[15] XENGEN was written by A. Howard and is available from Siemens Instruments. See Howard *et al.* (1987). *J. Appl. Crystalogr.* **20**, 383–387.

it usually has extra symmetry. With XtalView (discussed in Chapter 3) you can use XRSPACE to view equivalent sections by starting two copies of XRSPACE on both data sets. It is usually obvious what the problem is by comparing the patterns of intensities. When making these matrices, be careful to preserve the correct handedness. If you do one operation that changes the hand, then you must do another that changes the hand back. There must be an even number of hand changes. You may not care about the Friedel pairs for native data without an anomalous scatterer, but remember you will want to keep the Friedel pairs correct for heavy-atom derivative data later.

..... 2.7 .....
OVERVIEW OF OLDER FILM TECHNIQUES

X-ray film is still one of the best area detectors around. It is compact and easily stored, has very high pixel resolution, can be bent to form a curved surface, and is inexpensive. Its chief disadvantages are its high background, low efficiency, and low dynamic range. Films must be scanned by an optical scanner to be used as data, and in this step many of its advantages are lost. To increase its dynamic range, three films are usually used, which means that three files must be used to scan the data and three films must be handled. To take advantage of the high resolution, a fine pixel must be used on the scanner, increasing the scan time and producing three very large files. If the cost of the manual labor involved in a large film data set is factored in, it can be a very expensive method of data collection. However, for preliminary X-ray work it is hard to beat film for determining space groups and assessing crystal quality.

The rotation method is usually used with film data collection. Three films are placed in a pack, one behind the other. Each film partially absorbs the X-rays so that the exposure of each film in the pack is lowered by about a factor of three. The cassettes have small holes that are used to make fiducial marks on the films. Theses marks are needed to align the film precisely because the film can move around inside the cassette a small amount. The films are exposed while rotating the crystal through a degree or so (see the section on rotation photography). When the films are developed, it is important to use consistent times for each film. The film must not touch the sides of the developing tank or each other, and when they are wet care must be taken not to scratch the film. Typical development times are 5 min in the developer, 30 s in the stop bath, and 5 min in the fixer. When the developer becomes brown, it should be changed and the fixer changed at the same time. Wash the films for at least 5 min before exposing them to air to look at them or they will turn brown later. The total wash time should be a minimum of

30 min in clean, flowing water. Dry the films thoroughly before handling them. When the films are scanned, be sure to do it in a consistent manner so that the scanner axes are known relative to the camera axes for each film. If you forget to mark the fiducials on a film, often it can be carefully aligned with a rotation before or after and the fiducials marked with a pen.

..... 2.8 .....
FOUR-CIRCLE DIFFRACTOMETER DATA COLLECTION

Diffractometer is used in this book to mean a goniostat equipped with a single-counter detector (Fig. 2.33). This used to be a complicated topic, but with the advent of modern software, collecting data by diffractometer has

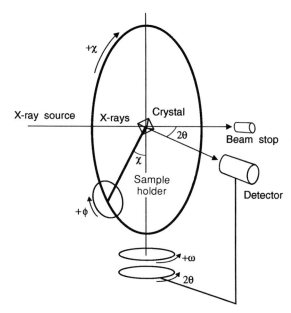

FIG. 2.33   Four-circle diffractometer. A 4-circle diffractometer consists of a 4-circle goniostat and a scintillation detector attached to an X-ray source. The four circles are ω, χ, ϕ, and 2θ. By moving the angles it is possible to bring every reflection into diffracting conditions such that the reflection can be counted with the detector. To integrate the intensity, the crystal is rotated on the ω axis and the profile of the counts used to determine the intensity. With an area detector the scintillation counter is replaced and a means of varying the distance from the crystal to detector is provided. The ω axis is then used to rotate the crystal to collect the data while holding ϕ and χ fixed. A 3-circle goniostat has a fixed χ angle (usually 45°) and a 2-circle would have $\chi = 0$ and $\phi = 0$.

become easier and faster. Many diffractometers are capable of auto-indexing the crystal. Check reflections are used to check the alignment as data collection proceeds, and more efficient peak scanning methods combined with faster motor slew rates have resulted in faster data collection. In fact, for collecting a low-resolution data set for the purpose of heavy atom scanning, the modern diffractometer can actually beat an area detector in speed and accuracy. The accuracy of diffractometer data properly scaled and corrected can be higher than any other data collection method. Diffractometers can be adjusted to cut down on the background reaching the detector through the use of slits and a long helium path. (Background falls off with square of the distance, while diffraction spots do not fall off at all if the path has no air.) The counter used in diffractometers has very high efficiency and virtually no background so that fewer counts are needed to achieve the same signal-to-noise ratio. Finally, an accurate absorption correction can be made on the area detector experimentally, while on area detectors and film this is done only approximately by scaling equivalent reflections. It is obvious that the diffractometer is a superior machine—so why isn't everyone using one?

There are a number of drawbacks to diffractometers that limit their usefulness in all cases. Depending upon the length of the detector arm on the diffractometer and the optics of the source, the individual reflections may not be resolved if the unit cell is too large, above 100–150 Å. Radiation damage must be relatively slow for a complete data set to be collected or else many crystals must be used with concomitant scaling problems. Finally, the diffractometer just takes too long to allow it to compete successfully given the scarcity of X-ray ports available in most labs. Far more data can be collected on an X-ray generator equipped with an area detector than on a diffractometer. This fact makes the area detector the method of choice for many groups where the speed with which data can be collected is the single limiting factor and space and money for more generators are limited. (The current detector fad has left many perfectly useful diffractometers languishing. You might be able to get your hands on one for heavy-atom screening, or you may have a crystal that can use the diffractometer to collect a highly accurate data set.)

An excellent discussion of the geometry of diffractometers and the factors affecting data collection parameters can be found in a discussion by Wyckoff.[16] For details of data collection, refer to the manual that comes with your diffractometer. In general, the data-collection strategy is to collect the data in shells of resolution starting with the highest resolution where radiation damage first shows up and proceeding to progressively lower resolution shells until all the data have been collected. Standards are measured periodi-

[16] Wyckoff, H. C. , in "Methods in Enzymology," Vol. 114, pp. 330–385.

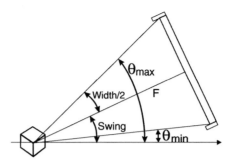

FIG. 2.34 Calculating resolutions for area detectors. The highest resolution (2θ) available for a given swing of an area detector is the swing plus half the angular width in degrees. The minimum resolution is the swing minus the half-width. The angular width of the detector is given by $\tan^{-1}$ (width of detector/crystal-to-film distance).

cally about every 100 reflections to monitor alignment and decay. Some of the standards should be in the resolution range you are measuring. A ϕ scan of a bright reflection on an axis is used to calculate a correction for absorption as a function of ϕ. If Bijvoet pairs are desired, they are collected in blocks of 25 reflections to minimize errors between pairs. First, a block of 25 reflections is collected and then the same 25 reflections are remeasured at $-h$, $-k$, $-l$. Some reflections that are symmetry mates can be collected to measure R-symm (see Chapter 3.1). If Bijvoet pairs are collected, then these can be the Bijvoet pairs of centric reflections [17] which cannot have an anomalous scattering component. R-symms, or the agreement between equivalent reflections, is the best indicator of overall data quality and a check on the accuracy of the data set.

····· 2.9 ·····
AREA-DETECTOR DATA COLLECTION

Area detectors are fast becoming the most common data collection technique. Since an area detector collects a larger volume of reciprocal space than a diffractometer, it is more efficent. The critical value for determining the speed of data collection is the angle subtended by the detector (Fig. 2.34). This angle will depend upon the width of the active surface of the detector

---

[17] Technically speaking, the difference in the distribution of intensities for centric and acentric reflections means that the centric reflections are not a perfect indicator of the symmetry errors in acentric reflections. Centric reflections tend to be either bright or weak, whereas acentrics have a more even distribution of intensities.

and the distance the detector is set back from the crystal. The width of the detector is fixed. The distance is dependent on the largest dimension of the unit cell and the size of a pixel on the area detector. Two adjacent diffraction spots must be separated by at least one pixel. For a Siemens area detector, Andy Howard's rule of thumb is

$$d(\text{cm}) \geq \frac{\text{longest cell edge } (\text{A}°)}{8.0}$$

In practice, spots have some width, so that the center-to-center distance that two reflections can have and still be resolved is also dependent upon the optics and the particular crystal. In practice, it is better to err on the safe side. Trying to squeeze too much data onto the detector and overlapping adjacent spots will lower data quality.

The easiest and best way to determine the d is to first do a rough back-of-the-envelope calculation and then put the crystal on the machine. Find an orientation with the closest spacing on the detector, collect a short data set with the longest axis on the face of the detector, and examine some of the frames (Figs. 2.35 and 2.36). The closest spots should be clearly separated. On the Hamlin detector this can be a single pixel, while on the Siemens detector there should be at least three pixels separation. If the spacing is too close, the detector must be moved back. Do not collect data with overlapping spots!

An area detector can be equipped with three types of goniostats, 2-circle, 3-circle, or 4-circle. The 2-circle is most limited, consisting of a rotation axis and a swing movement for the detector. Some improvement can be made by mounting crystals so that the rotation axis is a diagonal of the unit cell. With such a setup it is very hard to collect a single data set at high resolution from a single mount.

A 3-circle goniometer has a rotation axis and a ϕ rotation mounted with χ fixed at 45°. A swing angle is also provided for the detector. Data are usually collected by rotating around the rotation axis for as far as possible. The crystal can be rotated around the ϕ axis to collect new data. Usually rotating ϕ 90° will give the most new unique data.

A 4-circle goniostat will allow the most control over data collection. It has all the movements of the 3-circle, plus χ can be adjusted a full 360°. One disadvantage is that the size of the detector and the χ circle mean that they collide after ω moves over a limited range—usually about 60°. To overcome this, the crystal is ratcheted by advancing ϕ 60° and another ω sweep collected.

A useful recipe for data collection using a 4-circle goniostat with a Siemens area detector and an orthorhombic crystal is given in Table 2.4. It collects the unique data in a minimum amount of time at 2-Å resolution at a

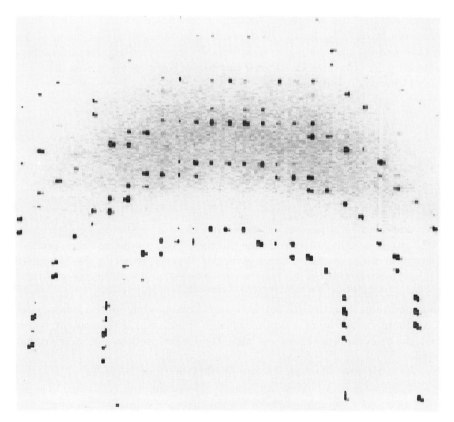

FIG. 2.35 Frame from San Diego Multiwire Systems (Hamlin) area detector. Frame from a multiwavelength data collection experiment at beam line I-5 at the Stanford Synchrotron Radiation Laboratories. Spots are well separated. Data collection geometry has been set up to allow Bijvoets to be collected simultaneously on the left and right halves of the detector (rotation axis is horizontal at this beam line) by taking advantage of the mirror symmetry of the samples space group ($P2_12_12_1$). Frame thanks to Brian Crane, The Scripps Research Institute.

Table 2.4

Example of Siemens Area Detector Data Collection for Orthorhombic Crystals

Run	ω, Start	ω, End	Swing	χ	ϕ	Oscillation (ω)	Number of frames
1	50	−10	22.5	15	0	0.25	240
2	50	−10	22.5	15	60	0.25	240
3	50	20	22.5	75	0	0.25	120

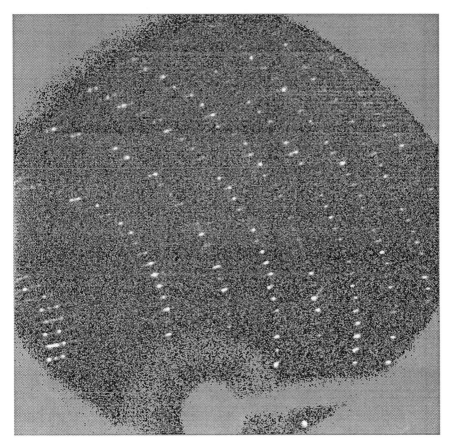

FIG. 2.36 Area detector frame. Shown is a frame of data collected on a Siemens area detector. The frame is 0.25° rotation of FBP. The detector was located at 17 cm. Note that the spots (see lower right) are just being separated.

crystal-to-detector distance of 12 cm. The crystal is mounted so that one axis is approximately along the capillary (i.e., at χ 0°, ω and this axis will be coincident). This needs to be only accurate to about +/− 5°. Optically center the crystal on the goniostat. Move χ to 0, ω to 50, and set the swing to 22.5°. Now rotate φ and take still frames until a 0-layer (the 0-layer is the one that passes through the beam-stop) is centered on the detector (Fig. 2.37) so that the outer edge of the circle is at the edge of the detector. Define this φ angle as 0°. As data are collected, the 0-layer circle will move from the center of the detector towards the beam stop. The χ is then offset to maximize the amount

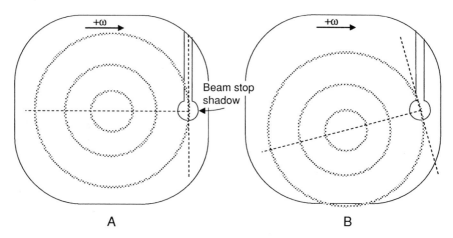

FIG. 2.37 Starting position for orthorhombic data collection. (A) The crystal has been aligned so that one axis is vertical, coincident with the rotation axis, and the crystal is rotated until the 0-layer stretches from the beam stop to the edge of the detector. The 0-layer always intersects the beam stop. (B) χ has been rotated 15° to maximize the unique data and to minimize the effect of the blind region around the rotation axis.

of unique data. Otherwise the data on the top and bottom of the detector will be related by mirror symmetry. Since not all of the data can be collected in one run (we need 90° + 22.5°), another run is done with ϕ rotated. To fill in the data that were missed by the limit of the detector height, a fill-in run of half the length at χ 90 − 15°, or 75°, is used. If Bijvoet pairs are desired, then interleave runs at $\chi = -\chi$, $\phi + 180$°, leaving the other angles the same, for a total of six runs (Table 2.5).

Table 2.5
Example of Siemens Area Detector Data Collection for Orthorhombic Crystals

Run	ω, Start	ω, End	Swing	χ	ϕ	Oscillation (ω)	Number of frames
1	50	− 10	22.5	15	0	0.25	240
2	50	− 10	22.5	− 15	180	0.25	240
3	50	− 10	22.5	15	60	0.25	240
4	50	− 10	22.5	− 15	240	0.25	240
5	50	20	22.5	75	0	0.25	120
6	50	20	22.5	− 75	180	0.25	120

Increasing Signal-to-Noise

Other than modifications to the optics and the beam stop, there are several easy ways to lower the background for marginal crystals. The first is to decrease the width of each frame so that each reflection takes about three frames to diffract completely. The background is a continuous value as the oscillation angle changes, whereas the spot is not. Taken to the extreme, it easy to see how this helps. If each frame is 1.0° wide and the spot diffracts for 0.25° of this oscillation, then in the pixels containing the reflection, background will have accumalated for four times longer than the reflection counts. This will greatly decrease the signal-to-noise ratio. In tests on our Siemens area detector we have found that collecting 0.1° frames instead of 0.25° frames increased the $I/\sigma(I)$ ratio for weak high-resolution reflections beyond 2.0 Å by two times. A second method of reducing background is to pull the detector back. The background falls off as a square of the distance from the crystal (ignoring air absorption for now). To a first approximation the diffracted rays are parallel and do not decrease in intensity with distance. So doubling the distance from the crystal will decrease the background four times. Of course, it will also decrease the amount of data that can be collected in a single frame. If distances greater than about 15 cm are used, a helium path is necessary or the gains in background will be lost to air absorption. If you have many small crystals or your crystals are not radiation sensitive, an increase in signal can be had by pulling the detector back and collecting more crystal positions to make up for the lower reflections per frame.

In tests done by us at Scripps using a Siemens area detector, we have found that I/σ increases about 1% per centimeter for helium versus air for distances greater than 10 cm. So at a d of 20 cm, an increase of 10% in I/σ is expected. Hamlin-style detectors require greater distances, so helium paths are a must.

..... 2.10 .....
IMAGE-PLATE DATA COLLECTION

Image plates are relatively new but have the potential of becoming the data-collection method of choice. They have a high spatial reolution of 100–150 μm, similiar to film, and subtend a large angle so that more data are collected at once.[18] Image plates can be used either as an alternative to film or as a replacement for the detector in an area detector. In the film mode they

---

[18] Miyahara, J., *et al.* (1986). *Nucl. Instrum. Methods* **A246**, pp. 572–578.

can be used in the same cassettes that X-ray film is used in and scanned off-line up to several hours later. In the area-detector mode they are automatically scanned after each exposure by apparatus built directly into the machine. The dynamic range of an image plate is much higher than that of film—it can reach 12 bits for image plates, whereas film is limited to 8 bits in practice.[19]

Image plates are exposed with X-rays, as with any other detector, and the X-ray photon causes a chemical change in the plate coating that releases a fluorescence that is detected by a photomultiplier when scanned with light of the proper wavelength. Image plates are read out by a laser beam on a scanner. The quality of this scanner largely determines the limits of the image plate. The construction of a high-quality scanner is a technically difficult feat because of the mechanical precision needed and the high quality of the electronics needed to take full advantage of the image plate's capability. The photomultiplier must have low noise and use a high-quality analog-to-digital converter, and the laser used for scanning must be stable and hit precisely when scanned. Image plates have a wider range of sensitivity with respect to X-ray wavelengths, which gives them higher counting efficiency at higher energies. This makes them the detector of choice for white-radiation Laue experiments that use very bright synchrotron light sources. Because only a few exposures are needed for Laue data sets, manual handling of the plates is not a great disadvantage. For collecting data sets with monochromatic radiation where hundreds of exposures are needed, an automated method of scanning the plates is a necessity, such as the MAR scanner (Fig. 2.38).

Image plates are erased by exposing to white light. This means they can be handled in the room light before they are exposed to X-rays. After exposure they must be protected from light. Cosmic background radiation will slowly expose the plate, so they need to be freshly erased before using. Exposure to very bright X-rays such as the direct beam will cause a spot that will take a long time to erase and can even show up for several exposure/erasure cycles.

With the use of an image plate as a film replacement on a monochromatic X-ray setup, the data are collected using the rotation method as previously described. Software used for the analysis of film can be adapted easily by removing the corrections for film sensitivity, because image plates are linear. Since image plates are becoming common for use in other applications to replace film, such as radiography of gels, an image plate scanner may be available at your institution. These scanners are perfectly adequate for replacing film in preliminary characterization of crystals.

---

[19] This is based upon practical experience and is not a theoretical limit in either case.

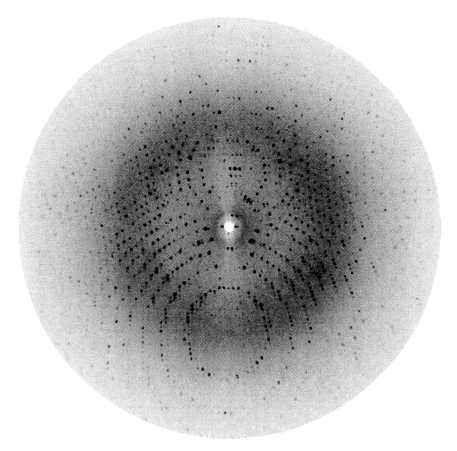

FIG. 2.38   Frame from a MAR research image plate. The crystal was rotated 1° about the horizontal axis for a 5-min exposure using a rotating anode source. The edge of the image is about 2.1-Å resolution.

····· 2.11 ·····
SYNCHROTRON-RADIATION LIGHT SOURCES

The term *synchrotron light* is misleading because the sources of synchrotron light are usually electron-storage rings. Synchrotrons are very different machines that are never used directly as X-ray sources. The first observation of synchrotron radiation was made using synchrotrons, and therefore the name. It is inaccurate to say "We are collecting data using a synchrotron," but the term has become so common that synchrotron is now synonymous with *synchrotron radiation light source* in protein crystallographic jargon. An excellent book on synchrotron sources and crystallography is by Helliwell (see Suggested Reading at end of the preface). There is only room here to touch on the subject and to point out areas of special interest.

Differences from Standard Sources

Synchrotron radiation as available at a storage ring has a continuous spectrum in the area of interest to protein crystallographers and is very bright (Fig. 2.39). Even after tight monochromatization where only a small fraction of the total energy is used, the sources are still up to two orders of magnitude brighter than the best rotating anodes. The tight monochromatization can mean lowered backgrounds and decreased radiation damage for the same exposure. Furthermore, the optics at storage rings is usually far superior

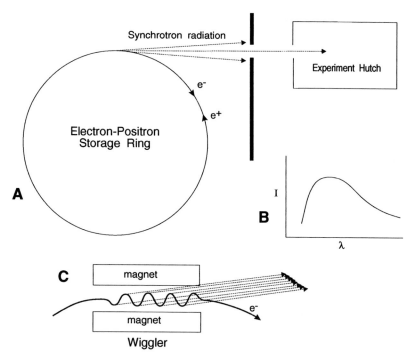

FIG. 2.39 Synchrotron radiation source. A storage ring has high-energy electrons held in orbit by bending magnets (A). As the electrons accelerate around the curve they emit synchrotron radiation (B). Because the beam is so intense, all experiments are done in shielded hutches that are interlocked so that personnel cannot be inside while the shutters are open. A wiggler (C) is a method of increasing the brilliance of the X-rays by combining several beams from local excursions of the electron path.

to anything used in laboratory sources providing tightly collimated high-intensity beams. One reason for the better optics is that the source is located meters away instead of within less than a meter, giving effectively a parallel source. The combination of brighter, tighter optics makes synchrotron sources the best for very large unit cells such as are found in viruses with cells from 300 to 1000 Å.

In our experience with many different crystals we have always found an increase in signal-to-noise at storage rings. The ability to tune the wavelength allows the use of more optimal energies. Wavelengths near 1.0 Å show very little absorption by the capillary and the solvent around the crystal, allowing wetter mounts while obtaining better signal-to-noise ratios. Corrections due to absorption are minimized. We have not found that harder radiation decreases lifetimes; in fact lifetimes are longer since the absorption that causes free-radical damage is more efficient at lower energies.

Special Synchrotron Techniques

The simutaneous availability of all wavelengths has led to the development of white-radiation Laue photography. Exposure times for Laue photographs can be very short—10 ms exposures for a typical lysozyme crystal at the best sources—and yet contain almost all the diffraction information in one or two photos. Furthermore, since most of the factors that need to be corrected for in reducing the data are a function of wavelength, especially absorption, the presence of the same reflection measured at different wavelengths in the same data set (Fig. 2.40) allows these parameters to be accurately accounted for by least-squares scaling. Moffat and co-workers have collected lysozyme data sets that compare favorably with data collected a diffractometer.[20]

The brightness of the source and high-quality optics make the storage ring an ideal place to collect data on very large unit cells as are found in viruses.

Time-resolved Data Collection

The short exposure times needed at the storage rings has allowed the collection of time-resolved protein crystallographic data. Reactions are initiated by laser flashing or in flow cells (for very slow reactions) and then data are collected by white-radiation Laue photography at appropiate time

[20] Temple, B., and Moffat, K. (1987). In "Computational Aspects of Protein Crystal Analysis. Proceeedings of the Daresbury Study Weekend, DL/SCI/R25," (Helliwell, J. R., Machin, P. A., and Papiz, M. Z., eds.). pp. 84–89. See also, Helliwell, J. R., *et al.* (1989). *J. Appl. Crystalogr.* **22**, 483–487.

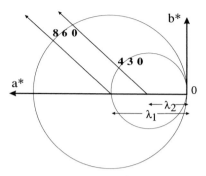

FIG. 2.40 How overlaps arise with white radiation. In white-radiation Laue photography a range of wavelengths is used simultaneously, and thus there are many Ewald spheres in diffracting conditions simultaneously. Two are shown here that differ in wavelength by a factor of two. The resulting diffraction exits the crystal in the same direction and is recorded on the detector in the same spot, leading to an energy overlap.

points. Using an undulator to intensify the beam and white-radiation Laue photography on storage phosphors, the exposure time can be as short as microseconds.

One of the chief difficulties can be the relatively high concentration of a protein crystal. This makes it difficult to deliver enough substrate, and the optical density can be very high. If light is to be used to start a photoreaction, high absorption necessitates intense light sources and causes gradients across the crystal. It is better to illuminate off of the absorbance peak at a position where the crystal is still transparent to the light so that light can get to all of the crystal volume. These experiments are, therefore, technically demanding and must be done carefully to ensure that most of the crystal is synchronized or the time resolution will be lost.

Reactions can also be started by diffusing in substrates using a flow-cell apparatus. In this case, the reaction must be very slow—on the order of hours—or else the diffusion time will be greater than the reaction time and the reaction will not be synchronized across the crystal.

····· 2.12 ·····
DATA REDUCTION

Integration of Intensity

Integration of the intensity in a spot is a matter of separating the background counts from the reflection. Two methods are in general use:

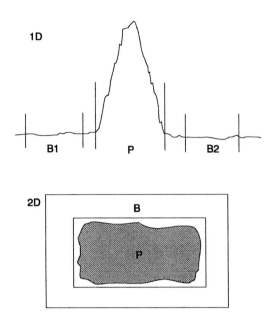

FIG. 2.41   Integration by masking in one and two dimensions.

1. Mask and Count.   In this method the region that is to be considered the spot is masked and the pixels within this region are summed (Fig. 2.41). The background is determined from the pixels adjacent to the spot and this value is subtracted to give the final intensity. The method works well when spots are well above background. The pixels can be counts in the case of counters, as in diffractometers and area detectors, or optical densities in the case of film.

$$Ihkl = \sum_{1}^{nP} counts P - \sum_{1}^{nB} counts B$$

2. Profile Fitting.   In profile fitting a curve is fit to the data and the area under the curve is taken to be the intensity (Fig. 2.42). The curve, or profile, can either be a geometric shape such as a guassian or it can be derived by averaging over the brighter spots. The advantage of the latter method is that bright reflections can be used to determine the profile, which is then applied to weak reflections. Different profiles are usually used depending upon the position of the spot on the detector. For example, the detector might be separated into a 4 × 4 array and a different profile used in each of the 16 areas. Then, in order to find the area of the spot, this curve is best-fit to the counts found in the area where the spot is predicted to be, and the area

FIG. 2.42   Profile fitting. Profile fitting can more accurately find the intensity of a peak—especially in the example on the right, where the background is sloped.

under the curve is then used to find the integrated intensity rather than the counts themselves.

Error Estimation

An accurate estimation of the error is important. The error of a single reflection is termed its σ.

Contributions to σ:

- Counting statistics: $\sigma_{counts} = \sqrt{NPeak + NBackground}$ (Note the inclusion of background counts)
- Instability of detector; usually a constant
- Profile fitting: deviation from observed and ideal shape
- Local variation in background

Other sources of errors are saturated pixels (photographic film is especially vulnerable to this), overlapped profiles, and errors in background models.

Merged multiple measurements of several reflections should be weighted by σ. Different data reduction packages will determine different σ, and the data are probably better averaged without σ weighting. In the author's experience the σ of some packages can differ by at least a factor of two. Reflections are often rejected by the ratio of intensity to σ, $I/\sigma(I)$.

Polarization Correction

This correction arises from the dependence of scattering efficency as a function of scattering angle. For polarized sources, the scattering efficency is also a function of the change of polarization direction with the angle of the scattering plane. Sources can be polarized by a monochromator, so this correction is dependent upon the optics of the source used. For unpolarized radiation,

$$p = 1/2(1 + \cos^2(2\theta))$$

Lorentz Correction

This correction accounts for the rate with which a reflection passes through the Ewald sphere. Reflections near the rotation axis remain in diffracting conditions for a longer time. At some point this correction becomes so large that the reflections very close to the rotation axis are rejected.

Decay or Radiation Damage

Prolonged irradiation of a sample induces radiation damage. Decay usually affects higher resolution reflections faster. If there is a choice, the higher-resolution data should be collected first. Decay should be monitored by collecting a set of standard reflections. If the decay exceeds about 80%, data collection should be halted. Although a decay correction can partially account for decay, different reflections can decay at different rates so that a single decay parameter cannot restore the accuracy of the data set. Radiation damage can be reduced by lowering the temperature. This slows down the free-radical chain reactions that are thought to induce radiation damage. Decay is a function of time and dose. However, it is not linear with dose, and brighter sources can collect more counts before the same amount of decay sets in. This is a great advantage of synchrotron sources. Also, once a sample is irradiated the free-radical chain reactions are initiated and will continue even after the beam has been off for some time. Irradiation affects samples at different rates, and some samples are very sensitive. The presence of a metal that absorbs X-rays more efficently, such as iron, platinum, or mercury, can speed up decay.

Absorption

This is probably the largest source of uncorrectable error in data sets. The path length of the diffracted X-rays through glass, crystal, solvent, and air determines the amount of absorption. This path length is different for each reflection. Unfortunately, there is no entirely accurate way to model this absorption. Two approaches are generally used. In the first, experimental measurements of the absorption in different directions through the sample are made and each reflection is corrected by these factors. In the second method, a least-squares fit is made to the differences between symmetry-related reflections as a function of some parameter believed to be a function of absorption. The experimental correction is easily calculated in the case of a diffractometer. In the case of two-dimensional detectors, the second method is normally used.

The overall error in a data set can be estimated by comparing symmetry-related reflections, which in the ideal case would be indentical. The reflections are calculated as

$$R_{symm} = \frac{\sum_{h} (I_h - \overline{I_h})}{\sum \overline{I_h}}.$$

3

COMPUTATIONAL
TECHNIQUES

There are several different crystallographic software packages available and it would be impossible to cover them all. The XtalView package is used for specific examples in this book. XtalView is a window-based visually oriented package that is especially easy for novices to learn. Options and commands are shown as buttons, sliders, and menus. All options are visible, making it easy to spot them and ideal for publishing in book form. You may not want to use XtalView—perhaps you already have a favorite package. In any case, most programs have similar options and features. For consistency, a single package is necessary for this book so that we can get right to explaining the methods and spend less time explaining the particular implementation.

XtalView was written at The Research Institute of Scripps Clinic by the author. It runs under X-windows, which is available on most workstations (Figs. 3.1 and 3.2). At present it has been ported to Sun workstations, (including the SparcStation series), Silicon Graphics, and DECstations running ULTRIX. Ports are underway for the Stardent machines and the Evans and Sutherland workstations. The package does not include data collection or protein refinement. For these XENGEN and XPLOR will be used as the primary examples.

A

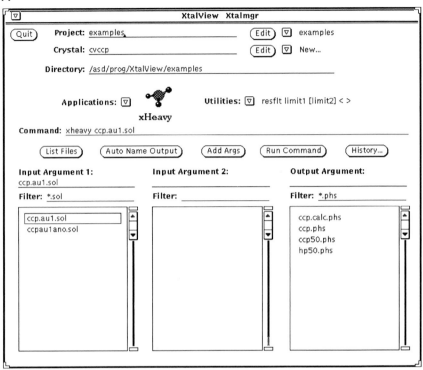

FIG. 3.1   XtalView Xtalmgr. (A) The XtalView Xtalmgr program is used to organize data and to start the individual applications. It has a graphical-user interface using buttons, pull-down menus, and scrolling lists to present a more intuitive interface. Data are organized into **Projects** which can be entered and edited using the field at the top of the window. The **Crystal** field is a key word used to access the parameters for a specific crystal type such as the unit-cell parameters and the the space-group symmetry operators. Other applications are selected from a pull-down menu (not shown) accessed from the **Applications** glyph. Selecting an application causes all files with the correct extension to be listed in one of the three file lists at the bottom. Files can be selected from these lists by clicking on them with the mouse. The command line is then built up using **Add Args** and then the application is started with **Run Command.** (B) The crystal editor is used to enter the unit-cell parameters, space-group information, and any other relevant information. The space-group symmetry operators for all space groups are kept in a table and can be accessed either by space-group number or by symbol as found in the *International Tables for Crystallography* Vol. 1. The information entered into the editor is then available to all XtalView programs by simply entering the crystal keyword (cvccp in this example).

B

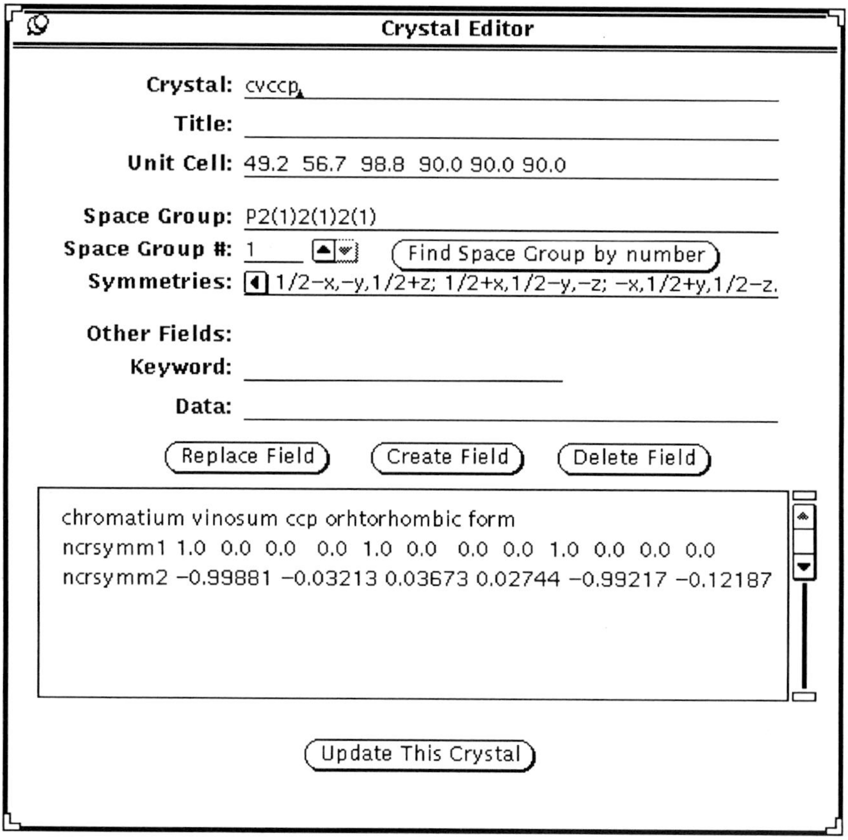

Crystal Editor

Crystal: cvccp

Title: _____

Unit Cell: 49.2 56.7 98.8 90.0 90.0 90.0

Space Group: P2(1)2(1)2(1)

Space Group #: 1   ▲▼   (Find Space Group by number)

Symmetries: ◀ 1/2−x,−y,1/2+z; 1/2+x,1/2−y,−z; −x,1/2+y,1/2−z.

Other Fields:

Keyword: _____

Data: _____

(Replace Field)   (Create Field)   (Delete Field)

chromatium vinosum ccp orhtorhombic form
ncrsymm1 1.0 0.0 0.0   0.0 1.0 0.0   0.0 0.0 1.0 0.0 0.0 0.0
ncrsymm2 −0.99881 −0.03213 0.03673 0.02744 −0.99217 −0.12187

(Update This Crystal)

FIG. 3.1—*Continued.*

····· 3.1 ·····
TERMINOLOGY

Reflection

A reflection is a single X-ray-diffraction vector that is the combined scattering resulting from the individual scattering of all of the electrons in the unit cell along a particular direction. It has a magnitude, $|F|$, that is referred to as F, a phase, α, and the Miller indices, h, k, l. The diffraction vector for protein is called F_P and the same diffraction vector with a heavy atom soaked in is F_{PH}. The diffraction vector for the heavy atom alone is fh. (The lower

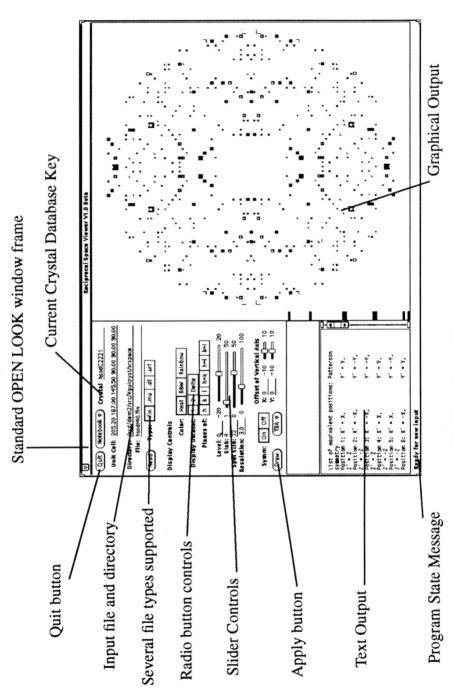

Standard OPEN LOOK window frame

Current Crystal Database Key

Graphical Output

Quit button

Input file and directory

Several file types supported

Radio button controls

Slider Controls

Apply button

Text Output

Program State Message

FIG. 3.2   XtalView application dissected.

case is to remind us that fh can never be directly measured but is always calculated.) Two separate observations of the same reflection will be called F_1 and F_2.

Resolution

Resolution is a loosely used term in protein crystallography. It usually refers to the minimum d-spacing (defined in the following) in the diffraction data set in question. It is usually assumed that the data are fairly complete to this resolution. A data set with one reflection at 1.7 Å and all the rest with d-spacings greater than 2.0 Å is really a 2.0-Å data set, not a 1.7-Å data set. In this book, high resolution is taken to mean a smaller number, usually in the range of 2.5 Å or less, low resolution is taken to be larger numbers, usually in the range of 100 to 4 Å, and medium resolution is usually around 3 Å. Not all crystallographers use these terms in a consistent way; it is important to keep the context in mind. Subjectively, a smaller minimum d-spacing implies more data and higher resolution and is, therefore, thought of as "better." Thus the phrase "this structure has been to determined to a better resolution" means a higher resolution or a lower minimum d-spacing.

Resolution is also commonly used to refer to the minimum distance that can be resolved. In a map with good phasing (low noise) the minimum distance that can be resolved is about 0.92 of the highest resolution. Thus, in a 1.6-Å map two carbon atoms 1.5 Å apart are just resolvable. Resolutions where individual atoms can be resolved are referred to as "atomic" resolution. Resolutions where not even the secondary structure can be seen, somewhere lower than 10 Å, are affectionately called "blobology," since that is the appearance that a typical protein has at very low resolutions.

Another important effect of resolution is that as the resolution increases so does the number of data. This is especially important in refinement as it determines the ratio of observed data to refinable parameters. Around 2.0 Å, depending upon the exact protein, it becomes possible to refine x, y, z and B values independently as the ratio of observations to refined parameters becomes greater than 1. Some properties of radial shells for a typical protein are listed in Table 3.1. The table shows the number of observed reflections over the number of parameters to refine the protein atom positions and the ratio of observed to parameters for a typical heavy atom refinement.

Resolution is often expressed in Å units, as mentioned above. Resolution is related to the scattering angle θ by the rearranged Bragg equation:

$$d = \frac{0.5}{\sin(\theta)/\lambda} \tag{3.1}$$

The equation is rearranged to show the relationship between resolution and $\sin(\theta)/\lambda$, which is often used. $\sin(\theta)/\lambda$ is a wavelength-independent quantity,

TABLE 3.1

Radial Shells of Resolution

Shell no.	d	2θ	No. reflections (s)	No. centric	Percent centric	N/22,500[a]	N/20[b]
0.5	7.92	11.17	500	229	46	0.02	25
1	6.28	14.10	1,000	365	37	0.04	50
2	4.99	17.78	2,000	577	29	0.09	100
3	4.36	20.38	3,000	756	25	0.13	150
4	3.96	22.46	4,000	917	23	0.18	200
5	3.67	24.21	5,000	1068	21	0.22	250
6	3.46	25.76	6,000	1201	20	0.27	300
7	3.28	27.14	7,000	1337	19	0.31	350
8	3.14	28.40	8,000	1459	18	0.36	400
9	3.02	29.57	9,000	1577	18	0.40	450
10	2.92	30.65	10,000	1687	17	0.44	500
12	2.75	32.62	12,000	1902	16	0.53	600
15	2.55	37.72	15,000	2212	15	0.67	750
20	2.32	38.90	20,000	2673	13	0.89	1000
25	2.15	42.04	25,000	3112	12	1.11	1250
30	2.02	44.81	30,000	3526	12	1.33	1500

Note. One thousand reflections per shell for a protein with an orthorhombic unit cell of 69.75 × 77.42 × 87.79 Å.

[a] Ratio of observed to parameters for protein refinement assuming 7500 atoms and three parameters: x, y, z.

[b] Ratio of observed parameters for a typical heavy-atom refinement assuming four sites with five refineable parameters: x, y, z, occupancy, B-value.

and its numerical value gets larger as the resolution increases. The wavelength, λ, is included in order to make the value wavelength independent. This makes statements like "a low-resolution 5-Å map is harder to interpret than a high-resolution 2-Å map" read as "a low-resolution $\sin(\theta)/\lambda = 0.1$ map is harder to interpret than a high-resolution $\sin(\theta)/\lambda = 0.25$ Å map." The ångstrom unit is a common measurement in chemistry and means something to crystallographers as well as non-crystallographers. However, it is not an approved international unit of measure, being equal to 10^{-10} m, or 0.1 nm, but since protein bonds are on the order 1 Å, it does make sense to use. This book will not break from tradition and will continue to use the ångstrom. Be aware, though, that some journals have banned it use.

Coordinate Systems

Two main coordinate systems are used in crystallography: fractional coordinates and Cartesian coordinates. Fractional coordinates are expressed as fractions of the unit cell in each of the three directions a, b, c separated by the angles α, β, γ. Fractional coordinates are parallel to the crystallographic axes and thus are not necessarily at right angles to each other. Thus the origin of the unit cell is (0, 0, 0), and the point one cell edge away in b is (0, 1.0, 0). The point (0.5, 0.5, 0.5) is in the center of the cell regardless of the cell dimensions and shape. In any space group, 1.0 or −1.0 can be added to any fractional coordinate without changing the meaning of the coordinate because of the infinitely repeating nature of crystal lattices. Thus, the point (−2.3, 1.4, 0.3) is the same as (0.7, 0.4, 0.3). Certain operations such as comparing symmetry relations are more easily done in fractional space. It is obvious in fractional space that (0.5, 0, 0) and (1.5, 0, 0) are the same point. Usually Cartesian space is used for atomic models, especially since this makes it easier to compare bond angles and lengths.

Cartesian coordinates are expressed as three mutually perpendicular axes, x, y and z, with ångstroms as the units. Coordinates in the Brookhaven Protein Data Bank (PDB) are reported in Cartesian coordinates and the files contain a 3 × 3 matrix for converting the Cartesian coordinates to fractional coordinates. The origins of both coordinate systems are the same point. Crystallographic calculations, such as Fourier transforms, use fractional coordinates, so it is necessary to convert from Cartesian coordinates, which are the best for display and geometry purposes. If the space group of the crystal is orthogonal, tetragonal, or cubic, then there is a unique way to superimpose the two coordinate systems: a along x, b along y, and c along z. When the space group has a non-90° angle, there can be several ways to superimpose axes. It is necessary then to use the matrix given in the PDB file to perform the Cartesian to fractional transformation. For instance, in monoclinic space groups, where γ is non-90°, there is the choice of aligning a with x or c with z, and b will always be along y. These two possibilities are referred to as abc* and a*bc, respectively (Fig. 3.3). Once the matrix for one of the trans-

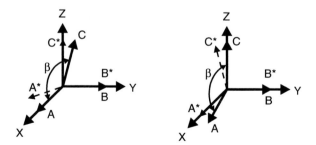

FIG. 3.3 Two ways to superimpose axes abc* and a*bc. On the left, the Cartesian coordinates x, y, z have superimposed upon the crystallographic axes abc such that a and b superimpose with x and y. Since β is not 90°, c cannot be superimposed but, instead, c* is superimposed on z, and thus this superposition is labeled abc*. On the right, the other possibility is shown with a*, b, and c superimposed on a*bc.

forms, Cartesian to fractional or fractional to Cartesian, is known, the matrix can be inverted to obtain the matrix for the opposite transformation. There is no standard method to relate fractional and Cartesian coordinate systems, and this is a common cause of confusion. For instance, if a rotation-translation solution is written out in $a*bc$, an attempt to refine it in a program that uses $abc*$ will give a random R-factor, leading the crystallographer to believe that he has the wrong solution when everything is fine!

Patterson space has its own coordinate system analogous to fractional coordinates except that the terms u, v, w are used for a, b, c. Reciprocal space has the coordinate system h, k, l, also referred to as the Miller indices. This coordinate system is related to the fractional system in an inverse manner.[1] In addition, there are many other coordinate systems: polar coordinates, Eularian angles, camera coordinates, and so forth. In this book we will avoid these additional systems.

R-Factor

R-factor is a formula for estimating the error in a data set. It is usually the sum of the absolute difference between observed and calculated over the sum of the observed:

$$R_{crystallographic} = \frac{\Sigma|F_o - F_c|}{\Sigma|F_o|} \qquad (3.2)$$

If two random data sets are scaled together, then the R-factor for acentric data is 0.59 and for centric data it is 0.83. The R-factor, often called just the R, is ubiquitous in protein crystallography and is probably given more weight than it deserves because it turns a rich wealth of detail in three dimensions sampled at thousands of points into a single number. It is dangerous to use a low R-factor as a guide to the correctness of a structure, although a high R-factor is a good guide of the incorrectness of a structure. One problem with using the R-factor as a guide is that it has led to a tendency to over-refine models and to "drive-in" phase bias by using the power of least-square algorithms. A good least-squares minimizer can always lower the R-factor by modeling the error as well as the data. Because of the nature of the Fourier transform, every point on one side of the transform contributes to every point on the other side so that it is possible to model the error on one side by subtly adjusting the points on the other side. It is also possible to lower the

[1] Stout and Jensen, pp. 26–33 (see Suggested Reading list in Preface for complete reference) has an informative discussion on reciprocal space and its relationship to real space. Especially nice are the figures showing the relationship of the real axes to their reciprocal counterparts.

R-factor by deleting observed data and by raising the number of parameters. For evaluation of an *R*-factor it is important to ask how many parameters were used and what percentage of the data were deleted and by what criteria. Knowing these, direct comparisons of R-factors can be a useful way to compare the relative accuracy of a model or data set. The relationship between R-factors and errors is discussed further in Section 3.10, Evaluating Errors.

Space Groups and Symmetries

Crystals are by definition symmetric arrays of molecules in three dimensions and must fall into one of the 230 known space groups that describe all the possible ways identical objects can pack in three dimensions. Of these 230 space groups, only 65 are possible for chiral objects, such as proteins, which are made up of L-amino acids. Crystals are built up from unit cells that are repeated by translation in each of three directions. The edges of this unit cell are denoted by *a, b, c,* and the angles between them are α, β, γ.[2] Note that the direction of translation does not have to be on a Cartesian coordinate system but can have any angular value. The unit cell is made up from one to several asymmetric units, arranged by symmetry elements. The symmetry elements are expressed mathematically as symmetry operators. By applying the symmetry operators to one asymmetric unit to form a unit cell and then repeating the unit cell in the three directions specified by α, β, γ, the crystal can be reconstructed. If a computer program is space group general, then it can read a list of the symmetry operators and perform the correct calculation. If a computer program is space group specific, then the symmetry operators of one or a few space groups are hardwired into the program and it can only be used for calculations in these space groups. XtalView is completely space group general, and it has a library file that contains all 230 space groups that can be accessed from the crystal-edit function of Xtalmgr. These can be overridden if desired to put in non-standard settings of known space groups, or even to enter impossible space groups, by entering the symmetry operators in algebraic form as found in the *International Tables for Crystallography,* Volume 1.

Most space groups have several different choices of origin. That is, several positions in the unit cell can be chosen as the point 0, 0, 0 to yield identical crystalline patterns that are shifted by a constant amount. In orthorhombic space groups, for instance, the origin can be placed at 0, 0, 0 or 1/2, 0, 0 or 0, 1/2, 0 or 0, 0, 1/2 or 1/2,1/2, 0 or 1/2, 0, 1/2 or 0, 1/2, 1/2, or

---

[2] For a complete description of space groups and symmetry see "The International Tables for Crystallography," Vol. 1 or Vol. A, (T. Hahn, ed.), D. Reidel Publishers, or any of a number of textbooks on crystallography.

1/2, 1/2, 1/2. One confusion that can arise from this is that two sets of coordinates can be on different origins and appear to be different solutions when in fact they are identical.

Matrices for Rotations and Translation

Symmetry operators can be expressed more generally in the form of matrices, as can any rotation and translation combination in the form

$$\mathbf{x'} = \mathbf{Rx} + \mathbf{t}, \tag{3.3}$$

where $\mathbf{R}$ is a 3×3 matrix and $\mathbf{x}$ and $\mathbf{t}$ are length-3 vectors. The vector x is transformed to form a new vector x'. In this way the arbitrary symmetry operator $x + y$, $2/3 + x$, $1/4 - z$ may be expressed as

$$\begin{pmatrix} x' \\ y' \\ z' \end{pmatrix} = \begin{pmatrix} x \\ y \\ z \end{pmatrix} \begin{pmatrix} -1 & 1 & 0 \\ 1 & 0 & 0 \\ 0 & 0 & -1 \end{pmatrix} + \begin{pmatrix} 0.0 \\ 2/3 \\ 1/4 \end{pmatrix} \tag{3.3a}$$

An example of a program segment applying the above matrix operation can be found in Appendix A, program 2.

A common problem in crystallography is to rotate from a starting position to a new position about one of the three laboratory axes, ϕx, ϕy, ϕz. The matrices for these transformations are:

$$\Phi_x = \begin{pmatrix} 1 & 0 & 0 \\ 0 & \cos \phi_x & -\sin \phi_x \\ 0 & \sin \phi_x & \cos \phi_x \end{pmatrix} \tag{3.4}$$

$$\Phi_y = \begin{pmatrix} \cos \phi_y & 0 & \sin \phi_y \\ 0 & 1 & 0 \\ -\sin \phi_y & 0 & \cos \phi_y \end{pmatrix} \tag{3.5}$$

$$\Phi_z = \begin{pmatrix} \cos \phi_z & -\sin \phi_z & 0 \\ \sin \phi_z & \cos \phi_z & 0 \\ 0 & 0 & 1 \end{pmatrix} \tag{3.6}$$

If the vector with the coordinates is $\mathbf{X}$, then the new coordinate $\mathbf{X'}$ is found by

$$\mathbf{X'} = \Phi_x \Phi_y \Phi_z \mathbf{X} \tag{3.7}$$

B-Value

The B-value is a measurement of the displacement of an atom from thermal motion, conformational disorder, and static lattice disorder. This vibration will smear out the electron density and will also decrease the scattering power of the atom as a function of resolution.[2a] The displacement for an isotropic B-value is related to the displacement u by the equation

$$B = 8\pi \overline{u^2}. \tag{3.32}$$

..... 3.2 .....
BASIC COMPUTER TECHNIQUES
File Systems

As your projects grow you will gain many files. It is important to keep these files organized in a sensible way or they will soon get out of hand. I have seen some users keep all of their files in their home directory until the list became longer than the directory command could process. A better approach is to put all the files for a given protein in a subdirectory. Appropriate subdirectories can also be made for groups of files with a common purpose such as phasing, refinement, or fitting. The use of subdirectories in a tree structure will allow all the files to be written off-line to a single tape to be stored. Even if you start off with several gigabytes of space, it will eventually all be filled if files are not stored off-line and deleted occasionally.

To keep track of files, a naming convention should be followed. Most files are given an extension to indicate the file type. The root of the file name should be a description of the contents. My convention is to use a short three- or four-letter descriptor of the protein; a one- or two-letter descriptor of the crystal, whether it is native or derivatized (n for native or the element name of the heavy atom); and a number designating the particular crystal. For instance, photoactive yellow protein was called yp and a file containing the native data from crystal number 5 in .fin format would be called ypn5.fin. If this information is merged with mercury derivative crystal number 1, yphg1.fin, to form a .df file, the new file would be ypn5hg1.df.

Portability Considerations
Avoid Space Group Specificity

While it may seem easier to write space-group-specific code, it usually turns out to be more work in the end. Quick-and-dirty programs have a way of growing beyond your original expectations and becoming a program you rely on daily. If a problem is worth solving, it is worth doing it right. Take

---

[2a] The equation for this falloff is $\exp(-Bs^2)$, where s is $\sin(\theta)/\lambda$ or $0.5/d$.

the extra time to make the program space group independent. It is much harder to modify a program later; when you look back at the code it can be difficult to decipher its meaning. Often space group specific code is faster. However, modern computers are so fast that the difference between no time and no time is still no time, whereas it takes humans relatively a much longer time to alter the code. For example, if you multiply a symmetry operator matrix, most of the elements will be zero. A small amount of time may be saved by writing the code assuming zeroes, but later the code will have to be changed to put these terms back in. The time it takes to rewrite the code later will probably add up to more time than the computer uses doing these extra multiplications in all of the runs of the program put together. (Actually, a smart math co-processor will recognize the zero and return zero without actually doing the multiplication).

Avoid Binary Files

At first glance, binary files seem to be a good idea. Instead of saving data as an ASCII equivalent, files are stored in their computer representation. This saves some space and speeds up I/O. However, binary files cannot be read by humans, so there is no way to tell what is in them. And computers can only read them if the order of bytes in the file is precisely known and whether they represent a float, an integer, characters, a double, etc. Unless the format is precisely specified, binary should not be used. Another problem with binary is that it is not portable between machines. Even worse, there is no universal standard for FORTRAN binary; different implementations use different headers and trailers on the records. Since the only constant in the world of computers is change, you may find in 2 years that your hot new machine cannot read binary files from your last machine.

Scratch files can be binary because they are not meant to stay around and binary I/O is faster than formatted. Also, for very large files, such as electron-density maps, binary may be the most practical alternative in terms of size and speed. Electron-density maps can always be recalculated from primary phases, so saving them is usually unnecessary. With the speed of modern computers, it is usually faster to Fast-Fourier-Transform a map than to read it in from a disk file, even if it is binary!

Language Extensions

Computer language extensions often seem to be temptations designed to make users dependent on a particular brand of computer. Avoid them. Since change the only constant, your extensions may become a headache when you have to update some old code that refuses to run on your new system.

Setting up Your Environment

UNIX users need to set up two files in their home directory: `.cshrc` and `.login`. The `.cshrc` file is executed every time a new shell is started and the `.login` file is executed once when you first log into the system. VMS users have a similar file called `login.com`. In order to use XtalView it is necessary to put a few lines into your `.cshrc` file, as outlined in Appendix B. Especially important is setting up your path variable, which determines where the operating system will look for commands. You can share executables with other members of your group by including the proper directories in your path. Also useful is the aliasing of common commands or groups of commands to a short word. For instance, XENGEN users will find the following alias useful: *alias ypset source ~dem/yp^$1/ process^$1.cmd*. This is used *ypset crystalname* to source the appropriate command file to set the environment variables for a particular crystal. Also, alias rm to `rm -i` and copy to `cp -i` to prevent overwriting files accidentally. Alias RM to `/bin/rm` and CP to `/bin/cp` to override the prompting when you are sure you want to delete the file(s). A related command is *set noclobber* that prevents overwriting of files with redirection commands.

····· 3.3 ·····
DATA REDUCTION AND STATISTICAL ANALYSIS

Evaluation of Data

R-Symm

R-symm is an internal measure of the accuracy of a data set. It compares the differences between symmetry-related reflections that should be identical in intensity. In the statistical world, these should be two independent measures of the same number, and the variance of symmetry-related reflections should reflect the variance of the entire data set. In the real world, symmetry-related reflections are usually not completely independent—they usually occur when the crystal is in similar geometry. However, R-symm does give a minimum variance—it is unlikely that the data are more accurate than the R-symm. In practical terms, data with an R-symm less than 0.05 are good data, less than 0.10 the data are probably usable, and above 0.20 the data are of questionable value. Two data sets that have no relationship to each other will have a theoretical R-symm of 0.57 simply by scaling them together. Random data scaled in bins with multiple scale factors can often reach an R-symm of about 0.35. By rejecting outliers, the R-symm can be lowered even further to the point where one might think the two data sets are related when they are from different crystals or the data were misindexed.

Ratio of Intensity to Sigma of Intensity

The sigma of a reflection intensity is an estimate of the accuracy of an individual measurement as opposed to the accuracy of the data set as a whole. It is a combination of counting statistics, background height, background variance, and the number of times the reflection was measured. In the author's experience, no two data-reduction packages will produce the same sigma, so direct comparison of data sets on the basis of sigma should be avoided unless some yardstick is available. The ratio of the amplitude of the reflection to the sigma of the amplitude ($F/\sigma(F)$) is often used as a rejection criterion. If this ratio is less than approximately 2/5, a reflection is often rejected as being unreliable.

Anomalous Scattering Signal

Many protein crystals contain a prosthetic group with anomalous scattering properties with CuK_α X-rays and, in addition, many heavy atoms used for derivitization have anomalous scattering. One test for the presence of an anomalous signal is to compare the R-symm of the centric data, in which the anomalous signal cancels, and the R-symm of the acentric data. The anomalous signal manifests itself as a breakdown of Friedel's law, F^+ is equal to F^-, so that a difference is found.[3] Even though the effect is often small, the Bijvoet pairs (Friedel pairs where the equality no longer holds) can be measured from the same crystal, and this allows measuring a smaller signal than if the measurements are from different crystals, as in the isomorphous-replacement case.

Anomalous scattering arises when the energy of the incident radiation is close to the resonant frequency of the tightly bound inner shell electrons of an atom. A simple experiment that demonstrates this effect is easily performed. Tape a lightweight Styrofoam ball onto the end of a 2.5-foot piece of yarn. With your arm outstretched, dangle the ball in front of you. Move your hand slowly back and forth and then gradually increase the speed. At slow speeds, where the driving energy (your hand) is well below the resonant frequency, the ball will follow your hand in phase. As the speed increases it will reach a speed where the motion of the ball is mostly in the vertical direction (the imaginary direction) and the sideways motion (the real direction) is minimal. As the speed increases further, the ball will become out of phase with your hand, being on the opposite side of each swing. The physical meaning of the term $\Delta f'$ (the change in the real part of the scattering) and $\Delta f''$ (the change in the imaginary part of the scattering) can be explained in terms of

---

[3] Stout and Jensen, pp. 218–222.

this mechanical model. The term $\Delta f''$ is the vertical movement of the ball at the point of resonance and $\Delta f'$ is the dampening of the range of oscillation in the horizontal movement. The sign of $\Delta f'$ is negative since the term represents an absorption of energy. The total scattering is then:

$$f_\lambda = f_0 + \Delta f'_\lambda + \Delta f''_\lambda \tag{3.8}$$

Percentage Completeness

This is the number of unique reflections measured over the number of unique reflections possible. The amount of completeness needed is dependent upon many factors, but above 80% complete, data are sufficient for almost all purposes. Data missing a single region are worse than data missing in random positions. If there are some data in all directions, then a lower completeness can be tolerated.

Filtering of Data

It is often important to filter out unreliable data and outliers that can skew results. The $F/\sigma(F)$ criterion has already been mentioned. If there are two measurements being compared, then the ratio of $(F1 - F2)/[(F1 + F2)/2.0]$ (the difference over the average) is very useful. If this ratio is greater than 1, then one of the measurements may be an outlier.

Merging and Scaling Data

Merging data refers to finding reflections with common indices and placing them together. Merging of data sets is desired for isomorphous phasing, for comparing two data sets, and for combining mutant and native data for difference-map purposes. *Scaling data* is the operation of setting the sum of one data set to the other. In the ideal case, a single overall scaling factor is sufficient to scale two data sets. Scaling can also be much more complicated. Systematic errors in the data due to absorption and decay mean that different parts of the data require different scale factors. Scaling is often done in bins based on resolution. Anisotropic scaling uses six scaling parameters derived by least-squares fitting. Local scaling scales together data from the same local region of reciprocal space and scales each reflection individually.

Resolution-Bin Scaling

The size of bins in scaling requires careful consideration. In the best circumstances only one scale factor is needed. To overcome systematic errors, the data are usually divided into small bins that are scaled individually. If the bins are too small, the scale factor becomes less accurate. At the extreme, if a separate bin is used for each reflection, then the two scaled data sets become equal and any information is lost. It is usually necessary to pick bins that have at least 100 reflections each. In single-parameter scaling, a single parameter is used to scale the data in each bin such that

$$\Sigma |F_1| = \Sigma |F_2| \times \text{scale} \tag{3.9}$$

A problem with bin scaling is that there is an abrupt transition at the zone between bins. To overcome this, a continuously varying scale factor can be used by fitting a line to the scale factor found in the bins or by interpolating between bins.

Anisotropic Scaling

Anisotropic scaling fits six parameters to the data to find a continuously varying scale factor in three dimensions along the reciprocal space lattice. The six directions are along the principal axes, h, k, and l, and along $h \times k$, $h \times l$, and $k \times l$ by fitting an equation of the form

$$s = h * h * a11 + k * k * a22 + l * l * a33$$
$$+ h * k * a12 + h * l * a13 + k * l * a23 \tag{3.10}$$

to the differences between the two data sets to be scaled. If a11, a22, and a33 are equal to 1.0 and a12, a13, and a23 are equal to 0.0, this is equivalent to a single uniform scaling parameter of 1.0. Finally, the two techniques, bin scaling and anisotropic scaling, can be combined. The data are broken up into large bins within which an anisotropic scaling is used. This is the method used by the XtalView program, Xmerge (Fig. 3.4). The data are broken up into bins of resolution and within each bin of resolution a six-parameter anisotropic scale factor is applied. This method is especially useful for heavy-atom data, although it is common for low-resolution, heavy-atom data to have larger differences due to the presence of the dissolved heavy atom in the solvent region. Thus, it is desirable to scale these data separately from the higher resolution data so that they are not adversely affected by the very large differences in the lower-resolution data.

The smallest number of reflections included in each bin should be about 100–200 reflections, with 500 being a good average number so that the six anisotropic scaling parameters are well over determined. Too few reflections will result in over fitting the data. To divide data into equal bins of resolution, the bins should be divided into equal segments of $(\sin(\theta)/\lambda)^3$ between the minimum and maximum values. This will put approximately the same num-

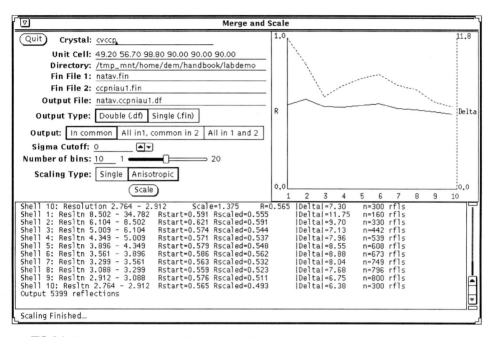

FIG. 3.4 Xmerge user interface. Xmerge is used to combine and scale two data sets together. The results are graphed on the screen and are listed. The user can change the settings and rescale until the best results are obtained.

ber of reflections in each bin. Another method is to sort the data on resolution and then divide the data into equal portions. Past experience with data scaling has shown that scale parameters change faster with lower-resolution data than at higher resolutions. Thus, the bins should include fewer reflections at low resolution than at high. In Xmerge this is handled by binning based on $(\sin(\theta)/\lambda)^2$ instead of $(\sin(\theta)/\lambda)^3$, which seems to be a good compromise.

Local Scaling

Another useful technique is local scaling. Some data sets vary too quickly to be handled by anisotropic scaling. In local scaling, a scale factor is computed for each individual reflection by considering only the data in a local block centered about the reflection. For example, the scale factor could be set by setting the sums of the data in $5 \times 5 \times 5$ blocks based upon h, k, l to be equal, and then applying that scale factor only to the reflection at the center of the block. The block is then moved over one row at a time. This method is used quite often to scale Bijvoet data together. Again it is important not to use too small a block size. Local scaling can smooth out the largest differences of any of the techniques discussed. Beware, however, for in isomorphous derivatives, where large differences are often real, local scaling can scale out a significant portion of the signal.

Multiple Data Set Scaling

So far, the scaling of a single pair of data sets together has been discussed. A more complicated situation is the scaling of several partially overlapping data sets simultaneously: for example, scaling together several different data-collection runs on the same or different crystals. This is usually handled in an iterative manner. Since the scaling of one data set affects the scaling of all the other data sets, it is not possible to derive the best scaling parameter for each data set independently. Therefore, the scaling parameters are set to initial values and each bin of data is rescaled until the process converges. If the scaling diverges rather than converges, this indicates serious errors in one or more of the data sets. Often the bad data can be detected by examining the individual scale factors. The bad data may be due to excessive radiation damage, the crystal may have suffered some catastrophe such as drying out, or a single run may have been misindexed when the others are correct. Throw these data away and continue the scaling with the rest.

Heavy-Atom Statistics

Several statistics are useful in determining the probability that a crystal has been derivatized and that the differences, if any, are isomorphous.

(1) The absolute size of the differences, $|F_P - F_{PH}|$ should fall off with resolution (Fig. 3.5). If they increase with resolution, then the derivative is non-isomorphous. This is because the heavy-atom scattering falls off with resolution due to its scattering and thermal factors. If the derivative is isomorphous, the total scattering is the vector sum of the derivative amplitudes and the original protein amplitudes. In a non-isomorphous derivative protein amplitudes are changed due to changes in the unit cell and heavy-atom-induced movements of the protein. Differences due to these effects increase with resolution.

(2) The root-mean-square differences should be larger for centric reflections than for acentric reflections. This is because the centric zones have phase restrictions so that the full heavy-atom magnitude is either added or subtracted. In the acentric case, the vectors can have any angular relationship so that at one extreme the heavy-atom vector produces no change in inten-

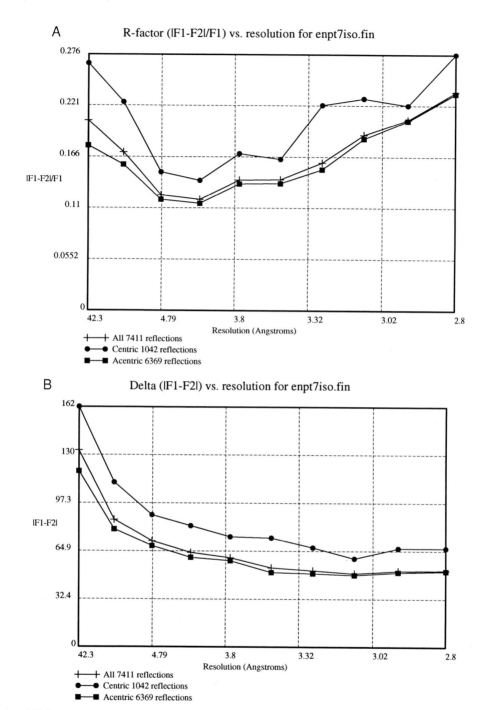

FIG. 3.5 Graphs of heavy-atom statistics. Graphs of statistics for a derivative with good phasing power are shown in (A), (B), and (C). Graphs from an unusable derivative are shown in (D), (E), and (F). See text for an explanation of the expected curves.

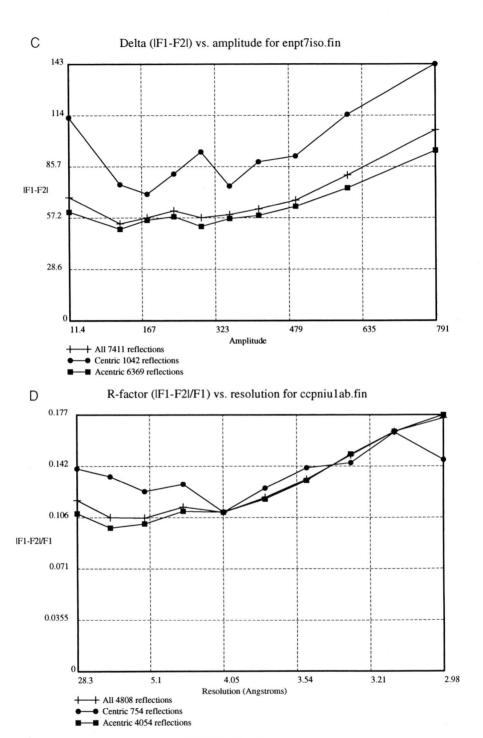

FIG. 3.5—*Continued.*

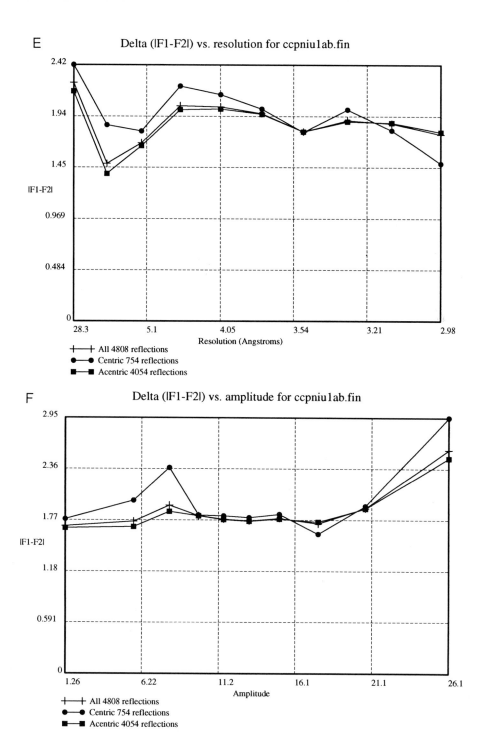

FIG. 3.5—*Continued.*

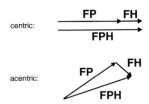

FIG. 3.6 Centric and acentric vector constructions. Where F_P and F_H are the same magnitude, in the acentric case $|F_{PH}|$ is considerably less than $|F_P| + |F_H|$. In general, centric reflections will have a difference equal to the amplitude of F_H (except in rare crossover cases), but an acentric difference is usually less than the amplitude of F_H (except, of course, when F_H and F_P have the same or opposite phases).

sity, and at the other extreme the vectors align to produce the maximal change. The average intensity change is $1/\sqrt{2}$ of the heavy-atom magnitude. This is illustrated in Fig. 3.6.

(3) The size of the differences should correlate with the size of F, which is partly a restatement of (1), since the mean size of F also falls off with resolution.

The overall size of the differences is important in order for the derivative to have any phasing power. In our lab we use some rough rules-of-thumb, which are fairly liberal to prevent throwing out a potentially useful derivative. If the root-mean-square differences on F (not I) are above 15% for the 5-Å data, there is definitely a signal present. If the differences are 9–15% and the data have been accurately collected, the derivative may be useful. Below this we have never found a useful derivative. We have solved the atom positions for weak derivatives with differences of about 7–8%, but when they are refined they have little to no phasing power.

The size of the isomorphous signal for a given heavy atom can be estimated from a simple formula derived here. The size of the resultant vector from adding a series of N random-walk vectors of length f is $f\sqrt{N}$. The maximum isomorphous difference is $f_H\sqrt{N_H}/f_P\sqrt{N_P}$, where f_H is the scattering power of the N_H heavy atoms and f_P is the average scattering power of the N_P protein atoms. Given the average ratios of carbon, nitrogen, oxygen, and sulfur atoms in a typical protein, the average scattering power of a protein atom is 6.7 e$^-$ and the average molecular weight of a protein atom is 13.4. Substituting these numbers into the equation, we get $f_H\sqrt{N_H}/6.7\sqrt{MW/13.4}$, or

$$\Delta F/F_{max} = 0.55\ f_H\sqrt{N_H}/\sqrt{MW}. \tag{3.11}$$

This gives the signal expected for centric reflections where the vectors are co-linear. For the acentric case the vectors are not correlated, and so the difference will be lowered on average by $1/\sqrt{2}$:

$$\Delta F/F_{\text{acentric}} = 0.39\, f_H\sqrt{N_H}/\sqrt{MW} \tag{3.12}$$

If we consider occupancy, this has the effect of lowering f_H by a proportionate amount, that is $f_H' = f_H^*$ occupancy. This formula gives an upper estimate since disorder will tend to lower this estimate. As an example, if we have a 32,000-Da protein with a one-site mercury ($f = 80$) derivative that is half occupied, the expected $\Delta F/_{\text{max}}$ is 12%. The expected $\Delta I/I$ will be twice this, or 24%. A second site of the same occupancy would raise the isomorphous signal by the square root of 2, or 1.4, to 17%.

••••• 3.4 •••••
THE PATTERSON SYNTHESIS

This old synthesis, named for the man who invented it, is still very useful today. Patterson's maps are obtained from a Fourier synthesis by assuming the phase is 0.0, squaring the amplitudes, and adding a center of symmetry. A Patterson map shows the vectors between atoms in the unit cell instead of the absolute positions of these atoms (Fig. 3.7). It is still the standard method of locating heavy atoms in the unit cell. A Patterson map of the difference between a data set with a heavy atom added and the data without ($F_{PH} - F_P$) gives the vectors between the heavy atoms.

Patterson Symmetry

The symmetry of a Patterson map is the symmetry of the space group in real space with a center of symmetry added and all translational elements set to 0.0. To add the center of symmetry, each operator, in turn, is copied and multiplied by $[-1, -1, -1]$ so that the number of operators is doubled.

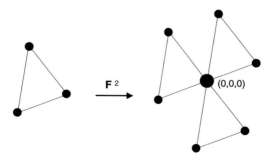

FIG. 3.7 Patterson vectors. On the left is a simple three-atom molecule. Its Patterson synthesis is shown on the right. The Patterson is derived by taking each atom in turn and placing it at the origin (0, 0, 0). Note that the Patterson pattern has an added center of symmetry.

The asymmetric unit is one-half the size of the real space asymmetric unit. Some symmetry elements are changed by this procedure into new ones. For example, a 2_1 axis turns into a mirror perpendicular to the original axis. The symmetry of Patterson space is similar to that of reciprocal space except for systematic absences and centering.

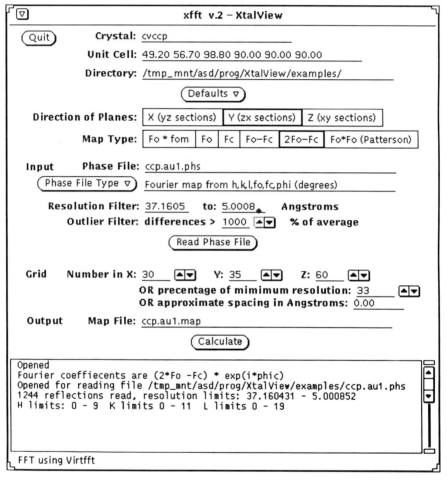

FIG. 3.8   xfft program interface. xfft is used to calculate electron-density maps by a Fast-Fourier-Transform method. The user interface is shown here, illustrating the options available. The user selects the desired options by selecting buttons and entering numbers in the appropriate fields and then pushes the Calculate button. The options can be saved and later loaded using the Defaults menu button.

Calculating Pattersons

Patterson maps are calculated as for a normal Fourier except that the Patterson symmetry operators are used, the amplitudes are squared, and the phases are all set to 0.0. If you are using XtalView, the proper symmetry is generated automatically and the amplitudes are squared when a Patterson synthesis is selected in the program xfft (Fig. 3.8). A difference Patterson is calculated by squaring the differences between two amplitudes and adding a phase of 0.0. Xfft will make difference Pattersons from `.fin` and `.df` files. Because the differences are squared, a few large differences can dominate the map. If these differences are caused by outliers, they can completely swamp the signal. Therefore, a filter has been added to xfft to detect obvious outliers and delete them. They are detected by rejecting any reflection where the $|F1 - F2| > [p * (F1 + F2)/2]$, or where the absolute value of the difference is greater than p times the average of the two reflections, and is usually set to 100% for isomorphous differences and about 30% for anomalous-difference Pattersons. If p is set greater than 200%, then no differences will be rejected. An example of the usefulness of this filter is shown in Fig. 3.9.

There are N^2 peaks in a Patterson map, where N is the number of atoms in the unit cell. Of these, N are vectors between the same atom and fall on the origin. This leaves $N(N - 1)$. The unique peaks are then $N(N - 1)/Z$, where Z is the number of asymmetric units in the Patterson space group. For example, if we have 3 atoms in the asymmetric unit of an orthorhombic space group, then there are 12 atoms in the unit cell. There are 144 vectors, of which 132 are not at the origin. Since the Patterson has a Z of 8, there are 132/8 unique vectors or 16.5. The fraction occurs because some peaks are on mirror planes and are shared by two adjacent asymmetric units.

Harker Sections

The peaks on a Patterson map result from all possible combinations of vectors between atoms in the unit cell including symmetry-related atoms. The symmetry-related peaks fall onto special positions, are called Harker peaks, and fall onto Harker sections. For example, in the space group P2, for every atom at x, y, z there is a symmetry mate at $-x, y, -z$ for which a vector will occur at $(x, y, z) - (-x, y, -z) = 2x, 0, 2z$ (Fig. 3.10). Thus, on the Harker section $y = 0$ peaks can be found at $2x, 2z$ for each atom in the asymmetric unit. However, while all Harker peaks fall on Harker sections, not all peaks on a Harker section are Harker peaks. If two atoms, not related by symmetry, happen to have the same y coordinate, they will produce a cross-vector on the section $y = 0$. The positions and relationships of Harker peaks (also

A

0.000 0.000   Y   0.500

Z

0.440

X = 0.500

B

0.000 0.000   Y   0.500

Z

0.440

X = 0.500

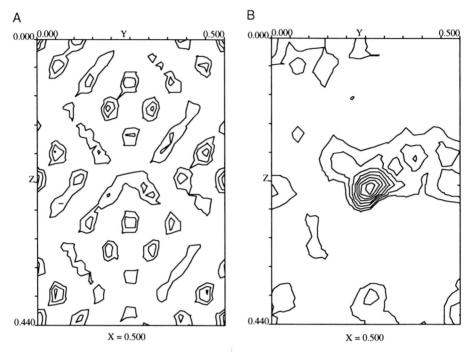

FIG. 3.9 Effect of filtering outliers. (A) Gold-derived Patterson map without filtering of outliers. (B) The same data filtered so that if $|F_P - F_{PH}|$ is greater than 100% of $\overline{F_P + F_{PH}}$, then the reflection is rejected. This filter rejects a handful of very large differences that were dominating the Patterson and making it uninterpretable. After removal of the outliers, the Patterson was interpretable, and this derivative turned out to have excellent phasing power.

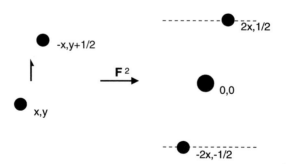

-x,y+1/2

F^2

2x,1/2

0,0

x,y

-2x,-1/2

FIG. 3.10 Origin of Harker planes. On the left are two atoms related by a 2-fold screw axis along y that displaces the second copy of the atom by 1/2 along y relative to the first. When a Patterson synthesis is constructed, the self-vectors fall along the planes $y = \pm 1/2$.

TABLE 3.2

Harker Vectors for Space Group P222$_1$

P222$_1$	x, y, z	$-x, -y, \frac{1}{2} + z$	$-x, y, \frac{1}{2} - z$	$x, -y, -z$
x, y, z	$0, 0, 0$	$2x, 2y, \frac{1}{2}$	$2x, 0, \frac{1}{2} - 2z$	$0, 2y, 2z$
$-x, -y, \frac{1}{2} + z$	$2x, 2y, \frac{1}{2}$	$0, 0, 0$	$0, 2y, \frac{1}{2} - 2z$	$2x, 0, \frac{1}{2} - 2z$
$-x, y, \frac{1}{2} - z$	$2x, 0, \frac{1}{2} - 2z$	$0, 2y, \frac{1}{2} - 2z$	$0, 0, 0$	$2x, 2y, \frac{1}{2}$
$x, -y, -z$	$0, 2y, 2z$	$2x, 0, \frac{1}{2} - 2z$	$2x, 2y, \frac{1}{2}$	$0, 0, 0$

called self-vectors) can be found by subtracting the possible combinations of symmetry operators for a given space group pair-wise as shown in Table 3.2.

The peaks for the example in Table 3.2, space group P222$_1$, fall onto three Harker sections: $x = 0$, $y = 0$, and $z = 1/2$. The Harker sections can be found by noting places in the table where one of the coordinates is a constant. An examination of the space group symmetry will also reveal the Harker sections, although not the full algebraic relationship. A 2-fold axis will have a corresponding Harker section at 0 in the plane perpendicular to the axis, a 2-fold screw at 1/2, a 3-fold screw at 1/3 and 2/3, and so forth.

Solving Heavy-Atom Difference Pattersons

Before you can use a heavy-atom derivative for phasing, the positions of the heavy atoms in the cell must be found. If no phasing information exists, the only map that can be made at this point is a difference Patterson map. The Patterson map must be solved for the x, y, z positions that produce this pattern of vectors. Note that the solution to the Patterson function is not unique; many different sets of positions can explain the same Patterson. The different solutions fall into two categories, origin shifts and opposite-hand choices. Fortunately, any self-consistent solution will give phases that will produce the same protein map. The choice of origin is arbitrary, and it will not matter which is chosen. The choice of hand is important because only a right-handed solution is correct. However, for isomorphous derivatives an incorrect choice of hand will produce a left-handed protein map that otherwise is identical to the right-handed map (this is *not* true for anomalous data). This can easily be remedied by inverting the signs of all of the heavy-atom positions and recalculating the phases. If two derivatives are solved from their Patterson maps, then there is no way to know if they have the same origin and hand. In order to solve this problem, difference Fouriers are used (see the following) to put both derivatives in the same framework. If this is not done, then the derivatives cannot be combined.

There are two ways to solve Patterson maps. One involves visual inspection and the calculation of heavy-atom positions by hand. This is usually not as difficult as it might seem at first because symmetry produces many special relationships between Patterson peaks that make it easy to find the solution. The other method involves using the computer, which, when it works, is obviously the easier method. But there is no guarantee that the computer can find the correct solution. Before trying any method, first evaluate the quality of the Patterson map to see if it is possible to solve it at all. Frequently, the first derivative has a Patterson that is not easily solved. It is sometimes more fruitful to keep looking for a derivative that produces an easily solved Patterson map that can be used to bootstrap the rest. In the author's experience, this simple strategy has almost always worked. Of course, if you have four protein molecules in the asymmetric unit, your chances of finding a single site derivative are pretty unlikely.

In order to improve your chances of solving a Patterson, either by hand or by computer, it helps enormously to make the best Patterson map you can. Because the differences are squared, Patterson maps are easily overwhelmed by a few large differences. Outlier differences should be filtered out as discussed earlier. A value of 100% is usually about right for this filter as differences larger than this are often incorrect. Deleting differences smaller than this can lower the signal-to-noise ratio. If the derivative has a high merging R-value (above approximately 0.20), then a larger percentage may be needed. The other filter to be set is resolution. Data below about 25 Å are often clipped by the beam stop on most data-collection systems. If the beam stop is not perfectly round, then the amount of clipping could be different between data sets. Also, very low resolution reflections will have strong differences that are due to the change in contrast from native to derivative mother liquor, either because the soaking mother liquor is higher in precipitant or because the dissolved heavy atoms in the soaking mother liquor interfere. Often, leaving out this data will improve the Patterson map. As resolution increases, data become noisier because the data are weaker, and often become less than perfectly isomorphous. This can give higher-resolution Patterson maps lower contrast. As a first approximation, 4–5 Å is a good upper resolution limit. The quality of the Patterson can be judged by looking at the peak-to-background ratio of the Patterson map. The background is taken to the root-mean-square, or sigma (σ), of the entire map. Peak heights are then expressed as ratios of the peak height over the sigma of the map. If you use xcontur to contour your Patterson map then it automatically sets the first contour level to one sigma, the second to two sigma, and so on.

Look first at the Harker sections for large peaks. If the Patterson is complex, then the Harker sections may be crowded and looking at general sections may be more fruitful. Try varying the resolution and outlier filter to produce the best contrast ratio. I cannot overemphasize the need to filter outliers and very low resolution data; otherwise interpretable Patterson maps may be overwhelmed by a few bad differences (see Fig. 3.9 for an example). The effect of changing the upper-resolution cutoff should be more subtle. At some point the contrast will be maximal, but the basic features of the Patterson map should not change with resolution. If they do, this is a bad sign and indicates that the differences are not due to the isomorphous addition of heavy atoms.

Classical Methods

In addition to this explanation of how to solve Patterson maps by hand, examples are given in Chapter 4. In most cases the strategy is first to find a Harker peak or set of Harker peaks that gives a first site and then to use cross-peaks to find additional sites. Make a table of symmetry operators and find the positions of the Harker peaks as is illustrated in Table 3.2. Sometimes special relationships can be found between Harker peaks that make it easy to find Harker peaks that arise from the same site (see the Patterson examples in Chapter 4, Section 4.1). This helps in identifying non-Harker peaks that happen to fall on a Harker section. Related sets of cross-peaks can also sometimes be found, and relationships between Harker and cross-peaks can often be identified that make it easy to find related sets of peaks. These relationships can be found with a little simple algebra using the symmetry operators of the space group.

As an example, say you have a Patterson in space group P222. The three sections $x = 0$, $y = 0$, $z = 0$ are Harker sections and the peaks on these are from vectors $2z$, $2y$, $2z$. You can look for single-site solutions that match between different Harker sections. In some cases a single site will explain all of the Patterson peaks, and there are no significant non-Harker peaks. Congratulations, you are done. If not, look for a cross-peak. You can find a second site by taking the position of site 1 and adding the cross-peak. However, remember that you do not know from which symmetry-related atom in the unit cell this peak arises. Also you do not know in which direction the vector goes: from atom A to B or from B to A. Therefore, you must try all of these possibilities. In this space group there are four symmetry-related atoms A_1, A_2, A_3, A_4. Given a cross-peak X we need to try $A_1 - X$, $A_2 - X$, $A_3 - X$, $A_4 - X$ and also $A_1 + X$, $A_2 + X$, $A_3 + X$, $A_4 + X$ in order to find atom B. We can confirm atom B by finding the Harker peaks predicted by atom B and also the other cross-peaks generated by site B and A symmetry-related atoms. These vector calculations are most easily done by a computer. Xpatpred in XtalView can generate all of the heavy-atom vectors given a list of sites. These can then be loaded into xcontur and compared against the Patterson map. Each of the possibilities can be tried and the results quickly scanned by looking at the agreement with the peaks in the Patterson map. You can maximize your chances of finding a match by starting with the highest Harker peaks and the highest cross-peak. This may fail, though, because the highest peaks may be due to two or more peaks that happen to fall in the same place, and the extra height is fortuitous. Many examples of this can be seen in the Pattersons for the heavy atoms of *C. vinosum* cytochrome *c'*, illustrated in Chapter 4. A non-crystallographic 2-fold causes many sites to be related, and the vectors fall in clumps. Also, it is possible that the cross-peak X is not between A and another atom, but between two other atoms. If you cannot solve a Patterson, put it aside and look for a derivative with one that is solvable. Later you can easily solve the difficult Patterson by cross-Fourier (see the following) with the solved derivative phases, so all is not lost.

Computer Methods

Several programs have been written that try to solve Patterson maps automatically. I know of three, HASSP, written by Tom Terwilliger, SHELXS, a commonly used small molecule crystallography program, and XtalView/xhercules, a correlation search method I wrote. HASSP and SHELXS first look for single-site solutions that explain the Harker vectors and then look at cross-peaks and try to find pairs of positions with the best match to the density in the Patterson map. This is very similar to the classical method. One problem I have seen is that the programs do not always account adequately for the fact that a peak has already been used and overlapping solutions can be found. Both programs work well on clean Patterson maps.

The XtalView program xhercules uses a different approach to automatic Patterson solution that works well but is very computer-intensive. A single atom is moved around the entire asymmetric unit on a grid, and at each position a correlation is calculated between the observed differences and the calculated heavy-atom amplitudes. (The use of a correlation function, rather than an R-factor, is important because the scale factor cannot be computed correctly and the correlation is independent of scale.) This atom is then placed at the position with the highest correlation (Fig. 3.11). A second atom is then moved about the asymmetric unit and the correlation calculated with the first atom held fixed. This atom is then fixed at its highest correlation. The relative occupancies are then refined by another correlation search. A third atom can then be searched for in the same manner, and so forth. Each correlation search takes a large amount of computer time—several hours on a typical workstation—which gets longer as more atoms are added. An intelligent choice of the asymmetric unit helps reduce time. Remember that the asymmetric unit is one-half the size in reciprocal space. The search grid needs

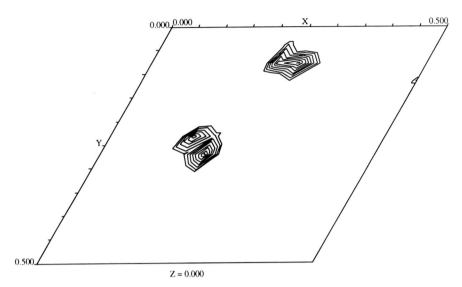

0.000 0.000   X   0.500

Y

0.500

Z = 0.000

FIG. 3.11 Hercules correlation map section of a single-site correlation search for a platinum derivative of photoactive yellow protein. The space group is $P6_3$—hexagonal with 6_3 screw axis along z direction. The single site is found at 0.25, 0.08, 0.0 (z is arbitrary because there is no orthogonal symmetry element). The second peak is related to the first by Patterson symmetry.

to be at least one-fourth of the minimum resolution and preferably one-sixth. The resolution cut off should be as low as 6 Å for large unit cells and as high as 4 Å for small unit cells. There must be a large ratio of differences to atom parameters, (x, y, z), because the differences are only an approximation of the heavy atom vectors. For most proteins at the resolutions suggested this can be kept at about 50 to 1.

Although the method can be used automatically, it is far better to check the results against the Patterson map. This can be done by writing out the solution, reading it into xpatpred, and displaying the output of predicted vectors on the Patterson map with xcontur. For best results, an idea of the relative occupancy of unfound sites is needed, although tests have shown it is not critical. The relative occupancy can easily be estimated by inspection of the remaining density in the Patterson map. How successful is the method? It seems very robust. A six-site solution to a complex Patterson with many overlapping vectors was correctly found for a $PtCl_4$ derivative of *C. vinosum* cytochrome c' in $P2_12_12_1$ (three Harker planes at $x = 0.5$, $y = 0.5$, $z = 0.5$). The solution was found before any other derivative and was later confirmed independently by cross-phasing from another derivative whose Patterson was solvable by inspection. The platinum derivative evaded a manual solution because the largest peaks on the Harker sections turned out to be due to a mixture of cross-peaks and Harker peaks.

Another successful method of heavy atom position determination is direct methods. The programs used by small molecule crystallographers, MOLTAN and SHELXS, have been used for this purpose by using the derivative differences as input.

Anomalous-Difference Pattersons

If there are atoms in the structure that have an anomalous scattering component (i.e., they have an absorption edge near the wavelength being used to collect data), then the differences between the Bijvoet pairs may be large enough to use for phasing. For proteins, the most likely naturally occurring anomalous scatterer is iron, and many of the heavy atom compounds used, such as mercury, platinum, uranium and the lanthanides, have significant anomalous scattering signals. Centric reflections do not have an anomalous component because the signal exactly cancels out in a centric projection. If you have collected symmetry mates of centric reflections, they can be used to estimate the anomalous signal by comparing them to the acentric reflections. The difference between the centrics is the "noise," and the "signal" can be found using the formula $\langle noise \rangle^2 + \langle signal \rangle^2 = \langle differences \rangle^2$. In practical terms this formula tells us that the signal will be larger than simply taking the difference between the centrics and the acentrics. The anomalous signal, in fact, can be detected even in noisy data because the pairs are usually collected from the same crystal near in time, which eliminates many of the scaling problems. Thus, random noise is less of a problem than systematic errors, such as those due to absorption and X-ray damage.

The expected anomalous difference $\langle F^+ - F^- \rangle / \langle F \rangle$ can be estimated using the same equation as for the isomorphous case (see preceding) except that f_H is replaced by $2f'$. Some expected anomalous differences are listed in Table 3.3 for atoms that give usable signals for CuKα radiation. If you can tune the wavelength as at a synchrotron, then the signal can be optimized and edges for more elements become available.

Before an anomalous-difference Patterson can be calculated, the data should be nearly complete to the resolution that you wish to use. This means complete in terms of Bijvoet pairs, which means collecting twice as many data in the correct positions.

An anomalous-difference Patterson (also known as a Bijvoet Patterson) is made using the differences between the acentric reflections as the Patterson coefficients (Fig. 3.12). Care must be taken not to use the centrics. In Xtal-View the difference between centrics will be 0, so even if they are left in, they make no contribution to the differences. Interpret an anomalous-difference

TABLE 3.3
Anaomalous Scattering Signals for CuKα (1.542Å) Radiation

Element	Electrons	Percentage $\Delta F/F$ (30K protein)	$\Delta f''$	Percentage F^+-F^- (10K protein)	Percentage F^+-F^- (32K protein)	Percentage F^+-F^- (100K protein)
S	16	5.1	0.6	0.46	0.26	0.14
Fe	26	8.3	3.2	2.46	1.38	0.77
Pd	46	15	3.9	3	1.68	0.94
Ag	47	15	4.3	3.31	1.85	1.03
I	53	17	6.8	5.24	2.92	1.63
Sm	62	20	12.3	9.47	5.29	2.95
Gd	64	20	11.9	9.16	5.12	2.86
Pt	78	25	6.9	5.31	2.97	1.66
Au	79	25	7.3	5.62	3.14	1.75
Hg	80	25	7.7	5.93	3.31	1.85
Pb	82	26	8.5	6.54	3.65	2.04
U	92	29	13.4	10.32	5.76	3.22

Patterson exactly as you would an isomorphous-difference Patterson. Because the centrics are left out, however, even a perfect set of anomalous differences will give series-termination errors that can lead to small peaks that are not due to scatterers. These can be detected by making a calculated Patterson using the same reflections list as the observed Patterson and coefficients calculated from the heavy-atom positions (with XtalView you can use STFACT for this purpose). If a peak appears that is not in the atom list used for the calculation, it must be a series-termination error. In fact, the lower the resolution the higher the percentage of centric reflections, so very low resolution anomalous-difference Pattersons have more noise due to series-termination.

One use of anomalous differences in heavy-atom work is to compare them to the isomorphous difference Patterson. An anomalous scatterer represents independent measures of the heavy-atom positions and, as such, comparison of the two Pattersons gives extra confidence in determining the heavy-atom positions. A peak on both Pattersons is more likely to be correct. It also possible to make a Patterson by combining the information from both sets of differences. Before doing this, it is worthwhile first to check that the anomalous Patterson actually has some signal. Adding noise to the isomorphous Patterson will not make it more interpretable. There are two methods

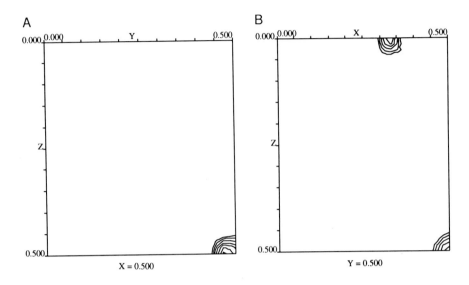

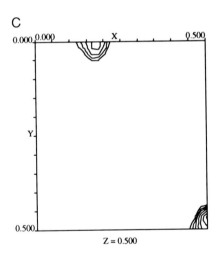

FIG. 3.12 Sulfite reductase Bijvoet difference Patterson. The anomalous scattering of sulfite reductase is due to the presence of a Fe_4S_4 cluster and a heme iron. At low resolution the individual scatterers are not resolvable and form a single large site. The space group is $P2_12_12_1$ and the coeficients are $(F_P^+ - F_P^-)^2$. (A–C) Harker sections $x = 0.5$, $y = 0.5$, and $z = 0.5$. Note that the peak on the section $x = 0.5$ overlaps onto the other Harker sections. At higher resolution the peaks are resolvable from the edges. (D) A three-dimensional stereo view of the Patterson, showing that there are just three peaks, not including the origin peak at 0, 0, 0.

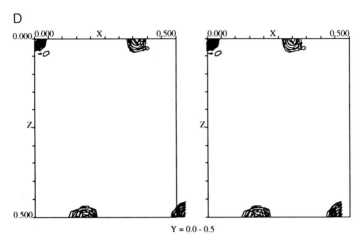

FIG. 3.12—*Continued.*

used to combine the signals. The simplest is to make a Patterson with the coefficients: $\Delta F_{iso}^2 + \Delta F_{ano}^2$. The other uses F_{HLE} coefficients,[4] which gives slightly higher peaks than the first. In either case, the improvement of the Patterson is better judged by the number of contours a peak of interest is above the root-mean-square density (or sigma) of the Patterson map. If the main peaks do not increase, then it is possible that the anomalous signal is too small to be of use.

If the anamolous data are too incomplete to give a Patterson map alone, it is still possible to use combined coefficients to augment the isomorphous data. Again, use the criterion of peak height to judge the effectiveness. Another criterion is the flatness of areas without peaks. They should become cleaner if the Patterson has been improved.

····· **3.5** ·····
FOURIER TECHNIQUES

It is a happy day in any structure determination when Fourier maps can be used instead of Pattersons, because they are very straightforward to interpret. On the other hand, a Fourier requires phases, and the quality of the phases very much determines the quality of the resulting map. For Pattersons,

----

[4]For a discussion and derivitization of F_{HLE} coefficients see Blundell and Johnson, pp. 338–340, as listed in Suggested Reading list in Preface.

A

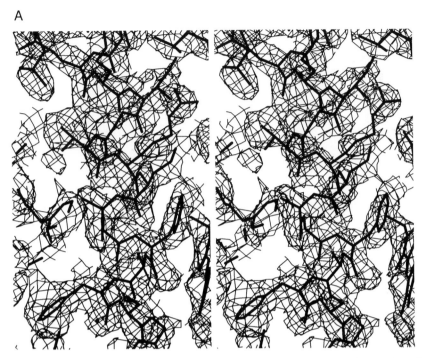

FIG. 3.13   Fouriers with random phases or amplitudes (A) and (B) the equivalent section of a Fourier synthesis with random phases. The thick lines represent the model that was used to calculate the correct amplitude and phases. Note that the map with random amplitudes is still interpretable, but the random phase map is uninterpretable. This apparently means that phases are more important than amplitudes. However, since phases are not directly measurable and must be determined from the measured amplitudes, accurate amplitudes are necessary to determine the phases accurately.

the quality is only dependent upon the accuracy of the amplitudes used. In fact, it has been shown that a while a Fourier map made with random amplitudes but correct phases is easily interpreted (Fig. 3.13), the opposite case, correct amplitudes and random phases, is not (which is the root of all this phasing trouble). This does not mean that the amplitudes can be ignored when there are phases. In crystallography we work in the gray area between the two extremes of correct and random phases. In this case, correct coefficients (e.g., 2Fo − Fc) *do* make an important difference in the quality of the resulting map. Also, do not take this to mean that the amplitudes do not need to be accurate. Without accurate amplitudes there is no way to derive accurate phases.

B

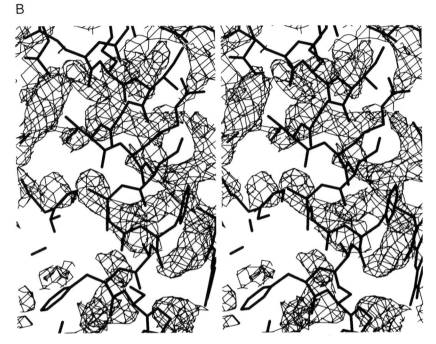

FIG. 3.13—*Continued.*

Types of Fouriers

Fo Map

This is the classic Fourier synthesis, the observed amplitudes with the most current phases Fo, α_{calc}, where Fo is the observed diffraction amplitude. It is not sufficient for all crystallographic needs and there are many other types. In particular, when the phases are calculated, this type of map is subject to model bias.

Fc Map

This is the least useful map for crystallographic purposes: the calculated amplitudes phased with the calculated phases and you get back exactly what you put in. Still, it can be used for checking what a map should look like, especially at lower resolutions, and to check for series-termination problems. For instance, even an F_c map in the resolution range 5–3 Å can be choppy and hard to interpret because of the missing low-resolution terms. An Fc map can also be used to check if programs are working correctly. If the Fc map does not look like what you put in or have the correct symmetry, then there is an error somewhere.

Fo − Fc, or Difference, Map

The difference Fourier, Fo − Fc, α_{calc}, is very useful in terms of information content but may be hard to interpret. In this map there are peaks where density is not accounted for in the model used to calculate Fc and holes where there is too much density in the model. This map is especially useful for finding corrections to the current model: for example, looking for missing waters, finding movements in mutants, misfittings, etc. Other types of differences maps are often used. An isomorphous difference map with $F_{PH} - F_P$, $\alpha_{protein}$ gives the positions of heavy atoms. Another difference map is $F_{mutant} - F_{wild-type}$, which can be used for looking at mutant protein structures (if the mutation crystallizes with the same unit cell). Note that a difference Fourier, $F_{mutant} - F_{wild-type}$, $\alpha_{wild-type}$, is not the same as the difference between two Fouriers, $(F_{mutant}, \alpha_{mutant}) - (F_{wild-type}, \alpha_{wild-type})$, which is equivalent to the difference between two electron-density maps.

There are three basic patterns of density found in a difference map. A peak indicates electron density in the Fo terms that is not accounted for in the Fc terms. A negative peak indicates a position where there is less density in the Fo terms than in the Fc terms. These differences can arise from movement of atoms, changes in B-values, or a change in occupancy. A third pattern is positive density paired with negative density. This indicates a shift in position from the negative to the positive density. The final position of the atom or group may be difficult to determine from the difference density. Its actual position is somewhat short of the positive peak because the negative hole next to the positive peak distorts its shape.

2Fo − Fc Map

This is the sum of an Fo map plus an Fo − Fc map (Fig. 3.14). It contains information from both the classic Fourier synthesis and a difference map and is easy to interpret since it looks like protein density. The quality of the 2Fo − Fc map depends upon the quality of the phases—a fact that often seems forgotten in the literature, where they are often used as "proof" of the correctness of a structure. At R-values above 0.25, they are of dubious value since they will look remarkably like whatever model was used to calculate the phases. In these cases a better map to use is the omit map, where the model in question is left out of the phase calculation (see below). Another problem with 2Fo − Fc maps is that they do not show the exact final position of the model in positions where there are errors, although they can come quite close. The 2Fo − Fc map shows the position where the model is and where the model should be. Usually the final position is a farther away than the map shows, but it is usually close enough to bring the model to within the radius of convergence for a refinement program. Variations on the 2Fo − Fc map, such as 3Fo − 2Fc (or 0.5 Fo + (Fo − Fc)) and 5Fo − 3Fc, are sometimes used and are claimed to alleviate the problem of phase bias.

Omit Map

In the omit map the portion of the model to be examined is left out of the phase calculation altogether, and the rest of the model is used to phase this portion of the map (Fig. 3.15). This powerful feature of a Fourier is pos-

A

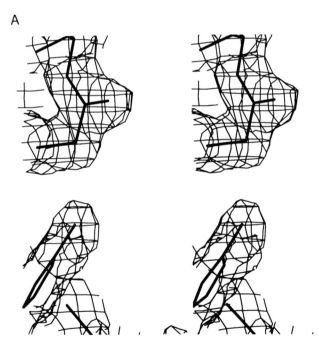

FIG. 3.14 Fo, Fo − Fc, and 2Fo − Fc Fourier Maps. The same section of map is shown using (A) Fo coefficients, (B) Fo − Fc, and (C) 2Fo − Fc. The Fo map does not contain any information that was not already in the model. The difference map shows some unexplained density in the solvent region that is taken to represent disordered solvent molecules. The 2Fo − Fc map is equivalent to adding the two maps together and shows the unexplained solvent density as well as the model density. This property makes it the most popular map type. (*Figure continues.*)

B

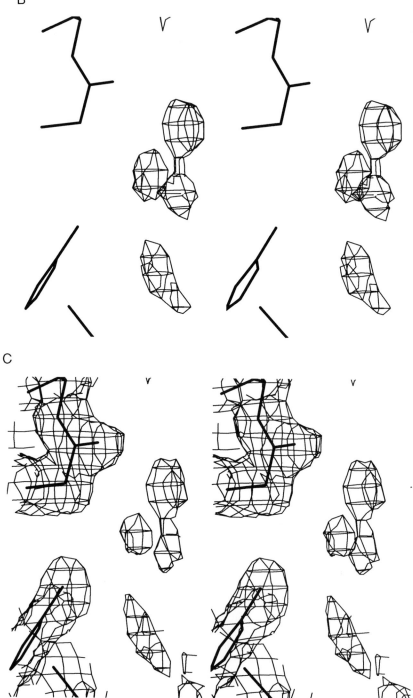

C

FIG. 3.14—*Continued.*

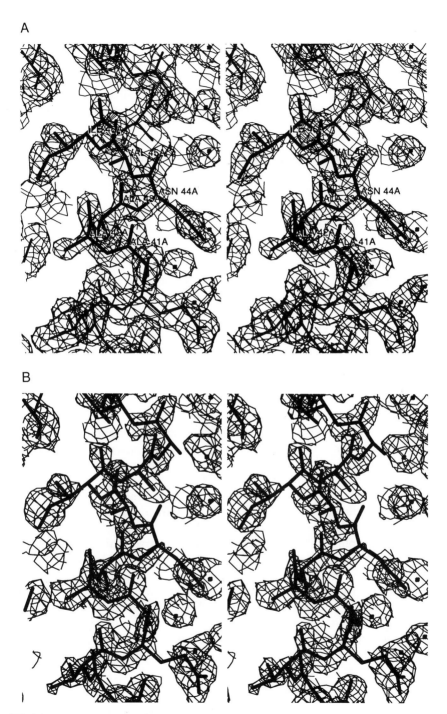

FIG. 3.15 Omit map. (A) The highlighted and labeled model in the center of the helix was omitted from the phase calculation and then an Fo, α_{omit} map was made. Note that the density for the omitted atoms comes back because of their presence in the amplitude information. In (B) 25% of the all the residues in the model were omitted. The density for the omitted residues, although noisy, still comes back. The maps are at 1.8 Å resolution and contoured at 1 σ.

sible because all parts of the model contribute to every reflection. The chief difficulty in an omit map is using the correct scale factor to scale Fo to Fc. The sum of Fc will be smaller than that of Fo if a portion of the model is omitted. The correct procedure is to use the proper scale before anything has been omitted and then to keep this same scale factor after the model has been omitted. Only about 10% of the model should be left out at any given time, so it is necessary to make many omit maps to examine the entire structure. Even omit maps can have some residual phase bias, as will be explained in the section "Phase Bias."

Figure-of-Merit Weighted Fo Map

This is the map used for MIR and is used to compensate for the error in the MIR phases. The coefficients used are Fo times the figure-of-merit. It can be thought of as a map where each reflection is weighted by the confidence in its correctness.

Fast-Fourier Transform

The Fast-Fourier Transform (FFT) is what its name implies: a quick Fourier synthesis. The FFT works by dividing the unit cell along the three principal directions by integer multiples, nx, ny, nz. Depending upon the particular FFT used, these must be multiples of small prime numbers. The FFT used in XtalView can use multiples of 2, 3, 5, and 7. This gives a wide range of possible integer values. In addition, the grid cannot be too coarse. The coarser the grid, the faster the FFT can be calculated and, also, the faster the resulting map can be displayed. The grid must be at least twice the maximum h, k, l values in the input-structure factors. This can be calculated from the formula $nx = 2 * a/d_{min}$, where a is the cell edge and d_{min} is the resolution limit of the input structure factors, both in ångstroms. If the grid is less than this, the FFT will return incorrect values because the cell will be undersampled. At a sampling of two times, a map will be coarse, and a better sampling is $3 * a/d_{min}$, which gives a smoother map that is easier to interpret. In cases where one cell edge is substantially longer than the other two, this edge can be sampled at two times and the others at three times. Such a map is hard to distinguish from one that is sampled finely in all three directions. Sampling on grids finer than three times will be slower and is usually not necessary. If you are not going to look at the map interactively, then sometimes oversampling is used to make smoother contour lines for displaying a static picture.

Solving Heavy Atoms with Fouriers

Once even a single derivative has been solved, the single isomorphous replacement (SIR) phases are usually of sufficient quality to solve the rest of the derivatives by difference Fourier using the coefficients $F_{PH} - F_P$, α_{SIR}. With XtalView, a derivative-difference Fourier, also called a cross-Fourier, is made by the following steps. First, the derivative data are merged with the native data using xmerge as previously described for difference Pattersons to produce a file with the derivative scaled to the native. This file is then phased using xmergephs to merge the fin file coefficients with the best available protein phases. The option to switch F_P and F_{PH} should be checked so that when the Fourier is made the peaks are positive. This new phase file with h, k, l, F_{PH}, F_P, $\alpha_{protein.}$ is then run through xfft with the Fo $-$ Fc option, which in this case will make a $F_{PH} - F_P$ map. Find the biggest peaks on the map using xcontur. The resulting coordinates are the correct coordinates on the same origin as the derivative(s) used to calculate the phases.

The peaks obtained this way should be checked against the Patterson map of the derivative to be sure that the heavy atoms positions are valid. The most common spurious peaks are ghost peaks, where a peak is present at the position of the heavy atom model for the derivative used to produce the phases. Less obvious is the possibility of ghost peaks at the opposite hand position. This can happen if the solution is on a special position, if it is centrosymmetric, or if it is pseudo-centrosymmetric. Particularly confusing is a centrosymmetric heavy-atom solution that gives rise to centric phases in an otherwise acentric space group.[5] Maps made with these phases will contain both left-handed and right-handed solutions to the new derivative superimposed. The map may still be useful if this can be resolved. An easy way to check this problem is to look at the distribution of phases for acentric reflections. If they fall on the cardinal points, 0, 90, 180, and 270, then the phases are centric; if they fall near but not exactly on them, then the phases are pseudo-centric.

····· 3.6 ·····
ISOMORPHOUS REPLACEMENT PHASING
Heavy-Atom Refinement

Having solved one or more derivative data sets for heavy atom positions, you next step is to refine the positions and search for minor sites and/or missing sites. The goal behind refinement is 2-fold: to improve the heavy-atom parameters and to get statistics that give information about the quality of the derivative and the highest resolution at which it can be used. Some guidelines about statistics will be given, but it must be remembered that they

---

[5] A single site in a polar space group is centrosymmetric, as is a multi-site solution where all the coordinates in the polar direction are the same.

A

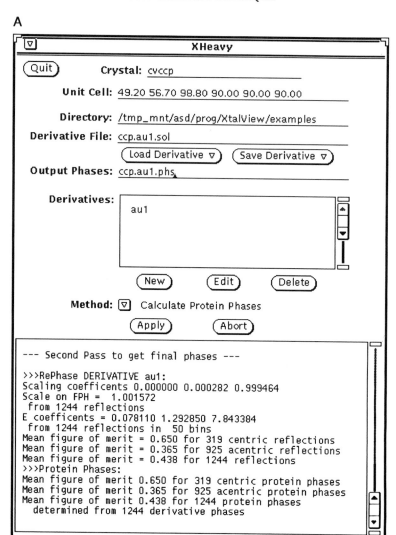

FIG. 3.16 Xheavy program. Xheavy is used to refine heavy-atom derivatives and to calculate isomorphous replacement phases. A derivative, or solution, file contains the information for one or more derivatives. This information can be edited using the **Derivative Edit** window (B). Xheavy refines heavy-atom positions by maximizing the correlation of F_H and $|F_P - F_{PH}|$ by moving each atom in turn until no further improvement can be found. The relative occupancies are then refined using the same correlation. To calculate protein phases, the single isomorphous replacement phases are first calculated for each derivative in turn. These are then combined to give an initial estimate of the protein phase. This protein phase is then used to get a better estimate of the SIR phases, and finally the new SIR phases are combined to give the final protein phases. The two-pass protein phasing method was adapted from a program written by William Furey of the University of Pittsburgh, PHASIT.

B

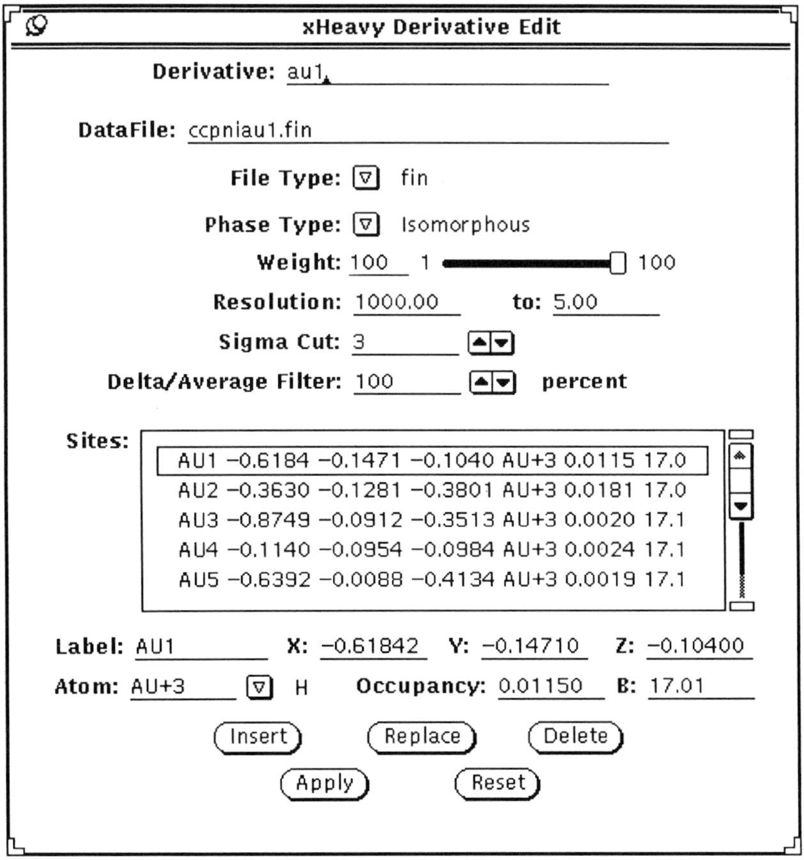

FIG. 3.16—*Continued.*

are only guidelines and that two heavy-atom derivatives with identical statistics can produce maps of different quality.

In order to calculate the most accurate phases, heavy-atom parameters are refined to improve the parameters and to get the best estimates of the errors. An accurate estimate of the errors is essential for good multiple isomorphous replacement phasing, since the errors control the figure-of-merits and the relative weighting of each reflection. In XtalView, the heavy-atom refinement is done using xheavy (Fig. 3.16). Xheavy takes a departure from the more traditional refinement programs by using a correlation search instead of least-squares refinement. The advantage is that it avoids local minima; the disadvantage is that it takes more computer time. With the faster computers available today, this is not a significant problem. The phases are calculated in two steps. First, each derivative is treated separately and estimates of the errors and phases are made without protein phasing information. Second, these phases are combined and a better estimate is made using these phases to produce a more accurate set of phases.

The correlation search in xheavy is done by moving the atom in a coarse box over a large range and then in progressively finer boxes and over a smaller range until no improvement is found. At each point the correlation is made between the observed differences and the calculated difference. The correlation used is

$$\frac{\Sigma \Delta_o \Delta_c}{\sqrt{\Sigma \Delta_o^2 \times \Sigma \Delta_c^2}},\qquad(3.13)$$

where Δ_o is the observed heavy-atom difference and Δ_c is the calculated difference. This function has the advantage of being immune to the scale between the calculated and observed differences.[6] Each atom is moved and then the occupancy and B-value are refined by a second correlation search.

Isomorphous Phasing[7]

Because of errors in the data and errors in the solutions and because true isomorphism is probably rare, isomorphous phasing must be done in terms of probabilities. What we know are the amplitudes $|F_P|$ and $|F_{PH}|$, and the vector F_H, which is calculated from the heavy-atom model. What we want to find out is the best phase of the vector F_P, given the errors in the data. The chief difficulty in this is estimating the errors in the data correctly. Blow and Crick assumed that all of the error lies in the magnitude of F_{PH} from which follows the equation[8]

$$P(\alpha) = \exp(-\epsilon(\alpha)^2/2E^2),\qquad(3.14)$$

where E is the estimate of the error and $\epsilon(\alpha)$ is the lack of closure error at a given value of α, the phase angle given by the expression

$$\epsilon(\alpha) = |F_{PH}| - |F_P + F_H|,\qquad(3.15)$$

which is the difference at a given phase angle between the the measured $|F_{PH}|$ and the amplitude of the sum of the vectors F_H and F_P.

---

[6] The scale factor cancels out. Say we have two quantities $F1$ and $F2$ related by the scale factor s such that $\Sigma F1 = s\Sigma F2$, then the correlation is

$$\frac{\Sigma F1 * sF2}{\sqrt{\Sigma F1^2 \ \Sigma (sF2)^2}} = \frac{s\Sigma F1F2}{s\sqrt{\Sigma F1^2 \Sigma F2^2}},$$

and the scale factor s cancels out.

[7] Be sure to read Watenpaugh, K. D. (1984), "Overview of Isomorphous Replacement Phasing," in "Methods in Enzymology," (Wyckoff, H., et al., eds.), Vol. 115, pp. 3–15. Academic Press, San Diego.

[8] Blow, D. M., and Crick, F. H. C. (1959). *Acta Crystalogr.* 12, 794–802.

The error estimate E is given by the equation

$$\langle E^2 \rangle = \langle (|\mathbf{F_{PH}} \pm \mathbf{F_P}| - |\mathbf{F_H}|) \rangle, \tag{3.16}$$

which is the difference between the calculated heavy-atom vector and the observed heavy-atom vector. If the reflection is centric, then it is possible to estimate the error by simply assuming that the observed difference $F_{PH} - F_P$ is the observed heavy-atom amplitude. For the acentric case, the difference $F_{PH} - F_P$ will be, on average, smaller than $F_{PH} - F_P$ by $1/\sqrt{2}$ (see above). Thus, the E can be estimated by

$$E \cong \sqrt{\frac{1}{n} \sum_{\text{centrics}} (|F_{PH} - F_P| - |F_H|)^2 + \sum_{\text{acentrics}} (|F_{PH} - F_P|\sqrt{2} - |F_H|)^2}$$

$$\tag{3.17}$$

In the general acentric case, there will be two maxima in Eq. 3.14 that are equally probable. The best phase is the weighted average of the two. In the centric case, the equations give one peak. In order to resolve the 2-fold ambiguity of acentric reflections and to overcome the errors, several derivatives are used.

For multiple isomorphous replacement, we repeat this process for each derivative in turn and multiply the phase probabilities. To simplify this process, we can use the Hendrickson–Lattman coefficients A, B, C, D to store each phase probability:

$$P(\alpha) = \exp(A \cos(\alpha) + B \sin(\alpha) + C \cos(2\alpha) + D \sin(2\alpha)) \tag{3.18}$$

This equation allows us to reconstruct the probability from the four coefficients and is a more compact form to store. To multiply two probabilities together, the coefficients are simply added, i.e., $A_{\text{MIR}} = A_{\text{DER1}} + A_{\text{DER2}} + A_{\text{DER3}}. \ldots$ Equivalent equations for the Blow–Crick equations above have been derived that give nearly identical results.[9] The equations for isomorphous replacement phasing are

$$A_{\text{iso}} = \frac{-2(F_P^2 + F_H^2 - F_{PH}^2)F_P a_H}{E^2}$$

$$B_{\text{iso}} = \frac{-2(F_P^2 + F_H^2 - F_{PH}^2)F_P b_H}{E^2}$$

$$\tag{3.19}$$

$$C_{\text{iso}} = \frac{-F_P^2(a_H^2 - b_H^2)}{E^2}$$

$$D_{\text{iso}} = \frac{-2F_P^2 a_H b_H}{E^2},$$

[9]Hendrickson, W. A., and Lattman, E. E. (1970). *Acta Crystalogr.* **B26**, 136.

where $a_H = \cos(\alpha_H)$ and $b_H = \sin(\alpha_H)$ with α_H as the calculated heavy-atom phase.

The best phase can then be found from the following procedure. The phase probability is calculated from the combined A_{MIR}, B_{MIR}, C_{MIR}, D_{MIR} for every $15°$, using Eq. 3.18 to generate the phase-probability ditribution. The phase is then calculated by integrating the probability distribution:

$$x = \frac{\sum_0^{2\pi} P(\alpha)\cos(\alpha)}{\sum_0^{2\pi} P(\alpha)}, \qquad y = \frac{\sum_0^{2\pi} P(\alpha)\sin(\alpha)}{\sum_0^{2\pi} P(\alpha)}$$

$$\alpha_{best} = \tan^{-1}(y/x), \tag{3.20}$$

and the figure-of-merit is given by

$$m = \sqrt{x^2 + y^2}. \tag{3.21}$$

The figure-of-merit, m, is the probability of α_{best} being correct. It ranges from a value of 0.0, where all phases are equally probable, to 1.0, where the phase is correct. The average figure-of-merit over all reflections gives us an estimate of the accuracy of the protein phase set.

Xheavy makes two passes through the phasing process in order to get a better estimate of E for each derivative. The initial set of protein phases uses centric data, if available, to calculate the initial estimate of E, the error. E is calculated as a function of the size of F_P by fitting a curve to the data. The initial Hendrickson–Lattman coefficients A, B, C, D are calculated with this estimate. If there is more than one derivative, then the coefficients of each derivative are summed and the protein phase and figure-of-merit are calculated. In the next pass, all of the data are used to calculate E, using the initial estimate of the protein. The new coefficients with the updated E values are again summed together and a final protein phase and figure-of-merit are calculated.

When calculating a map with isomorphous replacement data, it is important to calculate a map weighted by the figure-of-merit for each reflection. This is done by choosing the Fo*fom option in xfft. The effect of using figure-of-merit weighting is shown in Fig. 3.17.

The average figure-of-merit for the entire map gives an idea of the quality of the map. Figure-of-merits from different programs are not directly comparable unless they use comparable methods of calculating E. E is used in the denominator of the phasing equation and directly affects the figure-of-merits calculated. Also, phase sets run through solvent-flattening procedures

FIG. 3.17 Effect of figure-of-merit weighting. Two MIR phased maps are shown, (A) with no figure-of-merit weighting and (B), with figure-of-merit weighting. The weighted map is somewhat cleaner and easier to interpret, making it the map of choice to use for MIR phasing.

A

B

or density modification will have higher figure-of-merits, whether or not they are really improved. Nonetheless, there are some rules-of-thumb for figure-of-merit values and the corresponding map quality. If the figure-of-merit is less than 0.5 to 3.0 Å resolution, the map will be noisy and very difficult to interpret. Around 0.6 Å the map starts to be interpretable, and if above 0.75 Å, the map is almost certainly interpretable. If it is above 0.8 Å, without any modifications after the isomorphous replacement phase calculation, the map will be of excellent quality and a joy to interpret. In the author's experience, the increase in figure-of-merit that occurs with solvent flattening is of little use in judging the final map quality. It may be used as a relative number to judge the effect of using different solvent-flattening parameters. In the final conclusion, the best way to judge the quality of the phase set is actually to examine the map (Fig. 3.18).

To search for more heavy-atom sites, try to use a difference map or a residual map. In the difference map the coefficients used are $F_{PH} - F_P$, $\alpha_{protein}$, which produces a map that shows the positions of all of the heavy atoms (see the previous discussion on difference Fouriers). Any peaks on this map that

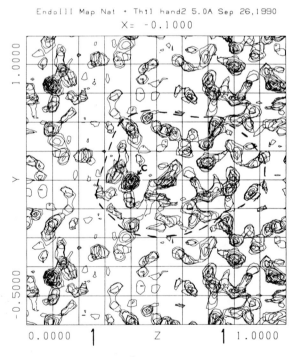

FIG. 3.18 Low-resolution MIR map. A 10-Å slab is shown of a typical low resolution MIR map at 5 Å. The clean solvent boundaries (indicated by dashed lines) indicate that this solution is on the right track.

are not already in the heavy-atom solution can be added and refined. Since it is possible for ghost peaks from other derivatives included in the phasing to show up, especially if the figure of merit is low, be careful not to add peaks that are in other derivatives used to calculate the phases. Whether the site is truly in common or just a ghost peak can be verified by examining the Patterson map. Add peaks in the order of size until the relative occupancy falls below 0.1–0.25 times the highest site. Adding too many minor sites may only model errors in the data and make the phases worse, not better. A residual map is similar except that the coefficients are $(F_{PH} - F_P) - (F_{PHCALC} - F_P) \ \alpha_{protein}$, which results in the peaks accounted for in the solution to be removed from the map so that only new positions are shown. These can be treated in the same manner as for the difference map.

As more derivatives are found and added, the resulting protein phases should improve (Fig. 3.19). With the improved protein phases, it is worth-

A

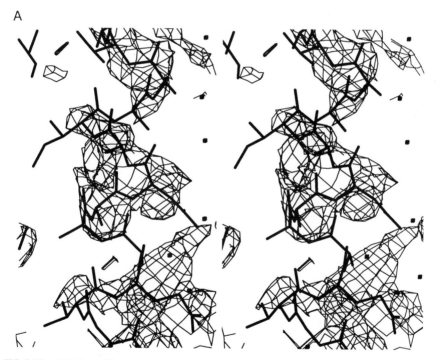

FIG. 3.19 Addition of heavy-atom derivatives. The same section of MIR map is shown, using one, two, three, and four derivatives to determine the phasing. (A) Notice that the single-derivative map is largely uninterpretable, with breaks in the main chain, and the effective resolution is quite low. (B) As a second derivative is added, things improve dramatically. (C) With the third derivative, the isoleucine side chain on the left is becoming visible. (D) The fourth derivative has little effect, and further addition of derivatives of this quality will probably not improve the map much. Compare these maps with Fig. 3.28.

B

C

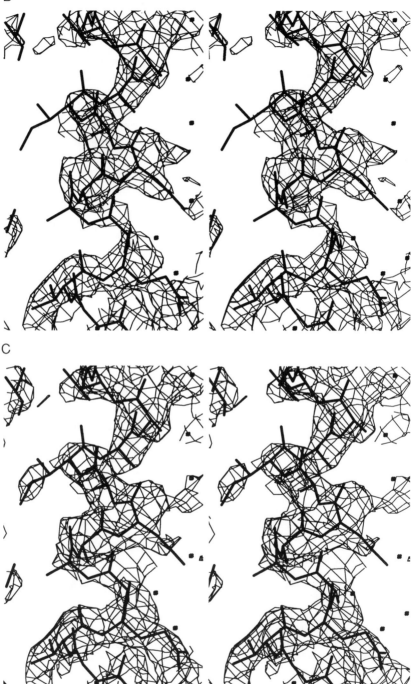

FIG. 3.19—*Continued.*

D

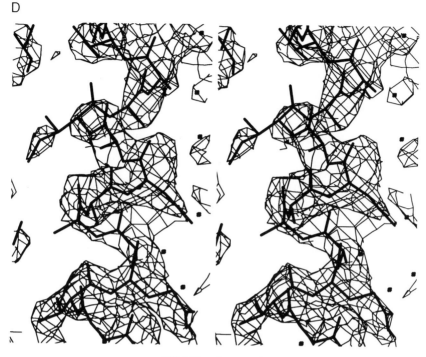

FIG. 3.19—*Continued.*

while repeating the difference Fouriers for all the derivatives to look for minor sites. Always make sure there is something in the Patterson map to justify the addition of the new site (especially check the cross-peaks). A site may not give all of the peaks on the Patterson, but if there is only one or none, then the site is not justified. If you use a heavy-atom refinement program that uses the lack-of-closure error refinement, then be wary of minor sites in one derivative that have the same position as a strong site in another derivative. This site is likely to refine to a good occupancy whether it is real or not.

Heavy-Atom Phasing Statistics

If you look at only one number in the phasing statistics, look at the phasing power or $\langle F_H \rangle / \langle E \rangle$, the size of the heavy-atom amplitudes over the error. This statistic takes into account the two factors that determine heavy-atom phasing power: the size of the differences and the size of the errors. This number should be above 1.0 for the derivative to contribute any to the phasing. Phasing powers above 2.0 are good, and the very best derivatives can sometimes reach 4.0. Since the sizes of the differences are fixed for a given data set to increase the phasing power, you must lower the errors by improving the solution or collecting better data. Otherwise, phasing power may be increased by longer soaking times and/or higher concentrations for a given derivative. More careful data collection and attention to scaling and merging may also lead to lowered errors. As resolution increases, the phasing power falls off, and this can be used as a guide of the maximum resolution at which the derivative can be used. F_H decreases with resolution because the scattering factors for the heavy atoms fall off with resolution. The errors increase with poor measurements, scaling errors, and non-isomorphism. Often a derivative is non-isomorphous above a certain resolution but can be used for phasing below this.

Another useful number is the centric R-value, for which the rule of thumb goes: above 0.70, the solution is wrong; 0.60–0.69, the derivative may be useful, but look for improvements to the solution; 0.50–0.59, the solution is definitely useful; below 0.5, this is an excellent derivative. You should find phasing power and the centric R correlated. There are lots of other possible phasing statistics, but the two that are mentioned have been the most used over the years.

Including Anomalous Scattering

There are two ways anomalous scattering can be included in the phasing. One is as an adjunct to the isomorphous phasing by measuring F_{PH}^+ and F_{PH}^-. Many heavy-atom compounds have a useful anomalous signal (Table 3.3), and provisions for including the anomalous signal are included in most isomorphous phasing programs. The equations for this have been worked out in terms of A_{ano}, B_{ano}, C_{ano}, D_{ano}, and these terms are simply summed, as for an isomorphous derivative. Also, inclusion of the anomalous scattering can improve Patterson maps. Since the anomalous portion of the heavy-atom structure factor is at right angles to the real part, the Bijvoet difference $F_{PH}^+ - F_{PH}^-$ will be largest when $F_P - F_{PH}$ is smallest, so that the phasing power of the two are complementary. One major difference holds when using the anomalous scattering. The hand of the heavy atom positions must be correct. Since there is no way to determine this a priori, both hands can be tried. In one direction the figure-of-merit should be slightly higher than in the other. The map can also be checked for clues from α-helices and β-turns (see the following), both of which are handed. Another strategy for dtermining the correct hand is to cross-phase a second derivative with SIRAS (Single Isomorphous Replacement with Anomolous Scattering) phases in both possibilities. In one map the peak height of the derivative should be higher than in the other map. All lines of evidence should point to the same answer. If there is a conflict (i.e., the map has left-handed helices even though the figure-of-merit is higher), then be sure that F_{PH}^+ and F_{PH}^- have not become switched somewhere. This can occur by using a left-handed description of the machine you collected data on (i.e., assigning the rotation angle to the opposite direction) or by transforming the indices that switched the handedness without also switching F_{PH}^{+} and F_{PH}^{-}, such as $h = -h$.

The other method is to use a native anomalous scatterer, that is, a scatterer in the native crystal such as an iron co-factor in a heme or iron sulfide, cluster. There are some extra difficulties in this case. There is no anomalous signal in the centric data, so these reflections must be left out of the syntheses. As can be seen in Table 3.1, centric reflections can represent a sizable fraction of the data in the low-resolution ranges. The centric isomorphous differences do not suffer from the approximations of the acentric ones, and their presence in Patterson maps, heavy-atom refinement, and error-estimation in phasing make a large contribution to the robustness of these methods. To get around this, use native anomalous Pattersons and refinements with the top 30%, or so, of the data (excluding outliers). The theory behind this is that the larger differences are more likely to appear when F_A is co-linear with the protein phase and, thus, these differences are better approximations of F_A. In practice, the author has found in at least two cases that the Patterson maps were little different when either all or 30% of the reflections were included. Since the terms are squared in a Patterson, the smaller differences do not add

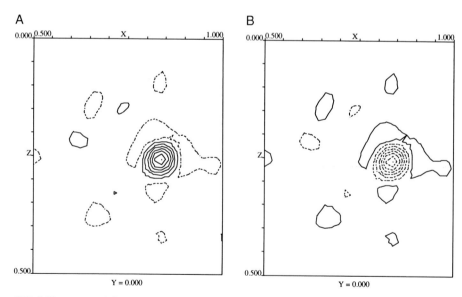

FIG. 3.20 Bijvoet difference Fourier. Sulfite reductase has a large, anomalous signal at 1.54 Å wavelength because of the presence of a Fe_4S_4 cluster and a siroheme iron. A Bijvoet difference Fourier using MIR phases from two derivatives is shown using data in left and right handedness. Solid contours are positive density, and dashed contours are negative. In the correct right-handed case, (A) a large peak is found at the site of the Fe_4S_4 cluster. In the incorrect left-handed case, (B), a large hole is found.

much to the map and their absence makes little difference. If a derivative exists, then a Bijvoet difference Fourier can be used to find the native anomalous scatterer (Fig. 3.20). This is the same as in the derivative case except that 90° is subtracted from the phase. If you are using xmergephs in XtalView, there is an option to do this built into the program. Refinement can be done using conventional heavy-atom refinement programs since $F_H \propto F_A$ using the reflections with the 25–30% largest differences. This approximation will be valid enough for refining the x, y, and z. A full refinement can be done with the computer program ANOLSQ written by Wayne Hendrickson (Columbia University). If the protein contains a cluster of known geometry such as an Fe_4S_4 cluster, then a rigid-body refinement of a standard cluster[10] will give improved accuracy by reducing the number of refinable parameters.

Finally, the native anomalous scatterers can be included in the phasing process. While only a few cases exist where native anomalous scattering has been sufficent to solve a protein without using multiple-wavelength methods (see the following) they can contribute substantially to an isomorphous solution. Hendrickson–Lattman coefficients can be derived for the native anomalous case:

$$A_{ano} = 2(F_P^+ - F_P^-)a_A/E_2$$

$$B_{ano} = 2(F_P^+ - F_P^-)b_A/E_2$$

$$C_{ano} = -(a_A^2 - b_A^2)/E^2 \qquad (3.22)$$

$$D_{ano} = -2a\,b_A/E^2,$$

where $a_A = F_A \cos(\alpha_A)$ and $b_A = F_A \sin(\alpha_A)$ where F_A, α_A is the calculated anomolous scatterer structure factor. Again, the model must be in the correct hand for the phases to be correct. This may be judged by examining maps in both hands and by looking at the figure-of-merit in both possibilities.

Fine-Tuning of Derivatives

Ideally, the computer should be able to refine all the derivatives to the best possible values and produce the best possible map. In reality, "fiddling" with the derivative parameters often leads to a better map. If you have a multiple-derivative solution, the first step is to see if removing a poor derivative improves the map. In order to tell if the map is better or worse, it helps if you can find a recognizable feature. A helix is usually the best for this, as the geometry of helices is tightly constrained. Another feature to look at is the solvent. As the phases improve, the contrast between the protein and solvent should improve and the solvent should become flatter. Once you have

[10] An example of this can be found in McRee, D. E., Richardson, D. C., Richardson, J. S., and Siegel, L. M. (1986). *J. Biol. Chem.* **261**, 10277–10281.

decided on a section of map to look at, you can adjust the parameters, recalculate the map, and try to decide if it is better. If you are using XtalView, you can set up all the windows you need and just keep reexecuting them to view the results. If removing a derivative seems to improve the map, then try lowering its resolution limit to see if it is still usable at a lower resolution.

Check for heavy atom ghost peaks and holes. These are large peaks or holes at the position of one of the heavy atoms. They can be removed by adjusting the occupancy of the site up or down. If you used lack-of-closure error, heavy-atom refinement, and refined occupancy often, this is almost sure to have happened.

Keep an accurate record of the changes you make so that you can return to a good solution later. Be objective. Increasing the resolution of your map beyond the phasing power of the derivatives will not make it easier to fit. Solvent flattening of very noisy data will only yield solvent-flattened noise. Beware of snake-oil and wooden nickels.

····· 3.7 ·····
MOLECULAR REPLACEMENT

Many structures can be phased by using a homologous structure and molecular replacement.[11] In this method, the homologous *probe* structure is fit into the unit cell of the *unknown* structure and the phases are used as an initial guess of the unknown structure phases. A six-dimensional search is required to find the best match of the probes transform to the observed transform—three angles and three translations. Fortunately, it is possible to split this search into two three-dimensional searches: a rotation search followed by a translation search. As an example of the amount of time this saves, suppose that a search of one dimension takes 10 sec of computer time. If we split up the search, the total time is $10^3 + 10^3 = 2,000$ sec as compared to $10^6 = 1,000,000$ sec, or more than 11 days. How identical does the probe need to be? This depends upon the structural identity of the two proteins as opposed to the sequence identity. Since we do not know the structural identity we are forced to use the sequence identity as a guide. A (very) rough rule-of-thumb is that above 50% sequence identity a molecular replacement solution should be straightforward since chances are that these two proteins are structurally very similar. The largest problem with searches using low homology probes is that once the solution is obtained, the phases will be poor estimates of the

[11] An excellent discussion of molecular replacement along with several examples can be found in an article by E. Lattman, "Use of the Rotation and Translation Functions," in "Methods in Enzymology," Vol. 115, pp. 55–77. The classic work on molecular replacement is a collection of articles edited by M. G. Rossman (1972), "The Molecular Replacement Method," *Int. Sci. Ser.,* **13**.

true phase, and there will be a high bias towards the probe structure, making it difficult to refine the correct structure. This bias is the main drawback of the molecular-replacement method. In many cases, a probe that represents only a portion of the structure is available, for example, in an antibody–protein complex. The method is robust enough that the probe can be accurately positioned in many of these cases.

Molecular replacement is often thought of as an easier alternative to multiple-isomorphous replacement. However, in practice the author has seen rotation–translation solutions take as long as, or longer than, heavy atom searches. The combination of both methods is more powerful than either alone. In the heavy-atom case, the phases are noisy but unbiased. In molecular replacement, the phases are heavily biased. A particularly successful strategy is to use a single derivative along with a molecular-replacement solution to cross-check each other and to combine the phases to produce a better map than either method can produce alone. The phases of the molecular replacement solution can be used to cross-Fourier the derivative differences in order to find the heavy atom solution. These single isomorphous replacement phases can then be combined with the molecular replacement phases to filter them and to remove the phase ambiguity, thus producing a map superior to what either can produce alone.

Rotation Methods

Rotation searches are actually done in Patterson space. Consider the Patterson of a protein molecule packed loosely in a lattice. In general, the short vectors will be *intra*-molecular vectors, and the longer ones will be *inter*-molecular. In a rotation function we want to consider only intramolecular vectors. Since all vectors in a Patterson start at the origin, the vectors closest to the origin will, in general, be intramolecular. Of course, closely spaced lattice contacts will also produce short intermolecular vectors, but they should be in the minority. By judiciously choosing a maximum Patterson radius, we can improve our chances of finding a strong rotation hit. The second choice to be made is the resolution range to use in calculating the Patterson. Higher-resolution reflections (above about 3.5 Å) will differ markedly even between homologous structures as they reflect the precise conformation of residues. Lower-resolution reflections reflect the grosser features of the structure, such as the relation of secondary structural elements. Very low resolution reflections, below about 10 Å, are heavily influenced by the crystal packing and the arrangement of solvent and protein, which is, of course, more dependent on the particular packing arrangement than on the structure of an individual protein molecule. Thus, the resolution range used for rotation searches is usually within 10–3.5 Å, with 8–4 Å being common. In practice, several ranges can be tried.

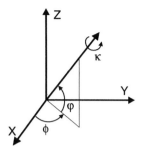

FIG. 3.21   Definition of Eularian self-rotation angles.

The first step is to calculate structure factors for the probe structure in an artificial P_1 cell. The cell should be about 30 Å larger than the probe in each direction so that there are no intermolecular vectors in the Patterson radius used for the rotation search. The probe is usually centered at the origin of this cell to simplify later steps. These calculated structure factors are then used in the rotation search. While the search is not actually done computationally in this manner, it can be conceptualized as the following. A Eularian angle system is used with three nested angles (Fig. 3.21). The range of the angles is chosen in order to cover the unique volume of the space group of the unknown structure. One angle is incremented 5° and the other angles are moved through all their values every 5° and at each point the match of the probe and observed Patterson functions is calculated and stored. The first angle is incremented 5° more and the process repeated. After all angles have been calculated, the list is sorted, grouped into peaks, and printed (Fig. 3.22). The peaks are usually reported in terms of their size relative to the root-mean-square peak size or sigma. If the probe structure is a good match for the unknown structure and the proper resolution ranges have been chosen, then there will be a single large hit at several sigma. Often, a decreasing series of peaks are found at slightly differing sigma. In such cases, it has been found that the correct peak may not be the first in the list. The order of the list may be altered by changing the resolution ranges slightly. Since the correct peak is unknown there is no a priori way to decide on the correct ranges. In these cases a new probe should be looked at if one is available. If not, there is the option of continuing the process, going down the list of rotation hits to see if a decision can be made based on the behavior of subsequent steps.

The asymmetric unit of the rotation function depends on the symmetry of both the probe's Patterson and the Patterson of the unknown. This problem has been examined and reported on in detail,[12] and Table 3.4 lists the

[12] Rao, S. N., Jih, J., and Hartsuck, J. A. (1980). *Acta Crystalogr.* **A36**, 878–884.

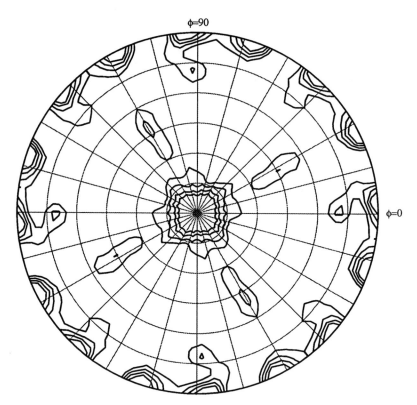

FIG. 3.22 Self-rotation function example. Peaks on this section ($\kappa = 180$) show positions of 2-fold symmetry in the diffraction pattern. The large peak in the center is due to the crystallographic 4-fold (which can be thought of as two 2-folds). The peaks around the outer edge are positions of non-crystallographic 2-folds.

more common cases when the probe is space group P1 for each of the 10 possible Patterson space groups for proteins. It makes a difference which Patterson is rotated and which is held still. This can be decided by a careful reading of the rotation program's documentation, because both conventions of rotating the probe or the unknown are used.

In Eularian space, the operator $\pi + \theta_1, -\theta_2, \pi + \theta_3$ is an identity operator. This should be kept in mind when comparing hits. Two rotation hits that look different may in fact be the same if the identity operator is applied.

Improving the Probe

It may be possible to improve the hit by systematically leaving out pieces of the probe and doing the rotation search again. As an example, every

3.7 Molecular Replacement

TABLE 3.4

Rotation Function

Unknown Patterson group	$\bar{1}$ Probe rotated		$\bar{1}$ Probe held still	
	$\theta_1, \theta_2, \theta_3$	$\theta_+, \theta_2, \theta_-$	$\theta_1, \theta_2, \theta_3$	$\theta_+, \theta_2, \theta_-$
$\bar{1}$	$0 \le \theta_1 < 2\pi$ $0 < \theta_2 \le \pi$ $0 \le \theta_3 < 2\pi$	$0 \le \theta_+ < 4\pi$ $0 \le \theta_- \le 2\pi$	$0 \le \theta_1 < 2\pi$ $0 < \theta_2 \le \pi$ $0 \le \theta_3 < 2\pi$	$0 \le \theta_+ < 4\pi$ $0 \le \theta_- \le 2\pi$
$2/m\ b$ unique	$0 \le \theta_1 < 2\pi$ $0 < \theta_2 \le \pi/2$ $0 \le \theta_3 < 2\pi$	$0 \le \theta_+ < 4\pi$ $0 \le \theta_- \le 2\pi$	$0 \le \theta_1 < 2\pi$ $0 < \theta_2 \le \pi/2$ $0 \le \theta_3 < 2\pi$	$0 \le \theta_+ < 4\pi$ $0 \le \theta_- \le 2\pi$
$2/m\ c$ unique	$0 \le \theta_1 < 2\pi$ $0 < \theta_2 \le \pi$ $0 \le \theta_3 < \pi$	$0 \le \theta_+ < 4\pi$ $0 \le \theta_- \le \pi$	$0 \le \theta_1 < \pi$ $0 < \theta_2 \le \pi$ $0 \le \theta_3 < 2\pi$	$0 \le \theta_+ < 4\pi$ $0 \le \theta_- \le \pi$
mmm	$0 \le \theta_1 < 2\pi$ $0 < \theta_2 \le \pi/2$ $0 \le \theta_3 < \pi$	$0 \le \theta_+ < 4\pi$ $0 \le \theta_- \le \pi$	$0 \le \theta_1 < \pi$ $0 < \theta_2 \le \pi/2$ $0 \le \theta_3 < 2\pi$	$0 \le \theta_+ < 4\pi$ $0 \le \theta_- \le \pi$
$4/m$	$0 \le \theta_1 < 2\pi$ $0 < \theta_2 \le \pi$ $0 \le \theta_3 < \pi/2$	$0 \le \theta_+ < 4\pi$ $0 \le \theta_- \le \pi/2$	$0 \le \theta_1 < \pi/2$ $0 < \theta_2 \le \pi$ $0 \le \theta_3 < 2\pi$	$0 \le \theta_+ < 4\pi$ $0 \le \theta_- \le \pi/2$
$4/mmm$	$0 \le \theta_1 < 2\pi$ $0 < \theta_2 \le \pi/2$ $0 \le \theta_3 < \pi/2$	$0 \le \theta_+ < 4\pi$ $0 \le \theta_- \le \pi/2$	$0 \le \theta_1 < \pi/2$ $0 < \theta_2 \le \pi/2$ $0 \le \theta_3 < 2\pi$	$0 \le \theta_+ < 4\pi$ $0 \le \theta_- \le \pi/2$
$\bar{3}$	$0 \le \theta_1 < 2\pi$ $0 < \theta_2 \le \pi$ $0 \le \theta_3 < 2\pi/3$	$0 \le \theta_+ < 4\pi$ $0 \le \theta_- \le 2\pi/3$	$0 \le \theta_1 < 2\pi/3$ $0 < \theta_2 \le \pi/2$ $0 \le \theta_3 < 2\pi$	$0 \le \theta_+ < 4\pi$ $0 \le \theta_- \le 2\pi/3$
$\bar{3}/m$	$0 \le \theta_1 < 2\pi$ $0 < \theta_2 \le \pi/2$ $0 \le \theta_3 < 2\pi/3$	$0 \le \theta_+ < 4\pi$ $0 \le \theta_- \le 2\pi/3$	$0 \le \theta_1 < 2\pi/3$ $0 < \theta_2 \le \pi/2$ $0 \le \theta_3 < 2\pi$	$0 \le \theta_+ < 4\pi$ $0 \le \theta_- \le 2\pi/3$
$6/m$	$0 \le \theta_1 < 2\pi$ $0 < \theta_2 \le \pi$ $0 \le \theta_3 < \pi/3$	$0 \le \theta_+ < 4\pi$ $0 \le \theta_- \le \pi/3$	$0 \le \theta_1 < \pi/3$ $0 < \theta_2 \le \pi$ $0 \le \theta_3 < 2\pi$	$0 \le \theta_+ < 4\pi$ $0 \le \theta_- \le \pi/3$
$6/mmm$	$0 \le \theta_1 < 2\pi$ $0 < \theta_2 \le \pi/2$ $0 \le \theta_3 < \pi/3$	$0 \le \theta_+ < 4\pi$ $0 \le \theta_- \le \pi/3$	$0 \le \theta_1 < \pi/3$ $0 < \theta_2 \le \pi/2$ $0 \le \theta_3 < 2\pi$	$0 \le \theta_+ < 4\pi$ $0 \le \theta_- \le \pi/3$

Note. Assymetric units in Eularian angles $\theta_1, \theta_2, \theta_3$ and pseudo-orthogonal Eularian angles $\theta_+, \theta_2, \theta_-$. The range of θ_2 is the same in both systems. The assymetric unit given is one of several possible choices in each case.

three residues can be deleted and the size of the rotation hit examined. The absolute sizes of the hits and not the peak/σ ratio are compared. If parts of the probe structure that contribute to the match and are thus likely homologous are removed, a lower hit will be found. If a portion is interfering, then removing it will result in a larger hit (Fig. 3.23). A probe can then be built,

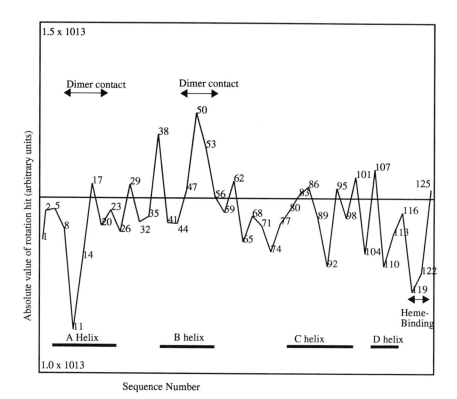

Sequence Number

FIG. 3.23 Systematic deletion of residues and the rotation function. Plotted on this graph is the absolute value of the rotation peak, using *R. molischianum* cytochrome *c'* as the probe against *C. vinosum* cytochrome *c'* data, with every three residues deleted throughout the entire probe molecule. The number indicates the first of the three residues omitted. The horizontal line across the entire graph indicates the value of the hit found using the entire model. Points above this line represent models that give larger hits when residues are removed, thus these residues are probably in a different conformation than those in the *C. vinosum* structure. Points below the line represent those that make the hit smaller when removed and are, thus, probably in similar positions in the two proteins. After the *C. vinosum* structure was solved (see Chapter 4), the models were compared by superposition of the coordinates. It was found that the best superposition is of the heme-binding pocket, which includes residue 12 and the residues indicated by the arrow and the legend "Heme-Binding." The biggest difference was at the dimer contact in helix B, where the last turn of the B helix is missing in *C. vinosum*. Thus, the rotation function used in this manner can give useful information about the unknown structure before a complete solution is available.

leaving out several of the residues that raise the hit and, thus, the overall rotation hit can be improved. A similar procedure can be done, leaving off side chains beyond the C_β atom (i.e., poly-alanine) on the hypothesis that the main chains follow the same path but the side chains differ considerably.

Refining the Rotation

The success of the translation search will depend on the quality of the rotation solution. As a minimum, the rotation search should be repeated in a limited range with a fine grid. In particular, if you used the Crowther fast-rotation method, then a finer search with the conventional rotation function should be done. Another alternative is to use a least-squares refinement method based on the Patterson correlations as implemented by XPLOR.[13] In this method, the probe is treated as a rigid body and its orientation refined to maximize the correlation between $|E_{obs}|^2$ and $|E_{model}|^2$ given by the function

$$\frac{\langle|E_{obs}|^2|E_m(\Omega,\Omega_i,t_i)|^2\rangle - \langle|E_{obs}|^2\rangle\langle|E_m(\Omega,\Omega_i,t_i)|^2\rangle\rangle}{\sqrt{\langle|E_{obs}|^4 - \langle|E_{obs}|^2\rangle^2\langle|E_m(\Omega,\Omega_i,t_i)|^4 - \langle|E_m(\Omega,\Omega_i,t_i)|^2\rangle^2\rangle\rangle}}, \quad (3.23)$$

where the overall orientation Ω is refined. It is also possible to break up the probe into two or more groups and to refine each orientation, Ω_i, and translation, t_i, individually. The data is normalized from F's to E's by binning the data in shells of resolution with equal numbers of reflections per bin and setting the average of each bin to the same value. In the normal case, where the amplitudes fall off with resolution, this will have the effect of emphasizing the shorter vectors. Another program that performs the same function of refinement of the rotation angles of unpositioned models is INTREF.[14] This program incorporates a radial weighting function to downweight longer vectors, which tend to be the undesired intergroup vectors. INTREF also permits refining an individual group orientation and the relative translations between groups.

Translation Methods

Translation Function

Once a rotation solution has been found, the translation function can be attempted. This step is not as robust as the rotation method. It is possible to get a correct rotation solution and still not find a translation solution for poor probe structures. Of course, it can be difficult to distinguish this from

---

[13] Brünger, A. T. (1990). *Acta Crystalogr.* **A46**, 46–57, and the XPLOR 3.0 manual, p. 278.
     [14] Yeates, T. O., and Rini, J. M. (1990). *Acta Crystalogr.* **A46**, 352–359.

the case where the rotation is incorrect. One reason for the low signal in the translation function is that the only difference between correctly and incorrectly positioned molecules is in the intermolecular vectors. The intramolecular vectors are identical, being independent of the translation position, and form a large background.

The simplest translation function is similar to a correlation function

$$T(x,y,z) = \sum_{h,k,l} F_c^2(h,k,l,x,y,z)F_o^2(h,k,l), \qquad (3.24)$$

where $T(x,y,z)$ should be at a maximum when the probe molecule is in best agreement with the observed data. Notice that F_c is calculated at every x,y,z and for every h,k,l, making this a large calculation. Because they are heavily influenced by the solvent, the lowest-order data below $10-25$ Å are left out of the calculation. On the other hand, the logic of this seems questionable because the intermolecular vectors are also of lower order since they represent the longest spacings in the crystal. The high-resolution cutoff is generally about $4-6$ Å because the probe is only an approximation, and the higher-order terms will not agree even at the correct position due to the differences in the precise conformations of the two structures. In order to remove the background intramolecular vectors, the formula can be recast to remove approximately the self-vectors:

$$T(\mathbf{x2} - \mathbf{x1}) = \sum_h (I_o - k(F_{M1}^2 + F_{M2}^2))F_{M1}F_{M2}^* \exp[$$
$$-2\pi i h(\mathbf{x_2} - \mathbf{x_1})], \quad (3.25)$$

which, although it looks complicated is now a Fourier summation, where F_{M1} and F_{M2} are the transforms of molecule 1 and molecule 2 and the function gives the vector between molecule 1 and molecule 2. The function is used to calculate Harker sections as in the heavy-atom case, and the problem is solved in a similar manner by considering the peaks on the Harker sections as self-vectors.

Thus, for the space group P222 the three Harker sections are $x = 0$, $y = 0$, and $z = 0$. The self-vectors are located at $0, 2y, 2z; 2x, 0, 2z;$ and $2x, 2y, 0$. There should be one unique peak on each section, and solving any two should give a solution that can be confirmed by the third.

Translation Search

An alternative to the translation function is an R-factor search where the molecule is moved on a grid and at each point the R-factor between the calculated probe amplitudes and the observed unknown structure amplitudes is calculated (Fig. 3.24). A more robust method is to calculate the correlation function since it is immune to scaling errors and an accurate scale

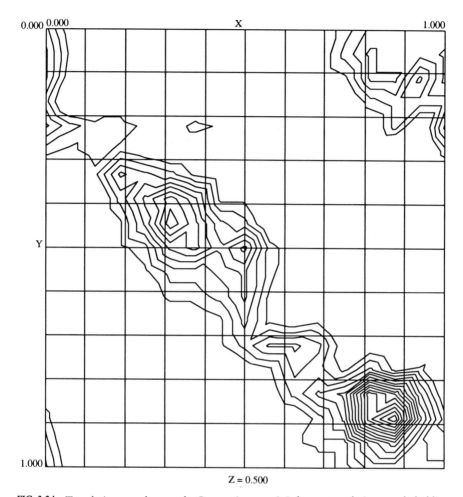

FIG. 3.24 Translation-search example. One section, $z = 0.5$, from a translation search, holding one-half of the C. *vinosum* cytochrome c' dimer fixed at a previously found position and searching with the second half. The second half was translated throughout the entire cell, and a correlation coefficient was calculated at each point. The map shows the correlation squared times 100. The largest peak in the map gives the correct translation of the second half of the dimer. The streakiness of the map is common in translation searches. Often the correct solution is where two or more streaks cross.

cannot be calculated at this point in the structure determination. The translation search takes more computer time than the translation function, but it is less prone to error and, as computers become faster, it will become the method of choice. Several proteins that could not be solved using the translation function were solved using the correlation search.

The volume that needs to be searched is dependent upon the space-group symmetry. Many space groups (those that belong to class 1, 2, 3, 4, and 6, the space groups that do not have orthogonal symmetry elements—see Table 2.2) are polar, and one direction is arbitrary. This direction need not be searched, and if it is the exact same hit will be found on all sections in this direction. For example, in $P2_1$ there is no need to search in the y direction.

Molecular Packing

There will only be a limited number of positions in the cell where the molecule can pack without overlapping one of its neighbors (Fig. 3.25). Since it is physically impossible for the molecules to overlap in three-dimensional space, these positions can be safely eliminated. The complication comes when one realizes that the probe structure may have a loop that is not in the unknown structure and, thus, a false overlap may be found. Conversely, the probe may not overlap when the unknown will. Thus, in practice, one needs to be fairly generous in deciding on the amount of overlap allowed. Also, since proteins are fairly symmetrical, one could imagine a perfectly good packing with one or more directions flipped. Although a unique solution may not be found, a large number of possibilities can be eliminated. Harada and co-workers have incorporated an interpenetration penalty into their translation function to incorporate molecular packing information.[15] Another approach is to run a packing search, generate a list of good packing positions, then calculate the correlation function for each of these positions to find the best translation.

[15] Y. Harada, Lifchitz, A., Berthou, J., and Jolles, P. (1981). *Acta Crystalogr.* A37, 273.

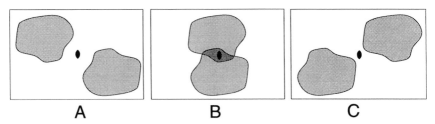

FIG. 3.25 Molecular-packing illustration. A hypothetical unit cell with a 2-fold in the center is shown. A molecule is moved across the cell and the symmetry mate is generated at each position. Note that in two positions the molecules do not overlap and represent allowable packings; however, in the middle position the molecules overlap. This packing is not allowed and need not be considered in other types of searches.

····· 3.8 ·····
NON-CRYSTALLOGRAPHIC SYMMETRY

Non-crystallographic symmetry, or local symmetry, is symmetry that exists locally within the asymmetric unit of the crystal (Fig. 3.26). For example, the protein may crystallize as a dimer in the asymmetric unit so that there is a dyad axis relating the halves of the dimer that is not coincident with a crystallographic 2-fold. This information can be used to produce averaged maps in which noise will tend to cancel out and can be used as a phase restriction to improve phasing. In the extreme case of viruses with 20- to 30-fold redundancy in the asymmetric unit, the phasing restrictions become so high they can be used to phase reflections de novo.

A symmetry operator for non-crystallographic (NCR) symmetry takes the same form of a 3 × 3 rotation matrix and a translation vector. Crystallographic symmetry operators can be expressed in the same form but can take only certain values. Non-crystallographic symmetry operators are not

A

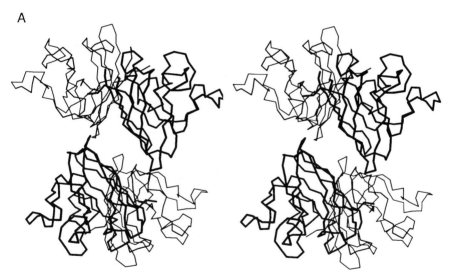

FIG. 3.26 Non-crystallographic symmetry averaging. (A) Stereo view of the contents of one asymmetric unit of bovine superoxide dismutase MPD crystal form. There are two dimers arranged head-to-head. Neither of the two dimer axes intersects a crystallographic axis and, thus, they are non-crystallographic. The densities of all four molecules were averaged in the original solution of the protein and aided in the tracing of the chain. (B) The medium-resolution density map of *C. vinosum* cytochrome *c'* shows an example of a non-crystallographic 2-fold. The axis of the 2-fold is indicated by the "X." The non-crystallographic dyad is nearly parallel to Z. (*Figure continues.*)

B

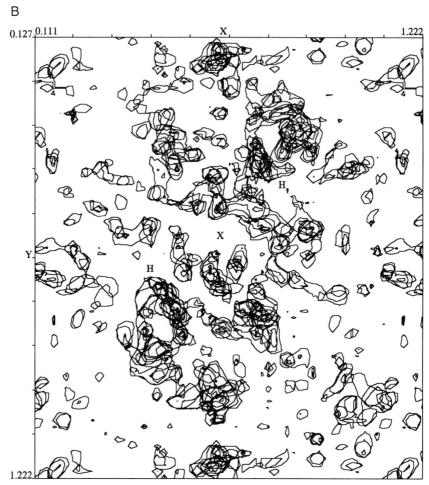

Z = 0.528; 0.537; 0.546; 0.556; 0.565; 0.574; 0.583; 0.593; 0.602; 0.611; 0.620; 0.630;

FIG. 3.26—*Continued.*

so constrained. It is usually customary to write crystallographic operators in algebraic form to show their generality, while NCR operators are usually written explicitly. Crystallographic symmetry operators can be looked up in tables, while NCR symmetry operators must be calculated on a case-by-case basis. It is necessary to have a list of equivalent points, either from a preliminary model or from equivalent points found in a map, to which a least-squares symmetry operator is calculated. XtalView has a program pdbfit that can calculate the matrix and vector that best fit two matched PDB models. Since pdbfit does not check the validity of the model, any arbitrary points could be input as fake atoms. It is only necessary that the equivalent pairs have the same atom type. Choose one molecule to be the standard reference position and calculate all of the pairs between this molecule as the target and the others as the source model. If you have solved the structure with heavy atoms, these pairs could be heavy-atom pairs. If you are only able to fit a portion of the unaveraged map, then these pairs can be partial models. Thus, if you have a dimer of A and B, there will be two operators A onto A, and B onto A. For four molecules A, B, C and D the four operators will be A onto A, B onto A, C onto A, and D onto A.

One big difference of non-crystallographic symmetry is its local nature. A crystallographic-symmetry operator can be applied to *any point* in the unit cell to find an equivalent point elsewhere. A local symmetry operation only relates a specific volume to another specific volume. There will be points in the unit cell where the local operation does not hold. Therefore, a knowledge of this volume is required to use non-crystallographic symmetry.

Usually this volume is one of the polypeptide chains, and the local symmetry operation describes how that chain is related to another. This description is useful in refinement programs. Before a model exists, this description is less useful. It is often desirable to use local symmetry to average out noise in electron-density maps. If there are three or more copies, this method becomes very powerful. An approximation may be useful where a box (or sphere) is used as the bounds and the operator relates this box to another, equivalent box. The box describes one of the equivalent molecules as can best be determined from the present map, and then there are one or more operators that relate how a point in other volumes (i.e., molecules) relates to this point. It is possible that the edges and corners of the box are not involved in local symmetry at all. This will cause a blurring. In most cases this blurring will be outside the bounds of the molecule and will not matter. Since every one of the equivalent molecules is packed differently, the crystal contact points will tend to be blurred. If you are using the averaging to help with fitting, this will not be a problem. If you are using it for phasing purposes, this will be more of a problem.

Another important difference of non-crystallographic symmetry is that it is not necessarily exact. Some parts of the molecule may be more similar than others. Usually the core and main chain of the molecule agree with the non-crystallographic operator to a close tolerance, but the surface side chains, which are more flexible, find themselves in different environments in each of the molecules and adopt different conformations (Fig. 3.27). When

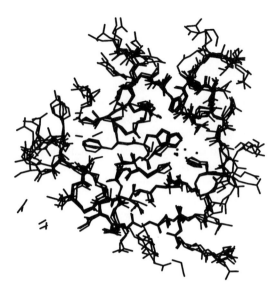

FIG. 3.27 Superposition of bovine SOD monomers. The models of all four momomers in the asymmetric unit of refined bovine SOD are superimposed for comparison. Shown is a narrow slab through the molecule. Note that in the center the four models superimpose closely, while at the solvent edges the side chains are in widely differing positions.

these multiple conformations are superimposed on one another with non-crystallographic symmetry operators, the density for them becomes blurred and confused. Outside loops of the protein may also adopt somewhat different or entirely different conformations. Still, non-crystallographic symmetry can be invaluable in tracing the main chain of most of the protein. With better phases you can then fit all of the molecules independently to improved maps to find the differences.

When using non-crystallographic symmetry during refinement, make some provision to allow for the different tolerances of fit. Since at the beginning of the refinement it is not known whether divergence in the model is error that you want to refine away or it is real, the main chain is usually tightly constrained but the side chains are not. A periodic check should be made to see if constraints should be loosened for a particular region. The constraints should permit the non-equivalent regions to drift apart. One strategy is to start the refinement with everything tightly constrained until convergence is reached. The constraints can then be removed and the model refined further. Regions of the model that have drifted apart can then be loosely constrained, or not at all, and the portions that have stayed close can be tightly constrained.

Self-Rotation Function

In order to discover non-crystallographic symmetry a self-rotation function is helpful. This is similar to the rotation method used previously, except that the probe amplitudes and the unknown amplitudes are the same. By looking on the sections of the outermost nested angle (usually κ), the position of these non-crystallographic symmetry elements can be found. For example, the section $\kappa = 180$ contains 2-folds. That is, when the second copy of the data is rotated 180° with respect to the first, near the direction of the 2-fold it will superimpose and give a peak in the function (Fig. 3.22). The position of the peak on the section gives the two angles that define the direction of the 2-fold. Similarly, peaks on the section $\kappa = 120$ are due to 3-folds; $\kappa = 90$, 4-folds; $\kappa = 72$, 5-folds; and $\kappa = 60$, 6-folds. All translation information is lost, so that if the symmetry element is a combination of a rotation and translation, then the self-rotation function will indicate the rotational portion only. The other difficulty with the self-rotation function occurs when the non-crystallographic symmetry is parallel to, but not coincident with, a crystallographic symmetry axis. In this case, the crystallographic symmetry will hide the non-crystallographic peak. For example, in the case of the *Chromatium vinosum* cytochrome c' structure discussed in Chapter 4, the direction of the 2-fold relating the halves of the dimer was very close to the c direction and could not be distinguished from the peak due to the crystallographic 2-fold screw along c. Also note that 4-folds are equivalent to two overlapping 2-folds 90° apart. Therefore, if you have a peak on the $\kappa = 90$ section, there will also be a peak on the $\kappa = 180$ section in the same direction. A similar relationship holds for 3-fold and 6-folds.

····· 3.9 ·····
DENSITY MODIFICATION

There are as many ways to handle density modification as there are programs written for it. They all seek to improve the phases by imposing restrictions on the density in real space and then using the phases of the modified map to filter or replace the experimental phases.[16] The weakness of this method is that there is no a priori way to judge the correctness of a density and, even if one were certain that a density is incorrect, determining the correct value is difficult. The success of the method may be highly problem and operator dependent. One assumption often made is non-negativity,

---

[16]Tulinsky, A. (1985). "Phase Refinement/Extension by Density Modification," in "Methods in Enzymology," Vol. 115, pp. 77–90.

which is only valid at very high resolutions above about 1.5 Å even with the F_{000} term included. Because lower-resolution maps have series termination effects, they will never be completely positive. A more rational approach is to assume that the most negative density is probably incorrect and that the largest peaks are also incorrect.

Solvent Flattening

Solvent flattening is a more straightforward technique and usually more successful. This method assumes that any density in the solvent region of the protein arises from noise fluctuations and that the solvent density should be flat everywhere throughout. (Actually, examinations of high-resolution maps with accurate phases still show some fluctuation in the solvent region even though there is no solvent model included. Still, the solvent is much closer to flat than the rest of the map.) The trick in this method is to find the boundary between the solvent and the protein. B. C. Wang has invented an averaging technique that has been successful at coming close to the correct boundary even in very noisy maps and has led to the development of automated solvent-flattening programs.[17] The algorithm is equivalent to a low-pass filtering of the data in reciprocal space. The lowest points in this smoothed map are then taken to be the solvent and the rest protein. Because the technique is not completely accurate, it may be wise to tell the program that the percent solvent of your crystal is about 5–10% lower than it really is. Another method is to consider this automated mask a first approximation and to edit the mask by hand to increase its accuracy. If a completely automated mask is used, then it should be periodically recomputed as the phases improve.

This mask is used to flatten the solvent regions of the map to produce a modified map. Other filtering may also be done at this point, such as removing excessively negative or positive density. This filtered map is inverted through a reverse Fourier transform to produce new structure factors. The new structure factors are then combined with the original MIR phase data to produced filtered structure factors. The filtered structure factors are used to produce a new map, and the process is repeated until the process converges. Convergence is reached when the phase change for a cycle is only a few degrees from the start of the cycle. These procedures have been implemented in a package of programs: PHASES.[18] In favorable cases the maps can show considerable improvement (Fig. 3.28).

[17]Wang, B. C. (1985). "Resolution of Phase Ambiguity in Macromolecular Crystallography," in "Methods in Enzymology," Vol. 115, pp. 90–112.

[18]The PHASES package is available from its author, Dr. William Furey, University of Pittsburgh.

Histogram Modification

A commonly used method for two-dimensional image processing is to modify the histogram of the image to match the expected histogram. The histogram in this case is the number of times a particular density value occurs. The densities are grouped into bins, and the number of densities that fall into that bin are recorded. This histogram is then compared to the expected histogram. In the case of a photograph, the expected histogram is computed from a similar image that has the characteristics desired. The histogram of the first image is then modified by multiplying each value in the image by a factor such that the final histogram is the same as the one expected.

A

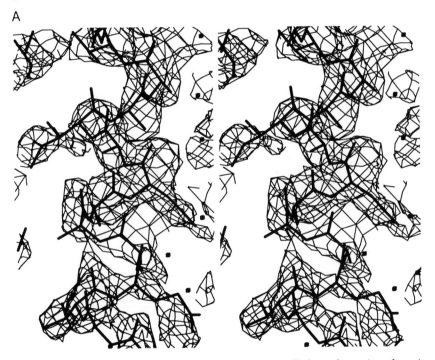

FIG. 3.28 Phase improvement by solvent flattening. A section of helix and a section of turn in a four-derivative MIR map are shown before and after solvent flattening and with the final refined phases at the same resolution. (A) Helix with starting MIR phases, (B) after solvent flattening, and (C) calculated phases from refined model for comparison. (D) Also shown is the turn region using MIR phases, (E) after solvent flattening, and (F) with refined phases. Note that the contrast of both maps is improved and some details are clearer after solvent flattening. In the turn region a false connection has become overemphasized and could lead to a false assignment. The original MIR map can be used to resolve the ambiguity. Compare the helix section with Fig. 3.19. (*Figure continues.*)

B

C

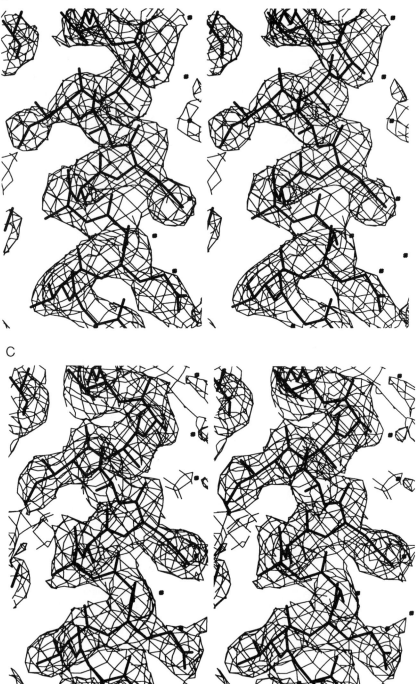

FIG. 3.28—*Continued.*

D

E

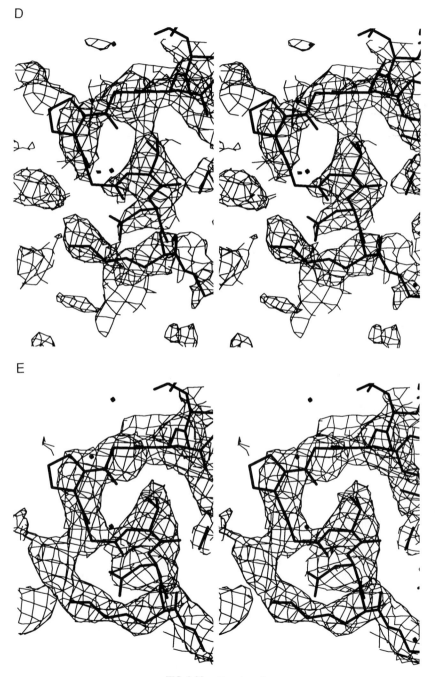

FIG. 3.28—*Continued.*

F

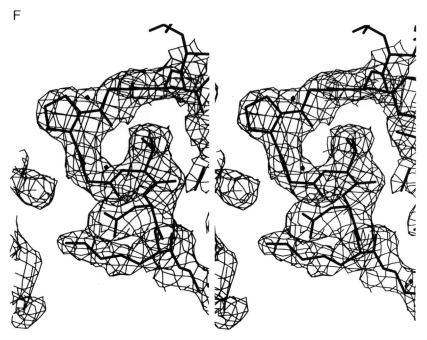

FIG. 3.28—*Continued.*

In the case of a protein, the expected histogram is that of the electron-density map of a well-refined protein. It is important that the electron-density maps having their histograms compared have the same resolution range and similar solvent contents. Tests have shown that it is not important that the proteins have the same fold. The histogram of mostly helical human hemoglobin is identical to that of mostly sheet bovine copper, zinc superoxide dismutase (Fig. 3.29). The presence of a metal is not a problem as long as it

---

FIG. 3.29   Protein histogram examples. Histograms from two proteins of radically different structure at 2.0 Å resolution are shown. (A) Superoxide dismutase is an eight-stranded β-barrel; (B) human hemoglobin is an all-helical protein. Both proteins were crystallized in cells with nearly identical solvent contents but with different space groups. Nevertheless, the histograms are nearly identical. The total histogram (solid lines) has been divided into the contributions from the protein (dotted lines) and the solvent (dashed lines). The protein portion contains a long tail of larger density values. These are the values we contour in order to interpret the structure. The negative values are smaller than positive ones but cluster more at an intermediate value. The solvent clusters tightly around 0 but is never completely 0. This is because there is always some weak density in the solvent region due to weak scattering of the solvent molecules, even after accounting for bound water molecules. The electron-density maps used for producing the histograms were calculated without including the F_{000} term. Including F_{000} would shift the histograms to the positive side by the amplitude of F_{000} but would not change the shape. In (C) and (D) the effect of resolution is shown on the histograms of copper, zinc superoxide dismutase and hemoglobin, respectively.

A

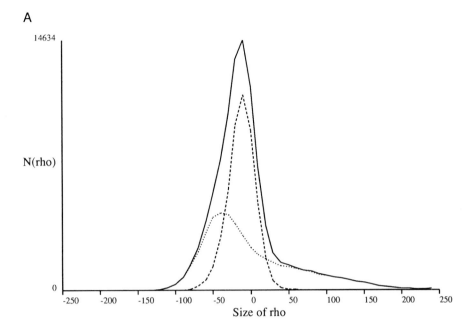

B

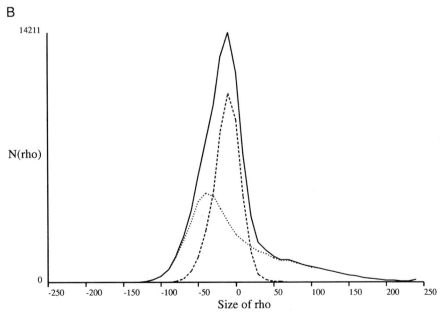

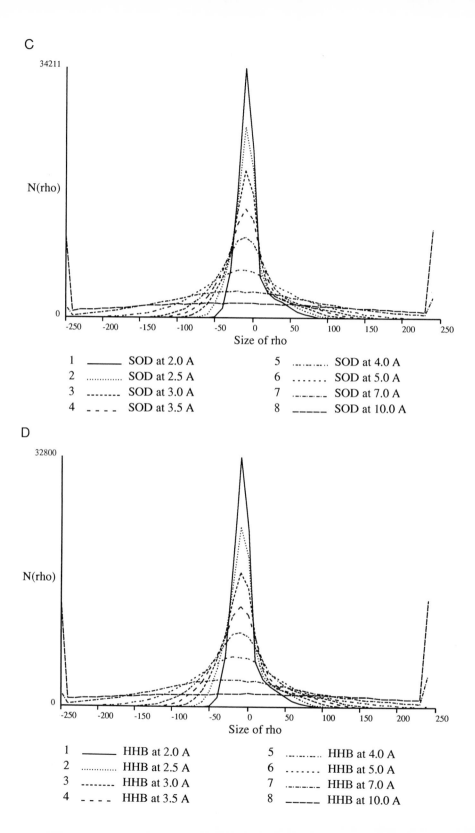

C

N(rho)

Size of rho

1	——————	SOD at 2.0 A	5	....	SOD at 4.0 A
2	....	SOD at 2.5 A	6	....	SOD at 5.0 A
3	---------	SOD at 3.0 A	7	.-.-.-.-	SOD at 7.0 A
4	- - - -	SOD at 3.5 A	8	_ _ _ _	SOD at 10.0 A

D

N(rho)

Size of rho

1	——————	HHB at 2.0 A	5	....	HHB at 4.0 A
2	....	HHB at 2.5 A	6	....	HHB at 5.0 A
3	---------	HHB at 3.0 A	7	.-.-.-.-	HHB at 7.0 A
4	- - - -	HHB at 3.5 A	8	_ _ _ _	HHB at 10.0 A

represents only a small portion of the total electron density. The method for modifying the histogram is the same as for a two-dimensional image. The electron-density values are binned by size, and the densities of each bin are scaled so that the final distribution is the same as that of the well-refined standard protein. This map will be easier to interpret, mostly because the contrast will be increased.

Histogram modification can also be used as a density-modification procedure to refine the protein phases. In this method the histogram-modified map is inverted to produce calculated phases that are combined with the original phases, and this produces a new phase set. The process is then recycled through the process until the phases converge. Convergence is slow in this process. In our tests we have used 20–40 cycles of cycling to improve MIR phases. The difference between the expected histogram and the one observed can be used as a guide to the progress of the method.

····· 3.10 ·····
MULTI-WAVELENGTH WITH ANOMALOUS SCATTERING

Recently a number of structures have been solved by this method, which takes advantage of the tunability of synchrotron-radiation X-ray sources and the presence of anomalous scatterers that have absorption edges in the range from 0.6 to 1.8 Å wavelength. In order for the method to work, there must be such anomalous scatterers in your molecule. Proteins often have these naturally in the form of metal cofactors. The method may also be used to solve a protein with a single derivative that contains an anomalous scatterer—most of the metals that are used for protein derivatives also have usable anomalous signals. In this case, isomorphism is not a consideration since the structure that is solved is the structure with the scatterer included. If the protein can be produced in *E. coli,* there is the possibility of substituting methionine with selenium–methionine, which gives a usable anomalous signal. The most complete discussion of the method can be found in Helliwell.[19]

Small blocks of data are collected, with each block being measured at three to four wavelengths. The wavelengths are picked to maximize the differences in the real and imaginary components of the anomalous scattering. This is done by picking wavelengths near the absorption edge of the anomalous scatterer. If possible, it is desirable to collect Bijvoet pairs close together in time. Often this can be arranged by aligning the crystal so that a mirror plane is present on the face of the detector. Otherwise, alternate blocks of

[19]Helliwell, p. 338–382 (see Suggested Reading list in Preface for complete reference). Computer programs for this method were written by Wayne Hendrickson, Columbia University, New York.

data can be collected such that the Bijvoets are collected in alternate blocks. Scaling is complicated by the necessity of keeping these pairs of data together. The phases are then computed by least-squares fitting of the terms to the multi-wavelength data. In order for this to work, the data must be of exceptional quality and accuracy, as the dispersive terms involved are usually small effects, on the order of 3–5%. Unlike isomorphous phasing, the phasing power actually increases at higher resolutions. The scattering factor for the protein atoms falls off rapidly with resolution, while the scattering factor for anomalous dipersion stays almost constant with resolution, so that the percentage difference increases. All data can be measured from a single cystal, alleviating scaling problems, and, consequently, there is no isomorphism problem.

The size of the expected differences can be estimated from formulas similar to those given for isomorphous differences (Eq. 3.12). For the Bijvoet differences the equation is

$$0.77 \Delta f'' \sqrt{N_A} / \sqrt{MW}, \tag{3.26}$$

and for the dispersive difference the equation is

$$0.39(^{\lambda_1} f' - {}^{\lambda_2} f') \sqrt{N_A} / \sqrt{MW}, \tag{3.27}$$

where MW is the molecular weight of the protein and N_A is the number of anomalous scatterers.

Choice of Wavelengths

A minimum of three wavelengths are recorded for each reflection. It is necessary to find the absorption edge of the crystal using an EXAFS scan. The wavelength is scanned through the range expected for the edge, and the X-ray flourescence is measured with a scintillation counter placed near the crystal. It is not sufficient just to look up this value in a table since the chemical environment of the anomalous scatterer shifts the edge (Fig. 3.30). The spectrum is then separated into the real and imaginary components to find $\Delta f'$ and $\Delta f''$ as a function of wavelength. The wavelengths chosen are at the f'' maximum, the f' minimum; these are found at the absorption peak and edge, respectively (Fig. 3.31). A third, remote, point is chosen to maximize the difference in f', $^{\lambda_1} f'_A - {}^{\lambda_2} f'_A$. In some cases a fourth point before the edge is chosen where f'' is at a minimum. This adds little phasing information but may help by increasing redundancy.

Collection of Data

In order to minimize effects of crystal decay and machine drift, the data for each wavelength should be collected as soon as possible in time. This

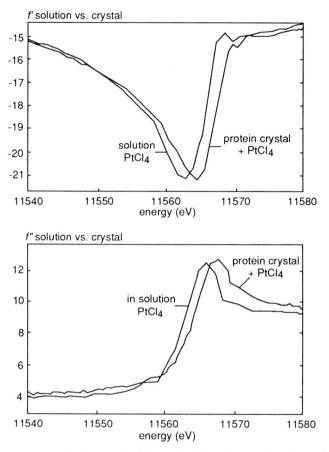

FIG. 3.30 EXAFS scan of platinum edge in bound and free forms. Plots of f' and f'' of $PtCl_4$ free in solution and bound to the protein are shown. The binding to the protein significantly alters the platinum edge, illustrating the necessity for a flourescence scan to determine the exact edge for each sample used in MAD phasing.

requires a computer-driven monochromator that can be accurately positioned many times. The monochromator should be calibrated frequently to check for drift by scanning the absorption edge of a piece of metal foil with an edge close to the wavelengths being collected. Ideally, if there are mirrors in the crystal, Bijvoet pairs should be arranged to occur on the same frame of data by causing the mirror plane to bisect the detector. Alternatively, two short runs can be done, one at a given χ and ϕ, and the next at $\chi = -\chi$ and $\phi + 180$, to collect the Bijvoets close to each other in time. In order to get the accuracy needed, use image plates, the FAST system, CCD detectors, or, if the flux is low enough, multiwire detectors. Film is not suitable for this method because of the low dynamic range and high background. The counts

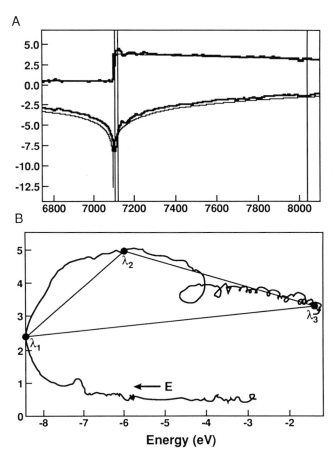

FIG. 3.31 Choice of wavelength for MAD phasing of a Fe_4S_4 protein. (A) Graph of energy versus absorption for a sulfite reductase crystal. The upper line represents the f'' component and the lower line is the f' component. The thick lines are the observed values and the thin lines are fitted. The change in absorption is caused by the anomalous scattering of the Fe_4S_4 cluster in the protein, which causes a step in f'' and a dip in the f' scattering components. The positions of the three wavelengths chosen for data collection are indicated by the vertical lines. If a fourth wavelength was chosen, it would be before the edge to the left. Scan done at the Stanford Synchrotron Radiation Laboratories by Brian Crane and Henry Bellamy. (B) Another representation of the same data, except that f'' is plotted against f'. The three wavelengths should be chosen to maximize the area of the triangle, which is proportional to the phasing power.

in each peak must be high in order to get accurate well-determined data. The data should be carefully processed. Scaling can be done with local scaling because the expected differences are usually small. Bijvoet pairs are usually kept together in the same run and not averaged across multiple runs. Each pair can be treated separately in subsequent steps. Even for the dispersive difference measurements between two wavelengths, the Bijvoet pairs are needed because the dispersive difference is the difference between pairs at different wavelengths,

$$[(^{\lambda_1}F^+ - {}^{\lambda_1}F^-)/2] - [(^{\lambda_2}F^+ - {}^{\lambda_2}F^-)/2], \tag{3.28}$$

and all four terms are needed to measure the difference.

Location of Anomalous Scatterers

The scatterers can be located with ΔF Pattersons as is standard for heavy-atom work. A better method is to use all of the wavelengths at once in a least-squares procedure to find the best value of F_A, the anomolous scatterer(s) structure factor.[20] In this way the method is different from most because quite accurate values of F_A can be found directly from the data rather than by approximations as in the isomorphous acentric case. Pattersons made with these values should be much cleaner and thus easier to interpret. If the Patterson is too complicated to solve, the multi-λ fit F_A values could also provide the basis set for the direct methods commonly used for small molecules. The positions of the anomalous scatterers are refined against the F_A values, as is done for isomorphous heavy atoms. If the Patterson is clean, this refinement should be straightforward.

Phasing of Data

The phase equations for MAD phasing are

$$|^{\lambda}F(\pm h)|^2 = |^0F_T|^2 + a(\lambda)|^0F_A|^2$$
$$+ b(\lambda)|^0F_T||^0F_A|\cos(^0\phi_T - {}^0\phi_A)$$
$$+ s(\mathbf{h})c(\lambda)|^0F_T||^0F_A|\sin(^0\phi_T - {}^0\phi_A), \tag{3.29}$$

with coefficients a, b, c defined as

$$a(\lambda) = [f'(\lambda)^2 + f''(\lambda)^2]/f^0)^2$$
$$b(\lambda) = 2[f'(\lambda)/f^0]$$
$$c(\lambda) = 2[f''(\lambda)/f^0]. \tag{3.30}$$

The subscript T designates the total structure, and the subscript A designates terms from the anomalous scatterers alone. The superscript 0 means the wavelength-independent value. The factor $s(\mathbf{h})$ is a sign and is positive for F^+ and negative for F^-. Once the locations of the anomalous scatterers are known, phases can be calculated by calculating the phase ϕ_A from these positions. The protein phase is then given by $\phi_T = \Delta\phi + \phi_A$ where $\Delta\phi = (\phi_T -$

---

[20]The least-squares procedure is described in detail in Hendrickson, W. A., *et al.*, (1988). *Proteins: Struct. Funct. Gene.* **4**, 77–88.

ϕ_A) is from the multi-λ fitting procedure. Equations have been derived by Hendrickson and co-workers[21] for calculating Hendrickson–Lattman coefficients A_{MAD}, B_{MAD}, C_{MAD}, D_{MAD} so that the phases can be combined with other phase sources such as an isomorphous derivative, if so desired. Of course, the correct hand must be used as for other anomalous dispersion techniques. This is done by trying both possibilities and examining the two maps. One should have clear solvent channels and the other should be less clear. Alternatively, the correct map could be selected by comparing the electron-density histograms (see above).

The other terms needed for the equations are the $^\lambda f''$ and $^\lambda f'$ values. The values can be extracted from the X-ray flourescence data. The values are often anisotropic along different crystal directions, which introduces a small error if a single value is assumed. Another method is to refine these values and treat them as variables. Weis *et al.*[22] have reported a method for doing this in the solution of the calcium-dependent lectin domain from a rat mannose-binding protein by MAD phasing using the L_{III} edge of holmium. The use of holmium gave large diffraction differences between Bijvoets and different wavelengths ranging from 10 to 27%—comparable to multiple isomorphous replacement—and the maps produced were of excellent quality.

····· 3.11 ·····
REFINEMENT OF COORDINATES

Available Software

The three most commonly quoted refinement programs are PROLSQ, XPLOR, and TNT. All three have been used with success, as have several others. PROLSQ was written originally by Hendrickson and Konnert and is a reciprocal-space least-squares refinement program. TNT is a similar program from Ten Eyck and Tronraud that uses a Fast-Fourier Transform and is faster than PROLSQ (although FFT versions of PROLSQ are available). XPLOR, written by Axel Brünger, is a very versatile program that has already been mentioned in the molecular replacement section. It can do conventional least-squares refinement as well as simulated annealing refinement where the protein is simultaneously subjected to molecular dynamics and crystallographic refinement in order to increase the radius of convergence.

Rigid-Body Refinement

Rigid-body refinement is the process of refining the positions of rigid groups of atoms against the observed amplitudes. The protein model is di-

[21] Pahler, A., *et al.* (1990). *Acta Crystalogr.* **A46**, 537–540.
[22] Weis, W., *et al.* (1991). *Science* **254**, 1608–1615.

vided into one or more groups. For each group it is possible to refine up to six parameters, three rotations and three translations. The shifts in positions for the groups are calculated by combining the shifts for all the atoms in the group. Rigid-body refinements should start at the lowest resolution range with enough reflections if large shifts are desired. The maximum size of a shift is limited to less than the smallest d-spacing included in the refinement. If too high a resolution is used, it is likely that the groups will be caught in a local minimum. On the other hand, the resolution must be high enough that the refinement is over-determined by a large margin. This is calculated by considering each group to be six parameters and then dividing the number of reflections in the resolution range being considered by the number of parameters. In practice, this number should be greater than approximately 10. The main trouble with rigid-body refinement by least-squares occurs if a local minimum is found before the correct minimum. Otherwise, rigid-body refinement is a robust technique because it is possible to over determine it greatly. Rigid-body refinement should be the first step for molecular-replacement models.

R-factor or Correlation Search for Rigid Groups

A method related to rigid-body refinement by least-squares is the R-factor or correlation search. The protein is divided into one or more rigid groups that are moved as a whole. These groups are then moved over a large range, such as the entire asymmetric unit on a grid of about one-third the minimum resolution, and at each point either an R-factor or a correlation is calculated. The correlation has the advantage that it is independent of the scale factor so it is more accurate if the model is incomplete. The obvious advantage of this method is that it can find the lowest minimum for the model because it checks all positions and does not get trapped in local minima. It is usually possible to do a three-parameter search, either the three translation axes or the three rotation axes, but for a full six-parameter search, where all rotations are tried at every translation, the amount of computer time needed becomes impractical. If packing is taken into consideration, then some of the positions will be clearly impossible because of overlaps in the model and these positions may be skipped. It may then become feasible to try all the remaining positions. Usually, a rotation solution is known, and instead of trying all six parameters, all the translations are tried, and the rotations are done over a small range at each translation to adjust for small angular errors in the rotation solution.

The other limiting factor, and a far more serious one, is inaccuracies in the search model. If you already knew the correct model it is unlikely you would be doing the search. The amount of inaccuracy in your model can make the correct minimum only slightly lower than other minima or perhaps not even the lowest one. To get around this, it is often desirable to examine the lowest 10 minima, or so, especially if there is not a minimum that is clearly lower. At each minimum further refinement may be tried to see if there is a clear winner. If the model is known to be close to the correct one, then even if a huge amount of computer time is needed, it is clearly worth doing a large search to find the correct position.

Protein Refinement

Unless one is working at very high resolutions, it is necessary to use stereochemically restrained refinement of proteins. This is because the ratio of observed (Fo) to refinable parameters (x, y, z, B) is below 1.0 and the problem is undetermined (see Table 3.1). Even at high resolution, the errors in the data, due to the relatively weak scattering of protein crystals, especially at higher resolutions, make restrained refinement a good idea at all resolutions. Different programs have different schemes for incorporating the stereochemical information, but for ease of conception we will use the energy model to explain the process. The different terms can be thought of as energy terms in an equation combining all of the information:

$$E_{\text{Total}} = w_A \Sigma E_{\text{crystallographic}} + w_B \Sigma E_{\text{bond distances}} + w_C \Sigma E_{\text{bond angles}}$$
$$+ w_D \Sigma E_{\text{torsion angles}} + w_E \Sigma E_{\text{non-bonded contacts}}$$
$$+ w_F \Sigma E_{\text{planar groups}} + w_G \Sigma E_{\text{chiral volumes}} \tag{3.31}$$

where the w terms are used for the relative weighting of the different terms. The energy for the crystallographic terms comes from the difference in Fo and Fc, and the energy for the stereochemical terms is evaluated by considering the difference between the actual value and the ideal value in such a way that, as the atoms deviate more from ideal geometry, the energy increases. The ideal values have been tabulated for proteins by examining highly accurate small-molecule crystallographic structures of amino acids, petides, and other similar compounds. The refinement program is set up to minimize the overall energy by calculating the shifts in coordinates that will give the lowest energy. The equations for the derivatives necessary for this minimization are non-linear. This means that the equations cannot be exactly solved but are, instead, approximated and a non-linear least-squares procedure is used to minimize the total energy. For this, the equations are evaluated and approximate shifts are calculated that should bring the total energy lower. The equations are then re-evaluated and more shifts applied. If the energy is lower in each cycle, then the process will eventually converge. However, because of the approximations that make the equations non-linear, the point of convergence may be a local minimum.

Some programs, such as XPLOR, also add an electrostatic term for the attraction of charged groups. Since the starting models for refinement are generally inaccurate, the use of the electrostatic term is dubious and probably should be turned off or at least down-weighted. It is also common to reduce the non-bonded weights at the beginning of refinements to allow the groups to move large distances while "slipping" past each other. Proper weighting is crucial to balance the stereochemical constraints with the derivative shifts. Otherwise, a low R-factor could be reached at the expense of distorted geometry, trapping the refinement in an incorrect minimum. Weighting the stereochemistry too high will slow the refinement and may even freeze the refinement by not allowing any positional shifts.

Isotropic temperature factors, or B-values, are also restrained during refinements.[23] Atoms that are bonded to each other influence each other's motion so that if one atom is undergoing large displacements, then atoms next to it must also be undergoing large displacements. B-values are restrained in such a manner that the average difference in B-values of bonded atoms is kept to a target value. That is, the B-values should vary smoothly along the protein chain and within a side chain. The usual target restraint for adjacent bonded main chain atoms is $\Delta = \langle B_1 - B_2 \rangle = 1.0$, where B_1 and B_2 are the B-values for two bonded atoms, and for side chains the target Δ value is raised to 1.5 to account for the higher gradient of side chains due to having one end free. In a similar manner, the B-values can be constrained for the 1–3 members of a bond angle. For main chain angles, the target Δ is usually 1.5 and for side chain angles Δ is 2.0.

Strategies

Unfortunately, a well-refined structure cannot be achieved by simply running a refinement with the highest resolution data until it converges. The refinement is almost certain to hang up in a local minimum. If the starting model may be fairly far from the final model, then start the refinement at a medium resolution and gradually add data by increasing the resolution. Refine at each stage until the R-value converges. Start with a single B-factor for the entire model. Keep adding data if the R-value is reasonable. If the R-value is high, then a round of fitting is needed to fix errors so that refinement can continue. Use difference maps to find these errors and carefully examine them with omit maps, omitting the residues in question and the near neighbors. If you have MIR phases, continue to use them throughout the refinement process in evaluating possible fittings. After refitting, the R-value will be higher, but it should drop to a value lower than before.

[23] Please see discussion of B-Values in Section 3.1.

At about 2.5–2.1 Å resolution there are enough data to allow individual B-factors to be refined. Examine the B-values carefully to detect portions of the model with errors. If the B-values are either very low (<2) or very high (>35), then this may indicate an error in this region of the model. The program should restrain B-values among neighboring atoms. In a well-refined structure, these will vary smoothly from atom to atom. Deviations from this pattern may give clues to possible errors.

At 2.0 Å with the R-value below 0.25 it is worthwhile to start adding waters. One common method is to use a difference map and put waters in all of the peaks in the difference map that do not collide with existing atoms. This practice is dangerous because it models the error with water atoms and can lead to serious phase-bias problems. While the R-value will drop due to the presence of more parameters, this will not help in the long run if the waters are in noise peaks. Instead, examine the neighbourhood of the water-proposed density peak and see if it has reasonable geometry for water. Is it in a crevice on the surface of the protein? Are there hydrogen-bonding partners at reasonable distances? If there really is a water in the peak under consideration, then adding it to the model will be beneficial, so it is worth being certain that the peak represents a water and not spurious noise. More waters will become apparent as the refinement continues. Weak density will often become stronger and a water can always be added later when it is justified by the density.

At some point the R-value will converge and difference maps will not reveal any features that indicate either refitting or solvent. If the model is well-refined then a plot of R-value versus amplitude should reveal good agreement for the strong well-measured data and the poorest agreement between weak, poorly measured data. Check a phi-psi plot for agreement with allowed values. There should be no, or only a few, outliers on this plot. Carefully check any outliers for errors.

Evaluating Errors

Stereochemistry

The examination of high-resolution, accurate crystallographic structures of peptides has provided the proper stereochemistry of protein groups. These ideal bond distances, bond angles, dihedrals, and van der Waals contact distances can then be compared to a protein model to calculate its deviation from correct stereochemistry. The average protein model does not approach ideal stereochemistry nearly as well as the average small molecule structure. First, the resolution of the maps is not as great. Even so, it should be possible to refine the protein structure to produce the correct stereochemistry. The interdependency of all of the linked groups in the protein then becomes apparent. If you push on one bond, it pulls on several others—all of the parameters are interdependent. A second, more subtle, reason may be that proteins are not static structures even when constrained by crystal contacts. The crystallographic map is a time-average of all structures in the crystal, and if the protein is in motion, the average may not be a true structure. Many cases are known of multiple positions of side chains and sometimes of small loops of protein where multiple positions can be seen in the electron-density map. Many cases of multiple positions cannot be distinguished in the electron density because they are too subtle. They may show up as small errors in the stereochemistry. In any case, the amount of error expected in a protein model is estimable from past experiences with refinement. The errors in bond lengths should be approximately 0.05–0.03 Å, and the error in bond angles about 3° on average.

Phi-psi (also known as Ramachandran) plots are especially good indicators of the accuracy of the protein model. The values of the main-chain dihedral angles in a protein can only take on a limited range of values. Glycines, not having a side chain, are the exception. For the other residues, their phi-psi values should all fall within the allowed values (Fig. 3.32). If there is a considerable number of outliers, then the structure is in need of improvement.

R-Factors

The R-factor, or the agreement between the observed and calculated amplitudes, is given by Eq. 3.2. Examination of the equation shows that as the agreement increases, this number becomes smaller until it reaches 0.0. This number can be taken as a guide of the progress of your model; it gives an estimate of the accuracy of the model. The R-factor must be taken with a grain of salt and not used as the sole criterion for accuracy. It is not possible to tell exactly how good your model is from this one number. The number is also dependent upon the completeness of the data, the resolution limits used, amd the accuracy of the observed data. This said, I will attempt to give some tentative guidelines for what the R-factor should do as the refinement progresses. Assume you have just fit an MIR or map or solved a molecular replacement problem and that you are refining, starting at 10–3 Å resolution with an overall B-factor. At the start of the refinement, the R-factor is usually around 0.40–0.45. This should fall rapidly as refinement progresses and then slow down until it stops falling at 0.25–0.35, depending on the quality of the starting model. Usually, the model must then be manually adjusted by examining difference maps to indicate regions that need adjusting. The R-factor will then slowly improve after the refitting. After refitting, the model

A

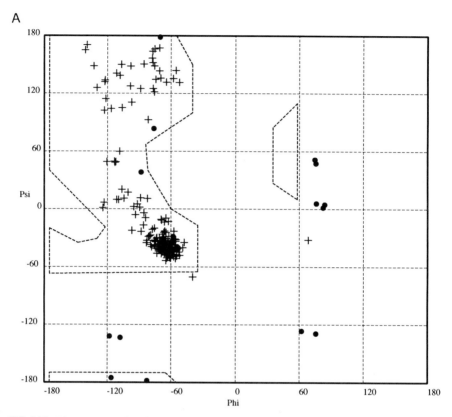

FIG. 3.32   Phi-Psi (Ramachandran) plots. (A) The plots are for mostly helical protein, (B) mostly sheet protein. Non-glycines are shown as plus signs. Glycine residues are shown as solid circles and can be found, having a hydrogen side chain, outside of the allowed regions shown by the dashed lines. Plot (C) is for a "poor" structure fit to an MIR map but not completely refined. While most residues are in allowed regions there are many that are in forbidden regions that need to be tweaked before the final model. Plot (D) is for a completely incorrect structure. The points fall everywhere on the graph. Plot (E) resulted when the structure in (D) was corrected. (*Figure continues.*)

is then refined again. As the R-factor improves, the resolution is increased. When the resolution is around 2.1 and the R-factor below 0.25, it is time to refine individual Bs for each atom. This will immediately improve the R-factor. The map is then refit and refined until the R-factor converges again. Waters are then added to the map and refinement continues. If the resolution is below 2.0 and the R-factor below 0.20, then the structure is probably essentially correct. Well-refined structures can have R-factors less than 0.17.

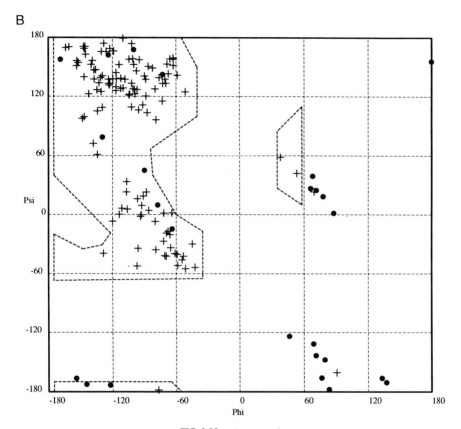

FIG. 3.32—*Continued.*

An R-factor below 0.12 is exceptional. And just to make you feel inadequate as a crystallographer, it should be noted that small molecules routinely refine to R-factors between 0.03 and 0.05. Why proteins cannot be refined to such low values is a topic of considerable debate.

It is also helpful to look at the R-factor as a function of resolution and amplitude. The R-factor with resolution should start out high at very low resolution, around 10 Å, where the solvent dominates the structure factors, then be lowest between 5 and 2.5 Å, and start steadily rising as the resolution increases. The last shell of resolution often has an R-value in the high twenties. A plot of R-factor versus amplitude should decrease steadily as the amplitudes are increased. For the weakest data, the R-factor will be very high, say around 0.40, and for the strongest data it is usually below 0.10. If there

C

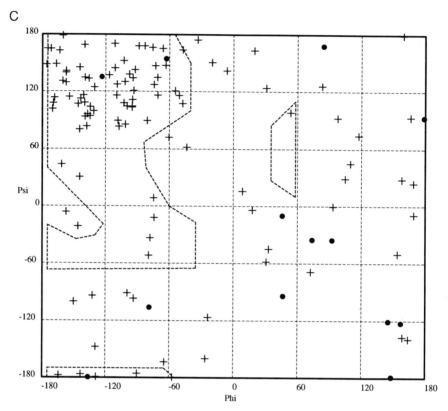

FIG. 3.32—*Continued.*

is no correlation of R-factor with amplitude or the brightest data do not have the lowest R-factor, then there are probably still some errors in the model.

After B-value refinement has been started, analyzing these values can provide an important clue to incorrect parts of the structure. Places with very low B-values (<2) are probably in need of attention, as are regions with very high B-values (>35). You might wonder how a B-value can go below 2 since such low values are probably meaningless. It probably arises from an atom that finds itself in density that is greater than can be accounted for by its scattering power. This can happen if the atom is in the density of a larger atom or if phase errors cause noise that makes the density at that atom higher than it should be. This also points out the problem with relying too heavily on the B-value to evaluate errors, as inaccurate phases caused by errors else-where in the structure may be the actual cause of the bad B-value and not necessarily a displacement of the atom in question.

D

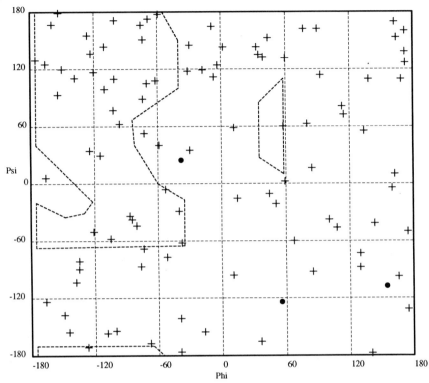

FIG. 3.32—*Continued.*

····· 3.12 ·····
FITTING OF MAPS

Calculating Electron Density Maps

Resolution Cutoffs

Normally, you can specify a minimum and a maximum resolution cutoff when calculating an electron density map (Fig. 3.33). Remember that the map is being calculated using a Fourier synthesis and that this causes series-termination errors for the higher resolution terms left out. These series-termination errors show up as ripples in the electron density with periods close to the highest resolution used in the map. The effect of leaving out low-resolution terms is to add low-period ripples that make the map look "choppy." Low-resolution terms are often left out of refinements and, thus,

E

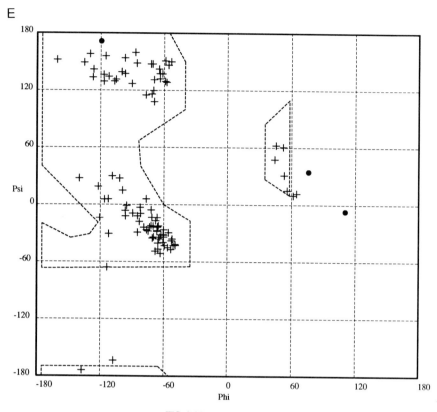

FIG. 3.32—*Continued.*

they often end up left out of map calculations. How much effect this has on the interpretability of the map depends upon how high the other limit is. For instance, a 4.0–3.0 Å map is hard to interpret even with perfect phases, while a 4.0–2.0 Å map is relatively straightforward. As a rule-of-thumb a 3-Å map should include data from 10–3 Å.

FIG. 3.33   Effect of different resolution cutoffs. The maps use the same data but differ in the resolution limits used. Thick lines are the model used to calculate the phases; thin lines represent the electron density contoured at 1 σ. (A) 37–5.0, (B) 37–4.5, (C) 37–4.0, (D) 37–3.7, (E) 37–3.3, (F) 37–3.0, (G) 37–2.0. The turns of the helix become apparent at 3.7, and the carbonyl bulges are apparent at 3.0. However, these maps are made with refined phases, and starting maps will appear to have lower resolution due to phase errors. Map (H), 5.0–3.7, shows the deleterious effect of truncating the low-resolution data too severely. Compare the density in (H) with that in (D).

A

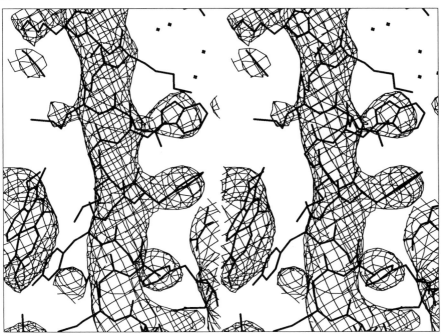

B

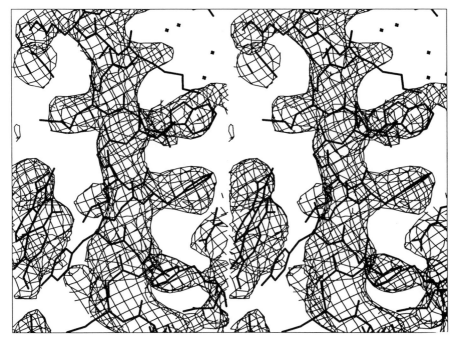

C

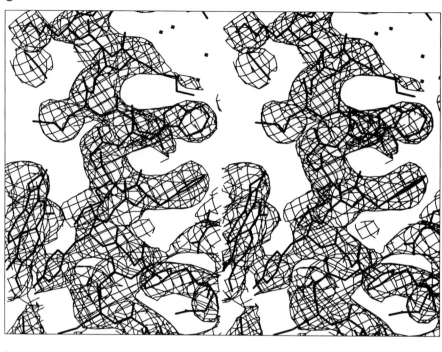

D

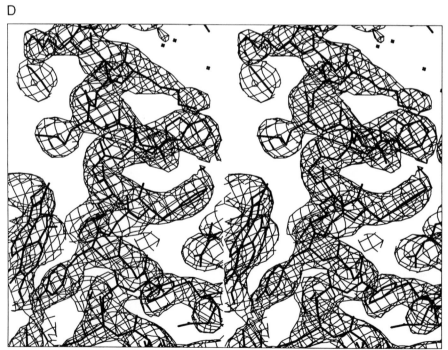

FIG. 3.33—*Continued.*

E

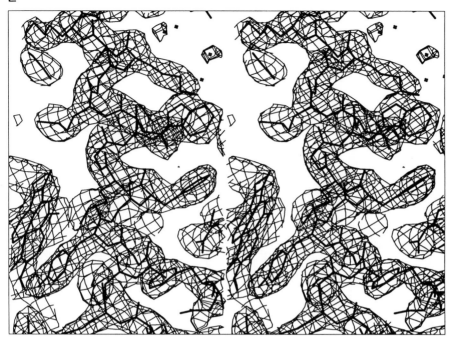

F

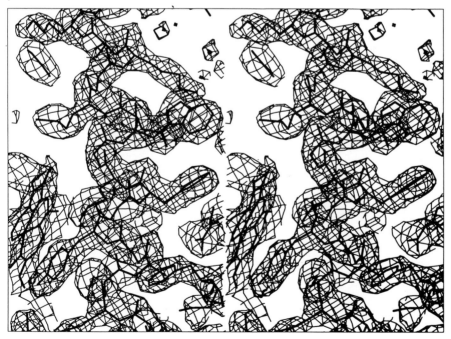

FIG. 3.33—*Continued.*

G

H

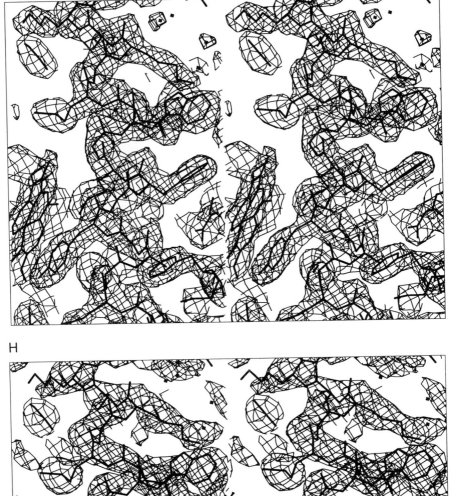

FIG. 3.33—*Continued.*

Non-Orthogonal Coordinates

The main problem with non-orthogonal coordinates is making certain that the map and the model are in the same coordinate frame as discussed previously under Coordinate Systems (see Section 3.1). It is also a common problem that after a model has been fit the cell transformation is incorrectly specified for the refinement program and the only indication this has happened may be a high R-factor. The XtalView system allows the user to specify a 3 × 3 matrix for the Cartesian-to-fractional coordinate transformation (for fractional-to-Cartesian the inverse matrix is computed), so that if another program is being used, the matrix from this other program can be entered. In XtalView the matrix used is the same as FRODO and XPLOR in the cases tested so far.

Map Boundaries

Many programs, such as FRODO, can only display as much of the unit cell as is stored in the precalculated map file. This means that the user must decide beforehand on the map boundaries that will cover an entire molecule. A mini-map can be used to determine boundaries that will cover an entire molecule. In xfit there is no need for this as the program is smart enough to know that the density at 1.1 is the same as at 0.1, and the maps always contain a full-unit cell.

Combined-Phase Coefficients

Phase bias is a serious problem in the early stages of fitting and refinement. One way to avoid phase bias is to use only the MIR phases to calculate the maps, this guarantees that the maps are unbiased with respect to the model being fit. However, MIR phases are noisy and usually limited in resolution. Several methods have been developed for combining calculated model phases with experimental phases to allow information from both to be used. This can be used to increase the resolution, to allow the use of partial models, and to help reduce model bias. All methods rely on the difference between Fo and Fc to weight the amount of calculated phase contribution. Combined phase coefficients allow the use of the low-resolution MIR phases, which are usually accurate, to be included with calculated phases from higher resolutions. The MIR phases alone should be used in the resolution range infinity to 5.0 Å, where model phases are inaccurate. At other resolutions both phases are used in a weighted manner. The phase combination program puts out a combined figure-of-merit that is used to weight the map, as in the MIR case. The combined maps are smoother, especially at resolutions between 3 and 2.5 Å, than maps made from an incomplete model for phasing.

The most common phase combination procedure is Bricogne's adaptation of Sim's weighting scheme.[24] Two-phase probability distributions are multiplied together by the following procedure. The phase probability for the MIR phases has been previously stored as the four Hendrickson–Lattman coefficents, A_{MIR}, B_{MIR}, C_{MIR}, D_{MIR}.[25] The phase probabilities for the calculated phases are calculated from the equation:

$$P_c(\phi) = \exp[2|F_{obs}||F_c|/\langle F_0^2 - F_c^2\rangle \cos(\phi - \phi_c)], \qquad (3.33)$$

where $\langle F_0^2 - F_c^2\rangle$ is the root-mean-square difference in intensities and is calculated in bins of resolution. It can be seen that when the R factor is large, that is *Fo* and *Fc* do not agree, then the contribution from the calculated phase is lowered because the denominator will be large. As the R-factor decreases, the calculated phase contribution will increase. This probability is expressed in terms of the terms *A* and *B*, where *A* is the cosine part of phase, *B* is sine part of the phase at the maximum value of the probability distribution and $\sqrt{A^2 + B^2} = mF^0$, where *m* is the figure-of-merit. These are then added to A_{MIR} and B_{MIR} with the relative weights set by w_{MIR} and w_{calc}:

$$A = w_{MIR} \times A_{MIR} + w_{calc} \times A_{calc}, \qquad (3.34)$$

with a similar equation for *B*. The new phase is found by evaluating:

$$P_j(\phi) = \exp(A \cos(\phi) + B \sin(\phi) + C \cos(2\phi) + D \sin(2\phi)), \quad (3.35)$$

which gives a new most-probable phase and figure-of-merit using Eqs. 3.19 and 3.20. The success of this process can partially be judged by an increase in the figure-of-merit. The weights should be adjusted so that the new phases are, on average, between the MIR and the calculated phase. Low-resolution phases will be closer to the MIR phase, and the higher-resolution phase will be closer to the calculated phases. The swing point, where more of the calculated phase is included than the MIR phase, should be located between 4.0 and 3.5 Å, based upon past experience. This seems to give maps that contain new information from the filtering effect of the calculated phases but is not overwhelmed by the them so that no information from the MIR is left.

Evaluating Map Quality

If an experienced crystallographer looks at a map, he or she can usually quickly assess the quality of the map. It is especially important for beginning

---

[24] Bricogne, G. (1976). *Acta Crystalogr.* A32, 832.
[25] Hendrickson, W. A., and Lattman, E. E. (1970). *Acta Crystalogr.* B26, 136.

crystallographers to try to learn this technique so that a lot of time is not wasted trying to fit poor maps.

Judging Electron Density

First look at the map on a large scale, where both protein and solvent should be visible. There should be a large contrast difference between the solvent and the protein. If the map is contoured at 1 σ, there should be few long connected regions in the solvent region (look at several sections). The protein region of the map should have connected densities that are cleanly separated. The heights of these ridges should be consistent over the protein region. Excessive "peakiness" is a bad sign. It might help at this point to consider that a protein is made up of a long polypeptide composed mostly of carbon, nitrogen, and oxygen, which all have about the same electron density. The exception is a small number of sulfur atoms. Any error in the phases will cause some volumes to have too little density, and another volume of the map will be correspondingly too high. With a practiced eye, you can quickly discern this.

Another feature to look for in order to judge the quality of the map is the contrast between protein and solvent regions. For this pupose, a slab of electron density over a large area is needed (Fig. 3.34). There should be a clear difference in level between the protein and solvent. The solvent should comprise a few large areas of low-level peaks rarely rising above 1–2 of the root-mean-square value (σ) of the map.

Electron-Density Histograms

The previous statements can be restated more precisely. In a correctly phased map, the distribution of the densities will be that for a protein molecule, which is independent of its fold or space group, and will have a characteristic histogram. In fact, the histogram can be used to compute the probable amount of phase error by comparing the histogram of an unknown with that of correctly phased maps of known structures (Fig. 3.35). The histogram is dependent upon the resolution range used and also on the percentage solvent in the crystal.

XtalView comes with a program, xedh (Fig. 3.36), which computes the histogram of a map and can be used to compare it with known histograms. This is easily enough done; except that, of course, there is a large gray area where a map may be interpretable in spite of the phase errors. However, the histogram-comparison method gives an objective method of estimating phase error that, with experience, will give a guide to the "interpretability"

A

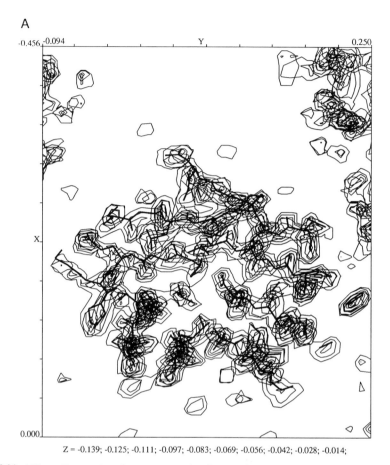

Z = -0.139; -0.125; -0.111; -0.097; -0.083; -0.069; -0.056; -0.042; -0.028; -0.014;

FIG. 3.34 Effect of increasing phase error on the electron denisity map. In (A) a section of bovine superoxide dismutase density at 2.0 Å resolution is shown using the final refined phases. In (B) an average of 22° of random error has been added to each phase. The solvent is becoming noisy but the map is still easily interpreted. In (C) an average of 45° phase error has been added to the model. The density is noisy but still interpretable; it is still possible to tell the solvent from protein. Finally, in (D) an average of 67° of phase error has been added, making the map virtually uninterpretable; some regions of solvent have as much density as the protein.

of a set of phases. The histogram method requires no model, and it does not matter how the phases are derived.

Fitting and Stereochemistry

Fitting is the process of making the model match the map while preserving the proper stereochemistry. This is done using an interactive computer-

B

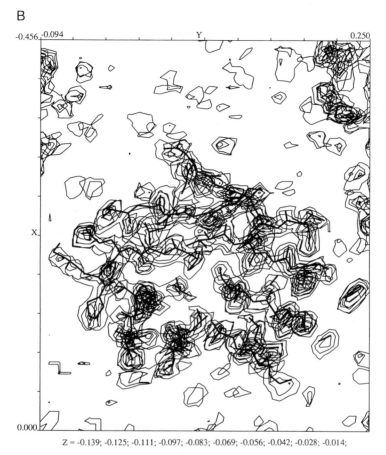

-0.456 -0.094   Y   0.250

X

0.000

Z = -0.139; -0.125; -0.111; -0.097; -0.083; -0.069; -0.056; -0.042; -0.028; -0.014;

FIG. 3.34—*Continued.*

graphics program, such as the XtalView program xfit (Fig. 3.37). If all structures could be fit to 2.0-Å maps, there would be little difficulty in doing this.
Most structures are first fit in the 3.0–2.5 Å resolution range. Stereochemistry is not as obvious at this resolution because the individual atoms cannot
be resolved. For instance, carbonyls either are small bumps or are not definable at 3.0 Å, and side chains often merge with the main chain for the smaller
amino acids.

Amino Acid Stereochemistry

Amino acids have a fixed stereochemistry, and this must be kept in
mind while fitting. The allowed bond distances and bond angles are fixed,

C

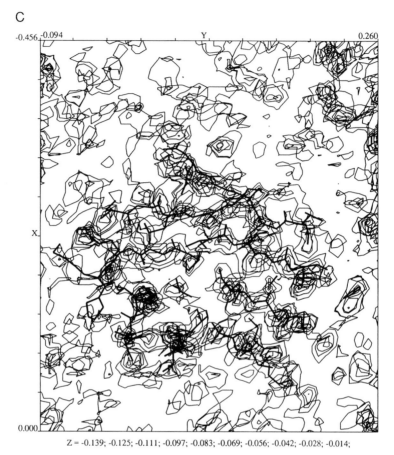

Z = -0.139; -0.125; -0.111; -0.097; -0.083; -0.069; -0.056; -0.042; -0.028; -0.014;

FIG. 3.34—*Continued.*

and all of the degrees of freedom are in rotations about bonds. Furthermore, the peptide bond is constrained to be planar.[26] Also, the dihedral angles of the main chain, phi and psi, have a limited range of allowed conformations (see above), and, if the positions of the residues before and after a given residue are taken into consideration, there is only a very small range left. In fitting, given the path of the polypeptide, and knowing the directions of the side chains, you do not have many choices left to make. Side chains are less

---

[26] A thorough discussion of peptide bond structure can be found in Chapter 2 of Schulz, G. E., and Schirmer, R. H. (1979). "Principles of Protein Structure." Springer-Verlag, New York.

D

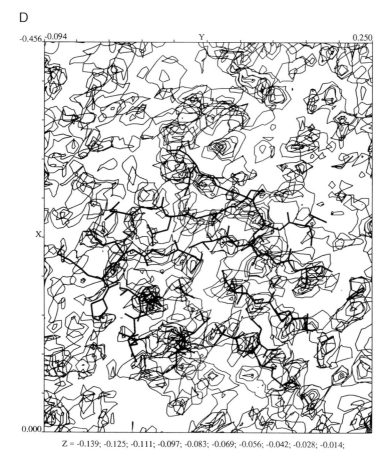

Z = -0.139; -0.125; -0.111; -0.097; -0.083; -0.069; -0.056; -0.042; -0.028; -0.014;

FIG. 3.34—*Continued.*

restricted than is the main chain. Side chains have preferred rotomer positions that should be kept in mind. Methylene, sp³, carbons, such as in the
side chain of lysine, prefer a staggered conformation with dihedral angles
near 120° and −120°. The guanidinium group at the end of arginine is planar. Until the maps are of hiqh quality, it is difficult to distinguish between
the two flipped conformations of the guanidinium head group since they
have similiar shapes. Rings, except prolines, are always tightly constrained
to be planar. This planarity extends to contain the attachment-point atom;
that is, in phenylalinine the Cβ atom is coplanar with phenyl ring. The car-
boxylate end groups of aspartate and glutamate also form a 4-membered

A

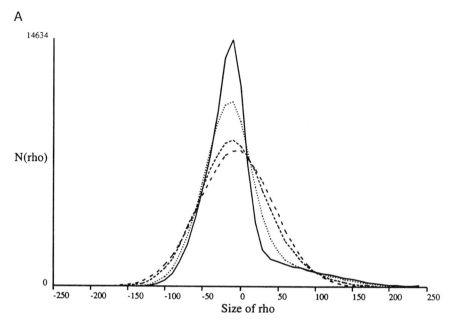

B

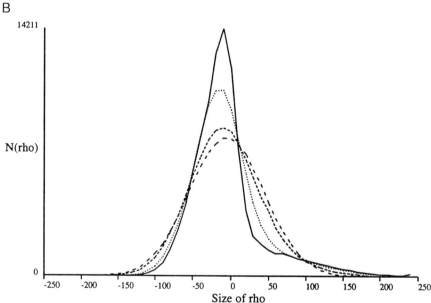

FIG. 3.35 Effect of phase errors on protein histograms. (A) Histograms of the maps in Fig. 3.34 are shown. The histogram without any error added is the solid curve. The histogram takes on a more gaussian shape as the error is added and the height decreases (dashed lines). The tail of points at higher rho values disappears, and the peak of 0 values found in the solvent decreases as noise enters the solvent region. In (B) the same figure for phase error was added to human hemoglobin. Note that although the two molecules have completely different structures and space groups, the histograms respond nearly identically to added phase errors.

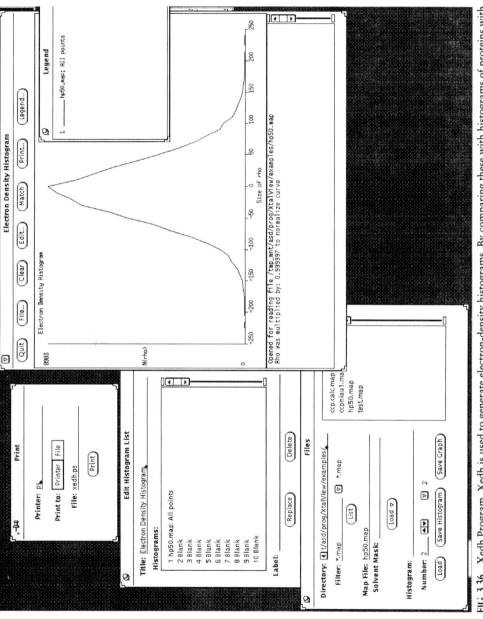

FIG. 3.36 Xedh Program. Xedh is used to generate electron-density histograms. By comparing these with histograms of proteins with similiar solvent content at the same resolution, an estimate of the phase error can be made.

195

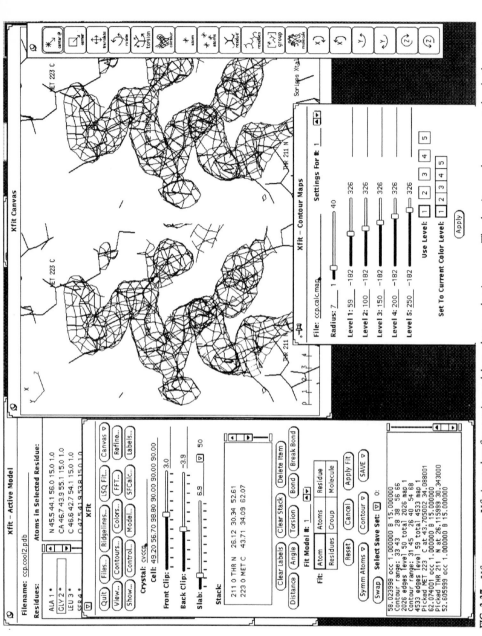

FIG. 3.37 Xfit program. Xfit is used to fit protein models to electron-density maps. The density can be represented as both contours and ridgelines. The program can also be used to compare models and to edit them. Using a macro language, complicated figures can be built (such as the ones in this book) and printed to any PostScript device. Most of the options available can be accessed through menus, buttons, or graphical aids.

planar group including, in the case of aspartate, Cβ, Cγ, Oδ1, and Oδ2. The same holds true for end groups of asparigine and glutamine.

Chain Tracing[27]

Mini-maps

A mini-map is mini relative to the large maps that formerly were used in a Richard's box.[28] The easiest way to make a mini-map is to plot out sections of the map and then copy these onto transparent sheets. The transparencies are then stacked up with pieces of Plexiglas separating them (Fig. 3.38). The mini-maps can still outperform the best graphics systems in terms of number of lines drawn, but they are awkward to manipulate. One

[27] An excellent discussion of practical issues involved in fitting to real maps is provided by Richardson, J. S., and Richardson, D. C. (1985) in "Interpretation of Electron Density Maps," in "Methods in Enzymology," Vol. 115, pp. 189–206.

[28] Richards, F. M. (1985). "Optical Matching of Physical Models and Electron Density Maps: Early Developments," in "Methods in Enzymology," Vol. 115, pp. 145–154.

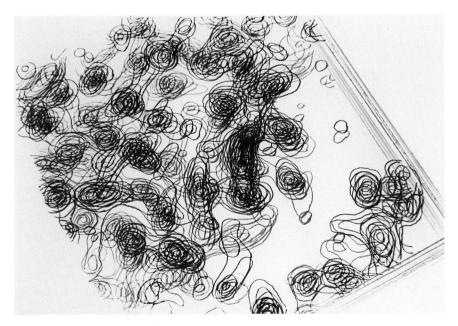

FIG. 3.38   Electron density mini-map.

great virtue of the mini-map is the ability to mark directly on the map with colored markers. Since the map is real and not a virtual image on the computer screen, it is easier to keep track of one's position. Symmetry is more easily recognized in a mini-map and symmetry axes can be marked directly on the map. A lot of people like to keep a mini-map near them as they fit on the graphics system. The mini-maps serve as a road map that can be viewed in detail on the graphics. Two final virtues of the mini-map are that it is portable and you do not have to sign up to use it.

Ridgelines

The traditional basket-weave representation of density requires too many lines to represent large portions of density, and it also has a "you can't see the forest for the trees" problem. An alternative method of viewing maps is to use ridgelines that are drawn along the three-dimensional ridges in the electron density map[29] (Fig. 3.39). This is a method of simplifying the map, or "skeletonizing" it. It requires far fewer lines to represent the density, so a much larger piece of map can be viewed on the graphics. Ridgelines also more closely resemble the stick figures commmonly used to represent chemical models.

The program GRINCH (Graphical Ridge lINes from Chapel-Hill) is a program meant to be used to perform the initial-chain trace of a map and to give an initial model. In this program, the first step in the interpretation is to color the density as represented by ridgelines, according to your interpretation. The map initially starts out all white to represent "unknown." As the map is interpreted, it is colored green for main chain, purple for side chains, and red for carbonyl oxygens. Density considered to be false connections are colored brown, and two colors, yellow and cyan, are left for special purposes such as cofactors. A special feature of the program is built-in heuristics that can do much of the coloring and fitting for you. For instance, to color a large stretch of chain green, you need only color one end green and then go to the other end and choose a ridgeline and the program finds the shortest path back to the first marked segment and colors the entire path green. Similarly, side chains are colored by picking one segment in the side chain and all the other segments connected to that are colored until the main chain or another

---

[29]The original author of GRINCH is Thomas Williams. His thesis contains a concise and thorough discussion of ridgelines. Williams, T. V. (1982). Ph.D. Dissertation, University of North Carolina, Chapel Hill. Another good discussion of an alternate method is: Greer, J. (1985). "Computer Skeletonization and Automatic Electron Density Map Analysis," in "Methods in Enzymology," Vol. 115, pp. 206–224.

color is found. Building a model is just as easy. A residue to be built is chosen, and two segments, one in the side chain, and one in the main chain, are picked and the program builds a residue to best fit the ridgelines. The chief drawback to the program is that a localized residue-by-residue fitting does not build good secondary structure. Fitting in larger pieces of three or four is more successful. As a quick way to trace a chain and to build a starting model, GRINCH is very fast. I have built a 120-residue protein in a single sitting using the system.

A

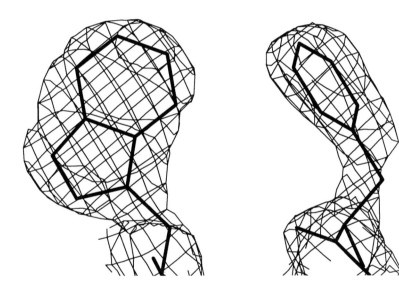

FIG. 3.39 Ridgeline representations of maps. (A) Conventional contour representation of electron density (thin lines) and superimposed stick-figure model (thick lines). (B) The same electron density was "skelotonized" to produce ridgelines and is compared with the model. Note that ridgelines look more like a stick-figure model, which can cause some confusion. However, they depict the density with fewer lines, which means that more map can be drawn on a graphics station before performance is slowed, and it is easier to see through foreground objects to the background. (C) Helix density with contours and ridgelines. The combination of the two is superior to either alone. (D) Helix density with ridgelines and model. (E) Looking down helix density represented by ridgelines. Note that a hole can be seen down the middle with side chains coming out radially, making helices particularly easy to identify with ridgelines. (F) Ridgelines are especially useful for looking at large views of the map. It easy to make out the large solvent holes between protein molecules. If this much map were contoured, the figure would have been black and the graphics system would have ground to a virtual halt.

B

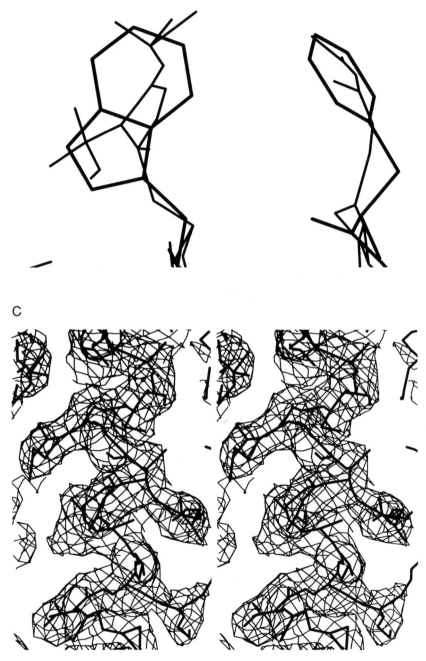

C

FIG. 3.39—*Continued.*

D

E

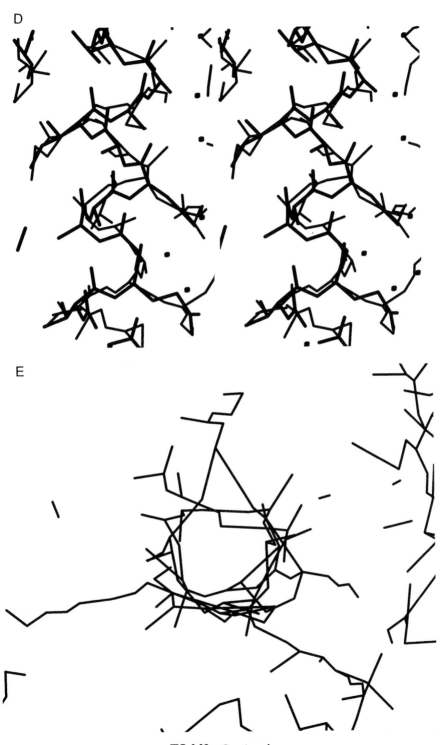

FIG. 3.39—*Continued*.

F

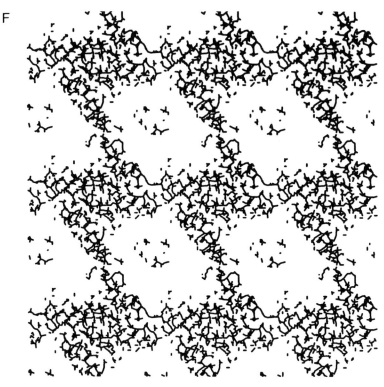

FIG. 3.39—*Continued.*

Recognizing Secondary Structure

Looking for large pieces of secondary structure can greatly speed map fitting. If the secondary structure is known, many of the breaks in the density can be confidently assigned as false because the constraints of the secondary structure dictate that the chain must bridge the gap. Helices are the easiest to recognize as long tubes of density disconnected from the rest of the structure. The side chains on a helix "point" to the nitrogen-terminal end (Fig. 3.40) so that it possible to discover the chain direction from a helix even if carbonyls are not visible. Helices are obvious in 5.5-Å maps. It is worthwhile searching a 5.5-Å map for long tubes of density. The trick to recognizing β-sheet structure is to find the correct viewpoint (Fig. 3.41). Sheets have a twist, and the easiest way to recognize a sheet is to sight down the length and to look for this twist. Then turning 90° to this, you can pick out the

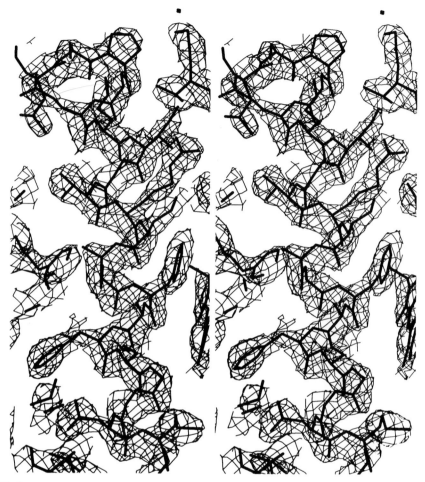

FIG. 3.40 Identifying chain direction in helices. The direction of helices can usually be determined from examining the direction the side chains "point." Shown is a 2.7 Å map with the final refined model superimposed. Most of the side chains point toward the N-terminal end of the helix, although some bend back and appear to point the other direction, especially in lower-resolution maps. An example of this can be seen with the Phe side chain in the lower left.

individual strands. Sheets often have regions with weaker density and gaps as well as false connections across the strands, usually at the position of the cross-strand hydrogen bonds. Sheets tend to break up at low resolution and are difficult to recognize at any resolution less than 3.0 Å. It is difficult

A

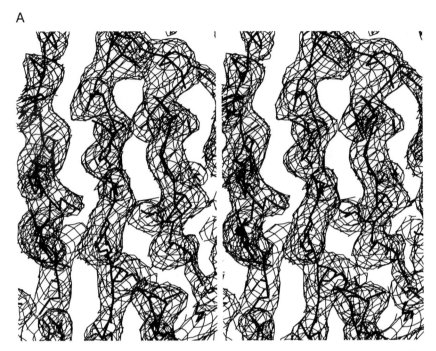

FIG. 3.41   Identifying sheet secondary structure in medium-resolution maps. Features of electron density of sheet in an MIR map are shown as contours with the derived ridgelines (thick
lines—not model!). (A) sheet front, (B) sheet top, (C) sheet side, (D) thin slab of sheet from the
side showing a single strand. The strands of a sheet run parallel, but the entire sheet is twisted.
Note in the top view (B) that the top and bottom strands are seen more edge-on, while the middle
strand is seen end-on because of sheet twist. The side chains of the sheet alternate back and
front. Adjacent strands are in register so that all of the side chains stick forward or back in rows
across the sheet. In MIR maps, breaks in the chain are common, and false crossovers often form
at hydrogen-bond positions. Another problem is that the the rows of side chains often "melt"
together, indicating a false connection and making side-chain assignment more difficult.

to decide on the chain direction in a sheet. If the sheet has reverse turns
(Fig. 3.42), the direction can be found by examining the reverse turn. If the
turn is traced in the wrong direction, the side chains will stick out on the
wrong side. By looking at which side the side chains come off the reverse
turn, the correct direction can be determined (Fig. 3.43). By locating as much
secondary structure as possible, it only remains to connect these pieces. By
making all the obvious connections, it may become possible to decide on all
of them by a process of elimination.

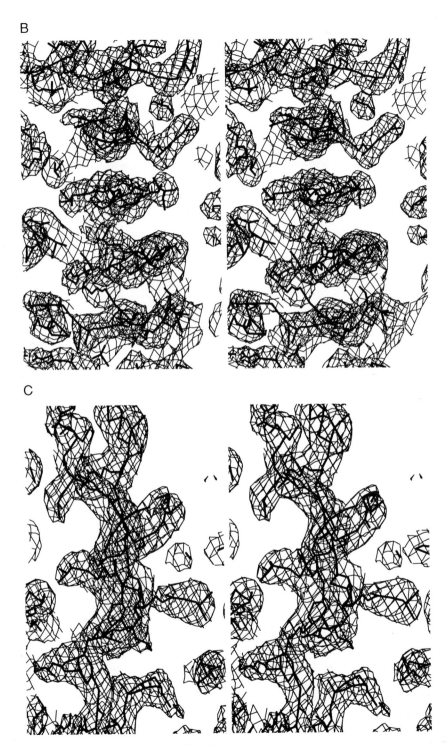

FIG. 3.41—*Continued.*

D

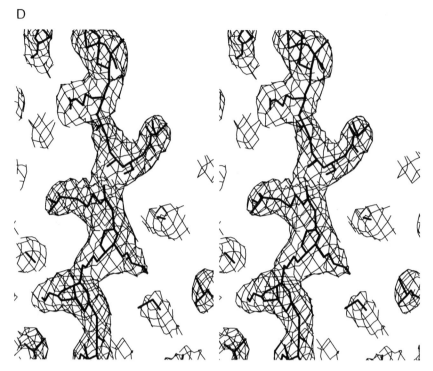

FIG. 3.41—*Continued.*

Sequence Identification

Another milestone in structure determination is finding the first match of sequence to the map. Since it is not possible to tell all the side chains apart from their density, especially at lower resolutions, the matching of sequence depends on finding a pattern of large and small side chains that is unique. Examples of side chains in medium-resolution electron density maps, where most initial chain tracing is done, are shown for an MIR map (Fig. 3.44) and for an Fo, α_{calc} map (Fig. 3.45). To illustrate how the density for side chains becomes clearer as the structural solution of a protein progresses, all tryptophan from aconitase are shown in Fig. 3.46 at three stages of its solution. Tryptophan is so much larger than all the other amino acids it can often be recognized. Reliably telling a tyrosine from a phenylalanine and an aspartate from a leucine is difficult, if not impossible, in the maps most of us have to work with. Keep in mind that hydrophilic side chains are often disordered, and their density prematurely ends at low resolution. It is rare, however, for

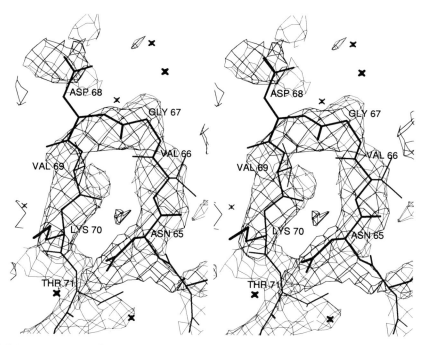

FIG. 3.42 Identifying tight-turn secondary structure in medium-resolution maps. It is common for the side chains of turns to be detached from the main-chain density.

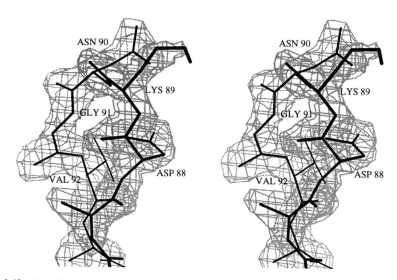

FIG. 3.43 Identifying direction by examining of β-turns. The direction of the chain through a tight turn can be determined by examining the direction that the side chains point. Rotate the density for the turn, as shown, so that the turn is upright and the side chains point to the right. The chain now runs from front to back.

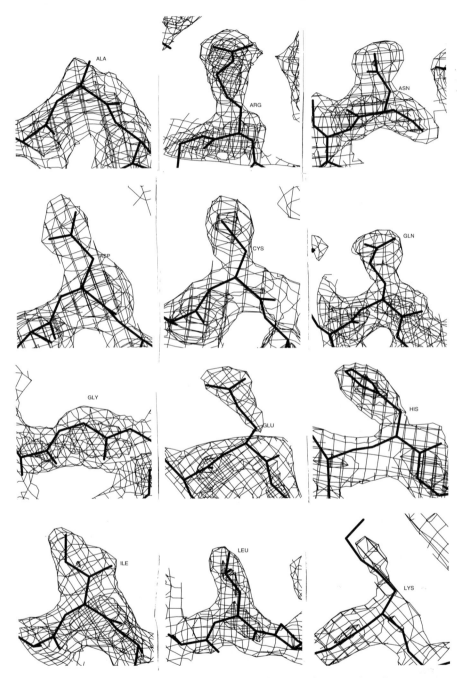

FIG. 3.44 Examples of all 20 residues in MIR maps. The maps shown are four-derivative MIR maps with an overall figure-of-merit of 0.7 at 3.0-Å resolution.

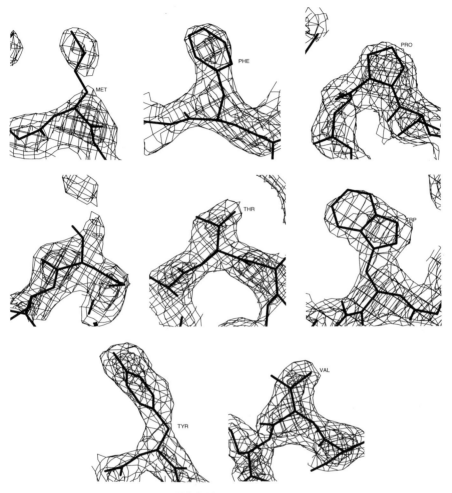

FIG. 3.44—*Continued.*

a hydrophobic side chain to be so disordered that its density is shortened. Thus, for example, looking for the pattern Lys-Arg-Glu is less likely to be successful than looking for Phe-Tyr-Leu. There are computer programs designed to look for matches, and I have tried them on occasion. In the end it seems that a human scanning a sequence can find a match more reliably and faster than entering the data into the computer. The true test of a sequence match is to continue the match in both directions to see if it holds up. After you have found several more amino acids in both directions, you can be confident that it is not an accident. Once one match is found, it limits the sequence that must be searched for other matches. After a while, all the se-

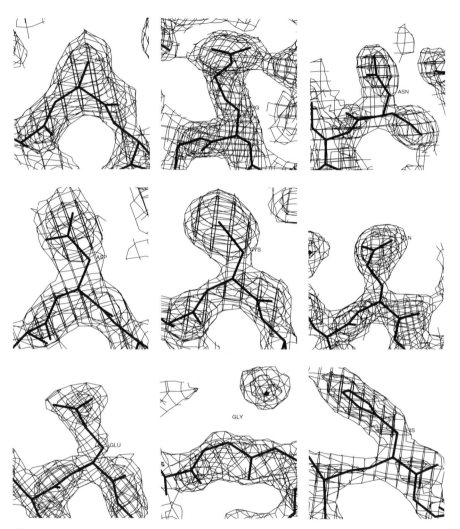

FIG. 3.45 Examples of all 20 residues in Fo maps. The map shown is a 3.0-Å resolution map using Fo, α_{calc} structure factors from a refined model.

quence can be found by building in both directions from matches and by a process of elimination.

Fitting by "Pieces"

Rather than fitting a model residue by residue, it is quicker and more accurate to add large pieces of secondary structure and then trim and tweak

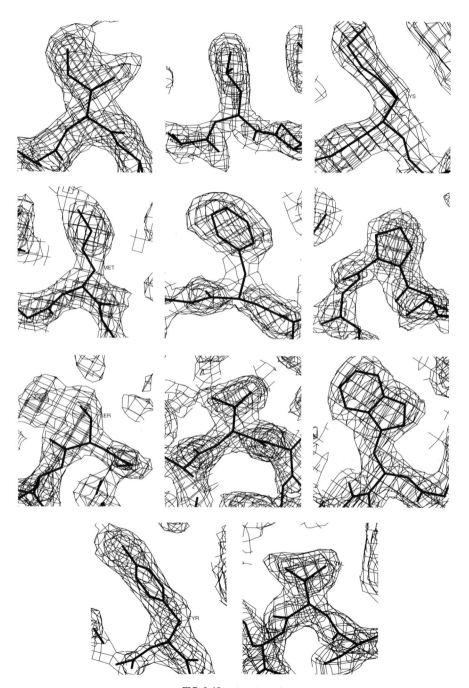

FIG. 3.45—*Continued.*

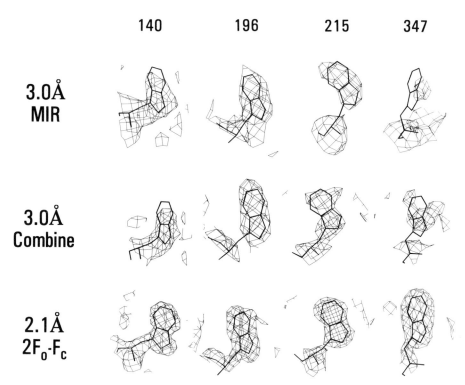

FIG. 3.46 Tryptophan's progress. The density of all of the tryptophans in aconitase are shown at three stages of the structure determination: MIR, MIR combined with a mostly complete model, and the final refined density. Figure courtesy of Drs. C. D Stout and A. H. Robbins.

these to match the map. It is difficult to build an α-helix accurately one residue at a time. Instead, a polyalanine helix made from another structure by truncating the side chains at the CB atom can be grafted into the map. This can be done by loading the new molecule in and then "flying" it into the correct position. With xfit an alternative method is to load the new piece in and then least-squares fit it to some marker residues. The marker residues are fake residues that have just a CA and CB atom that are placed where you think the helix residues should go. Then the helix can be least-squares fit to the marker residues. It is not necessary for the markers to be as long as the entire helix—a turn or two can be built, and then the new helix can be longer than this on either or both ends. The helix position can be further refined by real-space refinement. A similar method can be followed for β-sheet.

429   **548**   **577**   **631**   **739**

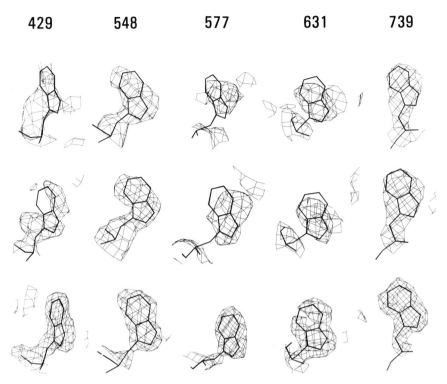

FIG. 3.46—*Continued.*

General Fitting

Fitting Main Chain

Fitting main chain is the trickiest part of fitting. The conformation is dependent on the previous and the next residue as well as the side chain (Fig. 3.47). Errors made in fitting one residue may propagate down the chain. Additionally, main-chain atoms are highly constrained. Knowing the position of three successive a-carbons highly constrains the phi-psi angles of the middle residue. In fact, proteins only have a few conformations available to the main chain. If the main chain is in an extended conformation, then the side chains alternate direction down the chain. In an extended chain it is not possible for two side chains to point in the same direction. In helices, the side-chain positions are so highly constrained that you can accurately predict the main chain and C_β atom positions with a refined α-helix from another pro-

A

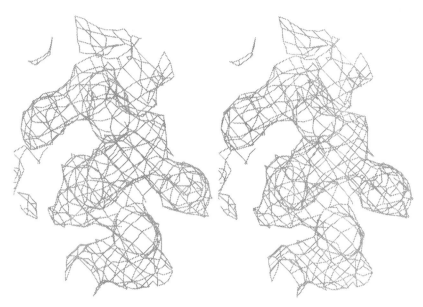

FIG. 3.47 Building model from scratch. These figures illustrate building a short section of helix to its density using best-fit pentamer peptides from well-refined structures. The model building was done using the xfit program XtalView. (A) The electron density (light gray) is contoured at 1σ. (B) Ridgelines (dark gray) have been added. (C) The electron density has been turned off for clarity in these figures, but on a color display it is not necessary. A marker residue consisting of a single Cα atom has been placed at the position of the first residue in the helix (black cross). (D) More Cα positions (black lines) have been added to build the protein backbone. (E) A search of the pentamer database is made for the best fit to the first five Cα positions and the pentamer thus found is shown in thin black lines. (F) The middle three residues of the pentamer have been inserted into the model to replace the marker residues. The ridgelines have been turned off for clarity. (G) The pentamer-fitting process is continued. The second pentamer overlaps by one residue because the first and last residues of the pentamer are not used. (H) A model built for the entire helix except for the first and last residues. The side chains now need to be replaced with the correct sequence and fit to the electron density. (I) The final model is compared to the ridgelines and in (J) the contours. The model includes polarizable hydrogens.

tein. Tight turns are also highly constrained. Memorization of the geometry of helices (α and 3_{10}), extended chains (anti-parallel and parallel), and tight turns will allow fitting 90% of the structure. The so-called "random-coil" portion of the protein is made up of short interlocked sections of helices, extended chain, and turns. Glycines and prolines are the complicating factors. Glycine is very flexible due to the absence of a side chain, and proline is usually found in "kinks."

B

C

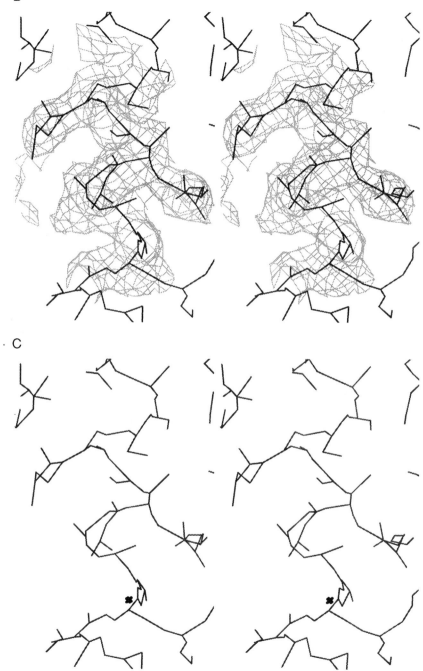

FIG. 3.47—*Continued.*

D

E

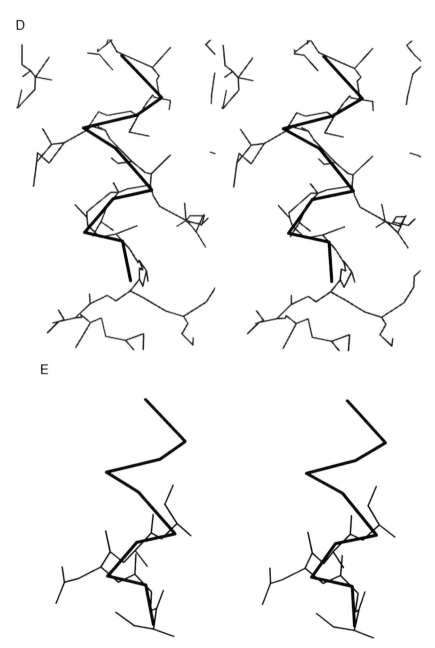

FIG. 3.47—*Continued.*

F

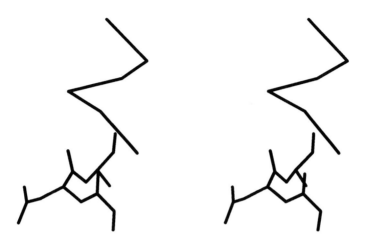

G

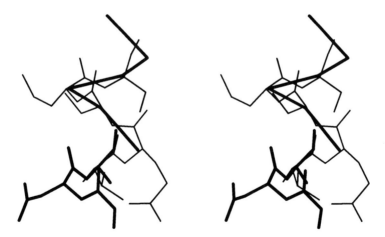

FIG. 3.47—*Continued.*

H

I

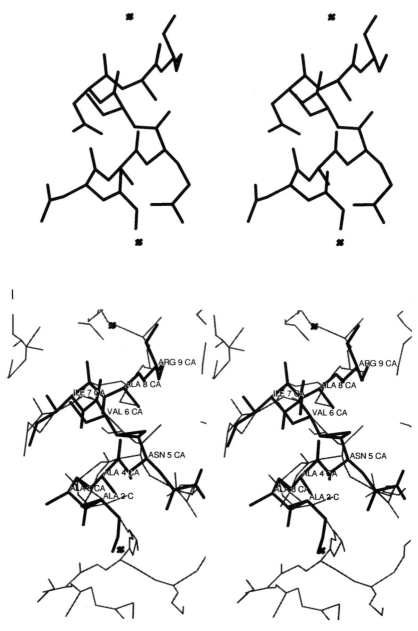

FIG. 3.47—Continued.

J

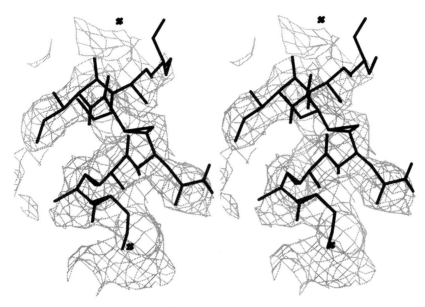

FIG. 3.47—*Continued.*

Because of the constraints on the main chain, if the path of the main chain is marked out, then a fairly accurate model can be built by finding a matching piece of chain from previously solved structures that follows that same path. In this manner, the entire main-chain model can be built by finding overlapping best-match polypetide pentamers from well-refined, high-resolution structures (Fig. 3.48). The match is performed by making a list of the difference vectors between the Cα atoms and comparing these to a previously computed vector list for a database of solved structures. Since the vectors between adjacent residues are always approximately 3.8 Å, there are only six unique vectors to consider. After the best-match is found, the pentamer is loaded and least-squares fit to the marker residues. Because the first and last residues are not well determined, only the middle three are kept. This process can be repeated until the entire structure has been built.

Remember that with all of the structures known to date, someone has already fit the conformation in front of you. If you find a new conformation never seen before, you will be hard-pressed to make anyone else believe you. These novel conformations seem to disappear with further refinement, never to be heard from again.

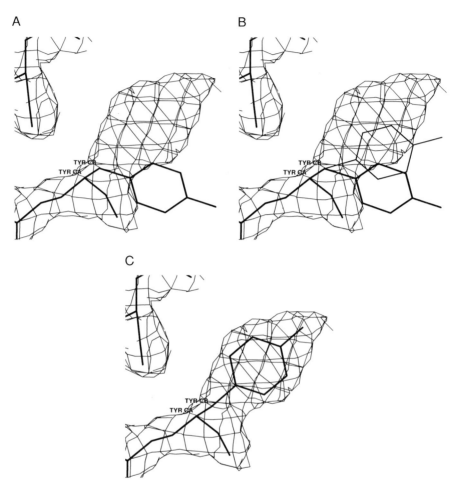

FIG. 3.48 Fitting a side chain by torsioning in xfit. A tyrosine residue that does not fit into its density is shown in (A). The bond to be torsioned is selected by picking the Cα atom followed by the Cβ atom and then choosing the Torsion menu item. This allows moving all the atoms connected to the Cβ atom to be rotated as a group about the vector defined by the Cα-Cβ bond. (B) The residue is shown being torsioned toward its density. The old position still shows as a guide. (C) After the bond was torsioned into position, the **Apply Fit** button was pushed and the residue was fixed into its new position. Alternatively, if you choose **Refine.Torsion Search** the program will find the best position by trying all 360° and choosing the position that maximizes the electron density at each atom.

Fitting Side Chains

Side chains have more degrees of freedom than the main chain. If you have fit the main chain accurately, then the C_β atom is accurately positioned. If at all possible, avoid breaking the side chain off from the main chain and fitting it separately. Instead, try to position the side chain by torsioning about the twistable bonds (Fig. 3.49). This is best done iteratively. First, try to get a rough position, then adjust the torsion angles to fit the side chain more accurately.

At medium resolutions the density is only a rough guide to the positions of the side-chain atoms. The density is smoothed and does not closely follow the path of the side chain. It can be difficult to tell the flat face of ring systems since they appear as a blob. The density at the C_β atom of many side chains is weak because it lies at the minimum of the ripple from the main chain and the larger side chains. In general, the C_β atom will not be centered in its density. One mistake that is easy to make in fitting medium-resolution maps is insisting on centering the C_β atom in its density. This will cause an error in the main chain because of the constraints on the angles around the C_α atom. Instead, let the C_β be defined by the main chain, and recognize that it will probably be somewhat off-center from its density. Small side chains, such as alanine and valine, tend to "melt" into the main chain and may appear as large bumps off the main chain. Because of the impossibility of accurately positioning the side chains, fitting must necessarily be an iterative process where the residues are roughed in and then further refined as more of the structure becomes apparent. In all fitting it is necessary to keep good stereochemistry in mind. If you fit with correct stereochemistry, the structure refinement will go much faster. When the position of a side chain is ambiguous, build it in a low-energy conformation. Again, novel conformations do occur but only if there is some reason.

Multiple Conformations

Side chains on the surface and, to a lesser extent, buried residues are likely to be in multiple conformations. The density for two (or more) conformations may be apparent in these cases if both have significant occupancies (Fig. 3.50). In some cases the conformations are close enough that densities overlap and cause a confusing widening of the density that no single conformation can adequately explain. In extreme cases a side chain just disappears after a certain point because it has many conformations, all of which have density below the noise level.

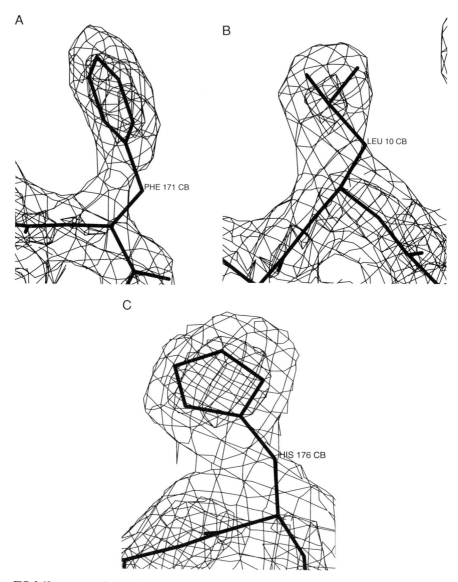

FIG. 3.49 Beware of weak Cβ density. At medium resolution the density at the C_β atom is weak and pinched off. (This is due to series-termination errors.) Three examples of the final refined model are shown superimposed on an excellent quality MIR map. (A) The density is almost gone at the C_β and positioned off to the side of the correct position. (B) The density is more even, but still the C_β would not be positioned in the center of the density. (C) The side-chain density is melting into the main-chain density because of the incomplete resolution. When fitting such maps, you must take into account the positions of adjacent residues and not force the C_β atom into the middle of its apparent density.

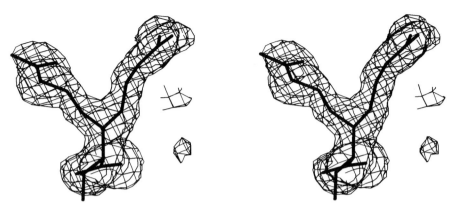

FIG. 3.50   Example of multiple conformations of side chains. This arginine residue from a β_4 hemoglobin shows two equally occupied positions hinged at the C_β carbon. The map is a 1.8 Å resolution Fo − Fc omit map contoured at 2σ. Figure kindly supplied by Dr. Gloria Borgstahl.

Phase Bias

The single biggest problem with fitting maps is avoiding phase bias. Phase bias results from the use of α_{calc} for calculating electron-density Fouriers. The phase portion of the Fourier transform tends to dominate the synthesis. This leads to a situation where whatever model you used to calculate the phases is the one seen in the map, even if it is incorrect. Phase bias is especially a problem in partially refined structures with higher R-values. If your R-factor is above 0.25, it is doubtful that a 2Fo − Fc map will give a true picture of the protein. Difference maps, Fo − Fc, are less prone to phase bias and are more reliable. It is wise to keep referring to the MIR map if you have one. The MIR map is probably noisy but is unbiased with respect to any given model and can be used to decide between models and conformations. If you have used molecular replacement, then overcoming phase bias can be a real problem. In this case there is no unbiased phase set to work with, and it can be a long uphill struggle to get the R-factor down to values low enough to get reliable maps. There are two useful methods of lowering phase bias: omit maps and phase combination, where the experimental phases are combined with the model phases (discussed previously).

Omit maps

A major strategy for overcoming phase bias is the use of omit maps. An omit map is made by removing the residues of interest from the model while calculating the phases (Fig. 3.51). In theory, this will allow the phases

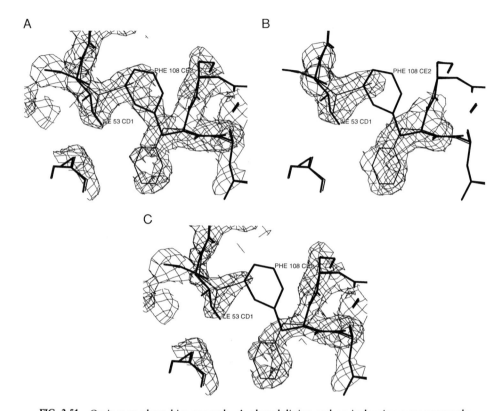

FIG. 3.51 Omit map phase bias example. A phenylalinine and an isoleucine were purposely misfit and then refined at 1.8-Å resolution for several cycles with the crystallographic terms weighted heavily over the geometry constraints. This had the effect of distorting the molecule and also building in some phase bias for the incorrect structure. (A) A map of the protein after refinement is shown. The thick lines are the incorrect model of the protein after refinement and the thin lines are the correct model. (B) The partial structure factors for the misfit Phe and Ile were calculated and subtracted using the omit current atoms option in xfit and an Fo − Fc omit map was calculated. Now only density for the correct residues shows up. (C) The omitted phases were used to calculate an Fo omit map. Note that in both types of omit maps the correct position shows up. In the Fo − Fc map the residues are cleaner. The Fo omit map is noisier, but it has the advantage of being independent of the relative scale of Fo to Fc and so can be advantageous if the scale is uncertain.

calculated from the rest of the model to phase the area of interest with no bias from the model left out. The method takes advantage of the Fourier-transform property that every point in real space is influenced by every point in reciprocal space, and vice versa. If the rest of the model is mostly correct, then the phases calculated for this portion will be close to the true phase and will produce a mostly correct image of the portion left out. About 10% of the total model can be left out without unduly affecting the phase accuracy. Omit maps can be calculated on the fly in Xfit if you have read in the map phases.

Unfortunately, phase bias can be spread throughout the entire model by the least-squares minimizer used in protein refinement. If the incorrect model has been included in the refinement, then the minimizer will adjust the coordinates of *all* of the model small amounts to fit the data to the model in question. If a portion of the model is in doubt, it is better not to include it at all than to put in a trial model. After refinement, the incorrect model will be "remembered" by the rest of the structure and will come back even in omit maps. This can be overcome partly by removing the piece in question and refining without it. XPLOR has a facility for doing molecular-dynamics-coupled refinement on the rest of the model to "shake it up" and remove this memory. A computationally less expensive method that achieves the same result is to add a small random number to all of the coordinates. A random number between 0 and 0.25 Å seems to work well. Omit maps made using the phases from the "shaken" coordinates show less bias in incorrect areas (see Chapter 4, Fig. 4.20).

Another difficulty encountered with omit maps is calculation of the proper scale factor. This can best be seen by writing out the coefficients used in a 2Fo − Fc map more fully to include the scale factor between Fo and Fc:

$$(2|F_{obs}| - s|F_{calc}|)e^{-i\alpha_{calc}}. \tag{3.36}$$

The scale factor s, is calculated in order to minimize the difference in the sums of F_{obs} and F_{calc}. If a portion of F_{calc} has been left out, then this scale factor will be incorrect—the scale factor s will be too large. This has the affect of subtracting too much of the calculated phases from the map. Now consider that the calculated coefficients have the omitted model subtracted from them. If the scale factor is too large, then we will add to the map $-(s - s_{correct})(-F_{omit})$, which is a positive image of the model we were trying to omit! If the scale factor is too small, then a negative image will be added to the map. Thus, making an omit map with 2Fo − Fc or Fo − Fc coefficients, where the scale factor is calculated in the normal way, will always bring back a calculated image of the omitted model. This is less than useful. This can partially be taken care of by first calculating the scale factor with the entire model and then using this scale factor instead of the one calculated for the incomplete model. Of course, your model may be incomplete without your realizing it. At early stages, solvent is not included and, usually, neither is an account of the disordered solvent. This raises the scale factor s, which, in turn, causes 2Fo − Fc omit maps to be biased unduly toward the model omitted. An easy way to avoid this is to use Fo coefficients in omit maps. The omitted portion comes back at about half the weight of the rest of the model but is not biased by scale factor problems.

Adding Waters and Substrates

A difference Fourier, Fo − Fc, will show unaccounted density. Often this density can be explained by water molecules. Protein molecules commonly have water bound in crevices and near hydrogen-bonding groups on the surface. If the potential water molecule does not make a bad contact with the protein and there are hydrogen-bonding groups at the correct distance, then a water molecule is warranted. Start adding water molecules at the largest peaks in the difference map and work your way down. Symmetry atoms should be generated periodically to ensure that a water position has not been accounted for already.

It is important to avoid adding too many waters too quickly and to avoid adding them where they are not justified. Because of the low number of constraints on water coordinates (i.e., they are not connected directly to the rest of the structure), they can model phase bias better than the highly constrained protein. The practice of putting waters in by fitting all the peaks in a difference map is dangerous. It can essentially freeze the refinement by modeling the noise with a large number of free parameters. This will allow the minimizer to fit the noise and lower the R-factor but not necessarily with any justification. On the other hand, adding water in correct positions will help the refinement.

When the waters are being added to density whose peak value is below 2 σ, or so, in a 2Fo − Fc electron density map, it may be worth checking their validity by protein refinement. Refine the B-values of the water molecules. If the B-values of the water molecules go above a conservative value of, say, 50.0, it is probable that the peaks are actually just noise.

A difficulty often encountered in adding water is partial disorder. In a crevice, for example, there is often a tube of density that represents the averaged position of several waters (see Fig. 3.14b). Most refinement programs will not let atoms get closer than their van der Waals contact distance, so it may not be possible to model the water with discrete atoms. Such density can be modeled with special water molecules that have no, or very small, van der Waals radii. Several waters, each with a partial occupancy, can be placed in the density.

Finally, not all unexplained density in protein maps is necessarily water. Buffer molecules and/or salt may also be bound by the protein. If a piece of density that needs several waters to explain it is encountered, it is worth carefully considering the exact contents of the crystal mother liquor. It is common for such density to be fit as water in early stages and then later, as the refinement progresses and the phases improve, to find the density revealing that another molecule is weakly bound there. It is worth reexamining all of the waters after the R-value has decreased to find such features. Try to explain the density in terms of chemistry as well as shape. If you have three negative side chains pointing toward one water molecule, it is worth considering if a cation is bound there.

..... **3.13** .....
ANALYSIS OF COORDINATES

Lattice Packing

There are several reasons for looking at the lattice contacts. First, the contact regions will have lowered B-values because of the increased resistance to movement. If you are doing a B-value analysis, this has to be taken into account. Another reason is to look for protein–protein interactions. In many cases, interactions that are important in the protein functioning may be revealed in the lattice packing. For example, the protein may act as a dimer in solution but have crystallized in such a way that the dimer was coincident with a crystallographic axis. The asymmetric unit would be a monomer, and lattice packing would reveal the dimer.

Many molecular-modeling programs have the ability to generate symmetry-related molecules, given the space group operators (Fig. 3.52). If not, they can always be generated in a separate step and read into the program. When generating contacts do not forget that translation by a cell constant in any of the three directions is a valid operator. That is, the operators listed in the *International Tables* assume $x+/-1.0$, $y+/-1.0$, $z+/-1.0$, since this is the fundamental relationship of crystals. Generating all possible contacts is done by generating a single cell by symmetry operators and then generating a $3 \times 3 \times 3$ block of unit cells around the molecule of interest by translating the single cell 0, $+1.0$, -1.0 in all three directions. Then a search is made to see if any atom within the molecule of interest is within some cutoff distance of each of the symmetry-related molecules.

The atoms at the contacts cannot come any closer than van der Waals packing distance. If they do, then something is wrong. Either the structure is incorrect, the unit cell is incorrect, or there is an error in one or more of the symmetry operators. Obviously, two atoms cannot exist in the same space.

Hydrogen Bonding

Hydrogen bonds are found in protein crystallography indirectly. If a proper hydrogen bond acceptor–donor pair is within the correct distance,

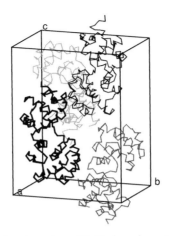

FIG. 3.52 Lattice packing of endonuclease III. Endonuclease III cystallizes in space group P2₁2₁2₁ (three perpendicular 2₁ screw axes) with one molecule per asymmetric unit. The box outlines one unit cell. The backbone of a single protein molecule is shown in thick black lines. The three other molecules that make up the unit cell are shown in shades of gray and are generated by application of the symmetry operators.

then the bond is taken to be a hydrogen bond. This distance is generally considered to be from 2.7 to 3.3 Å, with 3.0 Å being the most common value for protein and water hydrogen bonds.[30] The angle that the bond forms is also important in determining the strength of the hydrogen bond. The closer the hydrogen bond is to correct geometry, the stronger the bond. Hydrogen bonds often occur in networks—frequently with water mediating. Water is especially facile at hydrogen bonding since it is both an acceptor and a donor. Histidines can have various protonation states, and an analysis of the hydrogen bonds can allow a determination of the most likely protonation state by looking at whether it is hydrogen bonded to a donor or to a acceptor. In a similar manner, the orientations of histidines, threonine, glutamine, and asparagine are ambiguous in protein maps where the slight density difference between a carbon, oxygen, or nitrogen atom cannot be safely distinguished. An analysis of hydrogen bonding can give important clues in determining the correct orientation of an ambiguous side chain (Fig. 3.53).

[30] Stickle, D. F., Presta, L. G., Dill, K. A., and Rose G. D. (1992). *J. Mol. Biol.* **226**, 1143–1159.

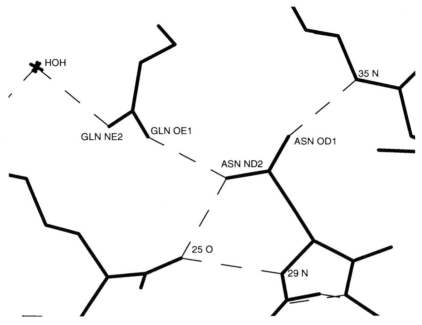

FIG. 3.53 Using H-bonding patterns to show Gln, Asn orientation. By examining the pattern of hydrogen-bond acceptors and donors, it is often possible to assign the N and O atoms of glutamine and asparagine side chains.

Solvent-Accessible Surfaces

Ultimately, it is the shape and chemical characteristics of the solvent-accessible surface that determine a protein's interaction with other proteins and substrates (Fig. 3.54). A solvent-accessible surface can be calculated and color coded by atom type to give some idea of the nature of this surface and of the solvent accessibility of surface groups.[31] Such a surface assumes a static view of the structure—which is an incorrect assumption. Another problem is dealing with hydrogens. Explicitly placed hydrogens are only part of the answer since they can be rapidly rotating. To overcome this, an average radius is often added to atoms containing hydrogens, and the hydrogens are left out of the model. For some residues the degree of protonation can also be ambiguous. Given these limitations, such surfaces are widely used in looking for substrate-binding sites and at protein interfaces.

[31] Richards, F. M. (1985). "Calculation of Molecular Volumes and Areas for Structures of Known Geometry," in "Methods in Enzymology," Vol. 115, pp. 440–464.

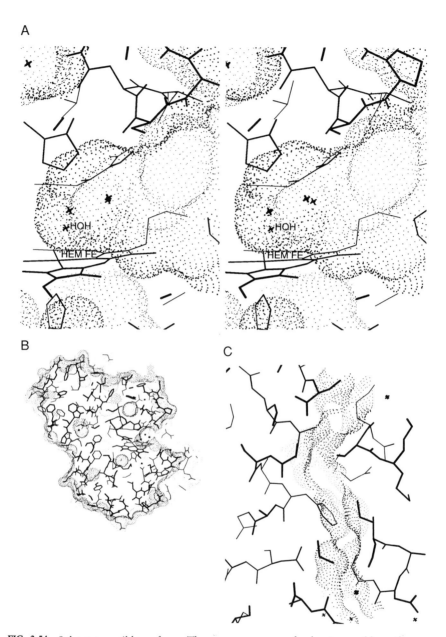

FIG. 3.54 Solvent-accessible surfaces. Three common uses of solvent accessible surfaces are shown. (A) The surface of an active site. Because it is bound to the Fe, the water, HOH, can be closer to the solvent-accessible surface. (B) A slice through a protein molecule, showing the shape of the outer surface and the presence of cavities in the interior of the protein. The active site of the protein is located at right center as a deep cleft in the protein. This is a common feature for active sites, and such a surface is a useful clue to the postion of an active site. (C) This surface shows the interface at a protein–protein contact. On color displays it is useful to color the surface according to charge to look for electrostatic interactions and hydrophobic contact surface.

Before calculating the surface, make a choice of the probe size. A water is generally taken to be a 1.4-Å sphere for calculating solvent-accessible surfaces. In many cases, a larger diameter of 1.6 Å is desirable. This helps account for small differences in atom positions and for thermal motion of side chains. The density of points to use in calculating the surface is also needed. A density a 10 points per angstrom^2 is common, while a lower density of 4 may be used to generate sparse surfaces. The higher density should be used if an accurate measure of the surface area is needed; the lower density will generate fewer points that will rotate faster when viewing with a graphics program. The surface is then calculated by running MS[32] (or another suitable program) with the chosen diameter. The output surface file can be viewed with xfit by running a simple awk command that extracts the x, y, z coordinates in the output surface file (see the xfit command in the Appendix B). It is also possible to color surfaces based on different parameters by using information in the surface file.

[32] Connolly, M. L. (1983). *Science* **221**, 709–713. Connolly, M. L. (1983). *J. Appl. Crystallogr.* **16**, 548–558.

4

PROTEIN CRYSTALLOGRAPHY COOKBOOK

..... 4.1 .....
MULTIPLE ISOMORPHOUS REPLACEMENT

This is the oldest method of phasing proteins and is still very successful. The basic method has changed little since myoglobin was solved, although the detailed implementation has.

The basic cycle of MIR phasing is diagrammed in Fig. 4.1.

1. Soak the crystals in heavy-atom solutions to scan for possible derivatives. Crystals that survive the soaking are tested to see if they still diffract. Those that do are scanned to see if they produce any changes in X-ray intensity.

2. If intensity changes are observed, a data set is collected. The first data should be collected quickly. If possible, a nearly complete data set to at least 5 Å should be collected within 24 h or even faster. Many heavy-atom derivatives are unstable in the X-ray beam, and the crystals quickly degrade or they change with time. Often the best data are from the first run, and even though the crystal still diffracts strongly, later data are found to have significantly lower phasing power. Continue collecting data if the crystal has not degraded.

3. Evaluate the heavy-atom statistics. By looking at the statistics of intensity changes it is possible to tell if the crystal is likely to be a good deriva-

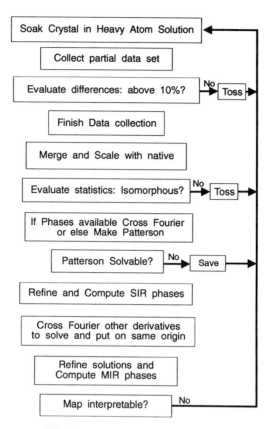

FIG. 4.1   Heavy-atom phasing scheme.

tive or not. The overall percentage difference should be above 10–12%. While you may solve and refine a derivative with weaker changes, it will have little phasing power; try resoaking to see if you cannot raise the percentage difference with longer soaks and/or higher soaking concentrations. The centric data should have larger differences than the acentric zones. The RMS magnitude of the differences should fall off with resolution in roughly the same proportion as the scattering factor for the heavy atom (including the temperature factor). If the differences do not fall off, it indicates noise or non-isomorphism.

4. Solve for the positions of heavy atoms. It is necessary to solve at least one derivative's Patterson map. This may not be the Patterson map of the first derivative you find. After solving one Patterson map, you can solve the others by cross-phasing with the SIR phases of the first. This also puts all the heavy atoms on the same origin.

5. Refine each heavy-atom solution. Look for additional sites. Be conservative about adding sites at this point.

6. With two or more derivatives you can co-refine the derivatives to improve the phases (if the derivatives share common sites this should be done cautiously). This gives you the first set of protein phases.

7. Make difference Fouriers of each derivative, preferably leaving this derivative out of the protein-phase calculation to reduce bias. Look for new heavy atom sites and confirm old ones.

8. Refine the updated solutions and co-refine to produce better protein phases.

9. Reiterate if necessary.

10. Calculate the protein electron-density map and evaluate its quality. If the map is difficult to interpret, go back to Step 1 and look for more derivatives or work to improve the ones that you already have. Be objective—you will not divine the structure from poor protein phases without considerable luck.

Example 1: Patterson from Endonuclease III

E. coli endonuclease III (Table 4.1) was solved at the Scripps Research Institute by MIR techniques.[1] The solution of a single-site derivative Patterson is presented here. Endonuclease III crystals were soaked in thiomersal, an organo-mercurial, at 1 mM for 2 days. Data were collected on the area detector and then merged with the native data. The isomorphous difference Patterson is shown in Fig. 4.2. The symmetry of the Patterson is *mmm*, orthogonal mirrors at 0 and 1/2 in all three directions. In space group P2$_1$2$_1$2$_1$, there are three Harker planes arising from the three 2$_1$ screw axes (Table 4.1) at $x = 1/2$, $y = 1/2$, and $z = 1/2$. A heavy-atom site will give rise to three unique self-vectors on these Harker sections (plus all the peaks related by

[1]Kuo, C. F, McRee, D. E., Fisher, C. L., O'Handley, S. F., Cunningham, R. P., and Tainer, J. A. (1992). *Science* **258**, 434–440.

TABLE 4.1

Endonuclease III Facts

Protein: *E. coli* endonuclease III

Unit Cell: 48.5, 65.8, 86.8, 90.0, 90.0, 90.0

Spacegroup: P2$_1$2$_1$2$_1$

Patterson symmetry: *mmm*

Harker planes: **1/2**, 1/2 +/− 2 y, +/− 2z
 +/− 2x, **1/2**, 1/2 +/− 2z
 1/2 +/− 2x, +/−2y, **1/2**

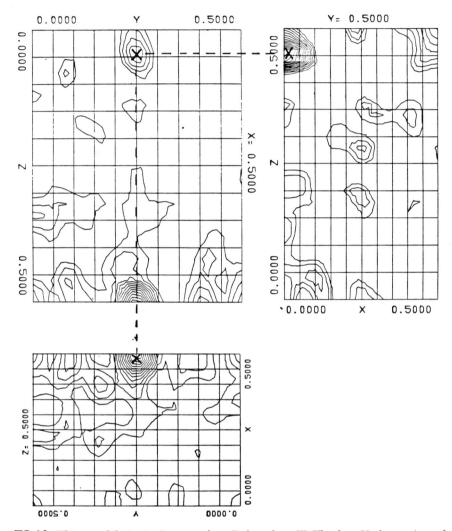

FIG. 4.2 Thiomersal derivative Patterson from Endonuclease III. The three Harker sections of an isomorphous derivative Patterson map at 5.0 Å resolution are shown for a thiomersal derivative of *E. coli* endonuclease III. The positions of three Harker vectors are marked with "X." The dashed lines illustrate how these peaks line up when the sections are aligned as shown. Notice that the peak on $z = 0.5$ overlaps onto the section $x = 0.5$. This Patterson shows no significant features on the non-Harker sections.

Patterson symmetry). Note in Table 4.1 that if there is a vector at $2x$ on one section it will be at $1/2-2x$ on the other. Thus, if we line up the Harker sections so that the common axes run in opposite directions and 0.0 is opposite 0.5, then the self-vectors on the two sections will line up (Fig. 4.2). On the Harker sections for this derivative there are just three unique large peaks, and they line up as shown by the dashed lines. The peak on $z = 1/2$ overlaps onto $x = 1/2$, making it appear as though there are two peaks on $x = 1/2$, but only one belongs solely to this section.

This Patterson can be explained by a single major site. The peak at $(1/2, 0.25, 0.05)$ is solved:[2]

$$1/2 - 2y = v,$$
$$y = -(v - 1/2)/2,$$
$$y = -(.25 - 0.5)/2 = 0.125;$$
$$2y = w,$$
$$y = w/2 = 0.05/2 = 0.025.$$

This gives us $(—, 0.125, 0.025)$ from this section. From the Harker peak at $(0.48, 0.25, 1/2)$ we can solve $(.01, 0.125, —)$. Combining these gives us a heavy-atom site at $(0.01, 0.125, 0.025)$. The $y = 1/2$ section can be used to confirm the other two.

From the *mmm* symmetry of the Patterson it follows that peaks occurring on the Harker sections lie on a mirror plane and will be doubly weighted, being the superposition of the peak and its mirror image. Peaks that occur on the edges are at the intersection of two mirror planes and will be weighted four times. Peaks in the corner are at the intersection of three mirrors and will be weighted eight times. This explains why the peak on $x = 1/2$ is about half the height of the peaks on the other two sections, which both occur at the intersection of two planes and are weighted four times versus two times. A single site thus accounts for most of this Patterson map, and the other features may be due to minor sites or perhaps just noise from the approximations used to derive the Patterson coefficients. In any case, this single-site solution was good enough to cross-phase other derivatives for this protein. Minor sites can be picked up more easily and accurately at a time later in the analysis, when other derivatives can be used to produce rough protein phases that can be used in a difference Fourier to give the positions

---

[2] Throughout this book we have used the coordinates x, y, z to indicate the three directions in a Patterson. Traditionally the indices u, v, w have been used to indicate the three axes so as to distinguish them from their real space counterparts, just as h, k, l are used for reciprocal space. I have used x, y, z mainly because the software makes no distinction between Patterson maps and other kinds. However, for the example of how to solve these equations it is useful to use the notation u, v, w for the Patterson coordinates, and x, y, z for real space to make the equations clearer.

of these minor sites. However, this particular derivative proved to be a true single-site derivative. In the solution of endonuclease III, this was the only Patterson that needed to be solved. All subsequent steps were done using difference Fouriers starting with the SIR (single isomorphous replacement) phases of this derivative.

Example 2: Single-Site Patterson from Photoactive Yellow Protein

Photoactive yellow protein is a 15 kDa protein that crystallizes in hexagonal space group $P6_3$ from ammonium sulfate with cell constants of $a = b = 66.9$, $c = 40.8$, 90, 90, 120. A $PtCl_4$ derivative can be made by soaking the crystals 7 h in a 0.1 mM solution of K_2PtCl_4 followed by back soaking the crystals in artificial mother liquor without any heavy-atom reagent. If the crystals are soaked longer, then they begin to deteriorate. The data for the derivative were collected by area detector. The data were merged and scaled to the native data using the anisotropic scaling method. A Patterson map was calculated using the coefficients $(F_{PtCl4} - F_{Native})^2$ at 5.0-Å resolution and throwing out large differences where $|F_1 - F_2|/((F_1 + F_2)/2)$ is greater than 1.0. A Patterson in space group $P6_3$ has two unique Harker sections at $z = 0$ and $z = 1/2$, which are shown in Fig. 4.3. The section $z = 1/2$ is especially useful to look at because peaks on this section occur at $x, y, 1/2$ and $2x, 2y, 1/2$ (Table 4.2). The peak at $x, y, 1/2$ is doubly weighted over the peak at $2x, 2y$ because it occurs twice as often in the vector table. This Patterson can be solved by simple inspection. The largest peak on the section $z = 0.5$ is at $(0.25, 0.08, 0.5)$, and a corresponding smaller peak is at $(0.5, 0.16, 0.5)$. We can, therefore, assign a heavy-atom position at $(0.25, 0.08, 0.0)$. In this space group z is arbitrary because there is no symmetry element perpendicular to z to fix the origin, so z is arbitrarily assigned to 0.0. If a second derivative is found, the relative z will need to be found by other means, usually by cross-Fouriers. We can confirm our solution looking at the section $z = 0.0$. The peak here arises from the vector $x + y, 2y - x$, and a cross labeled A-A shows the calculated position where our site will fall on this section. Most of the other peaks on the sections are related by the 6-fold to the peaks we have already used. However, an astute observer will notice that there are still some peaks of heights two contour levels and less unaccounted for. No other site could be found to account for these peaks, and they probably arise from one of three considerations: (1) The synthesis was artificially terminated at 5.0 Å. In this small cell there are not very many reflections, and we may have series-termination errors. To test this, the Patterson can be made at a higher resolution to see if it becomes less noisy. (2) An approximation to the heavy-atom vector was used and not the true heavy-atom vector, which invariably causes

A

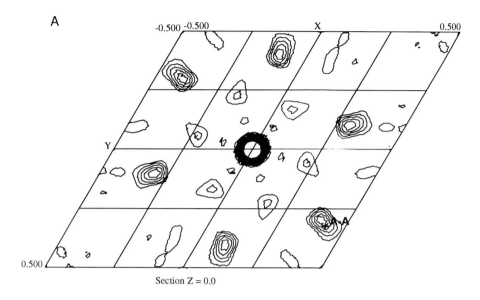

Section Z = 0.0

B

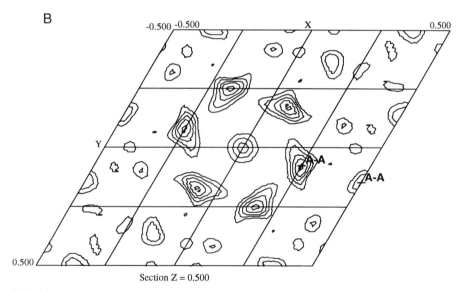

Section Z = 0.500

FIG. 4.3 Photoactive yellow protein PtCl$_4$ derivative Patterson map. Two sections are shown from the Patterson map of the PtCl$_4$ derivative of photoactive yellow protein, space group P6$_3$, a single 6$_3$-screw axis parallel to z: (A) Section $z = 0.0$ and (B) section $z = 0.5$. On each section the unique vectors are marked A-A. The rest of the peaks are generated by 6-fold symmetry about the z axis.

TABLE 4.2
Patterson Vectors for Space Group $P6_3$

$P6_3$	x, y, z	$-y, x - y, z$	$y - x, -x, z$	$-x, -y, 1/2 + z$	$y, y - x, 1/2 + z$	$x - y, x, 1/2 + z$
x, y, z	0, 0, 0	$-x - y, x - 2y, 0$	$y - 2x, -x - y, 0$	$-2x, -2y, 1/2$	$y - x, -x, 1/2$	$-y, x - y, 1/2$
$-y, x - y, z$	$x + y, 2y - x, 0$	0, 0, 0	$2y - x, y - 2x, 0$	$y - x, -x, 1/2$	$2y, 2y - 2x, 1/2$	$x, y, 1/2$
$y - x, -x, z$	$2x - y, x + y, 0$	$x - 2y, 2x - y, 0$	0, 0, 0	$-y, x - y, 1/2$	$x, y, 1/2$	$2x - 2y, 2x, 1/2$
$-x, -y, 1/2 + z$	$2x, 2y, -1/2$	$x - y, x, -1/2$	$y, y - x, -1/2$	0, 0, 0	$y + x, 2y - x, 0$	$2x - y, x + y, 0$
$y, y - x, 1/2 + z$	$x - y, x, -1/2$	$-2y, 2y - 2x, -1/2$	$-x, -y, -1/2$	$-x - y, x - 2y, 0$	0, 0, 0	$x - 2y, 2x - y, 0$
$x - y, x, 1/2 + z$	$y, y - x, -1/2$	$-x, -y, -1/2$	$2y - 2x, -2x, -1/2$	$y - 2x, -x - y, 0$	$2y - x, y - 2x, 0$	0, 0, 0

some noise in the Patterson. (3) Inaccuracies in the data sets, scaling errors, and non-isomorphism also contribute.

Example 3: Two-Site Patterson from Photoactive Yellow Protein

In another heavy-atom trial, a photoactive yellow protein crystal was soaked for several days in 50 mM GdSO$_4$ and then in 0.1 mM K$_2$PtCl$_4$ for 7 h to produce a double derivative. It was already known that both GdSO$_4$ and K$_2$PtCl$_4$ produce single-site derivatives, so a double-site derivative was expected. After collecting the data and merging with the native (Fig. 4.4), we produced the Patterson in Fig. 4.5. On the Harker section $z = 0.5$ there are the vectors for both the platinum (Pt) and gadolinium (Gd) site. By chance, on the $z = 0$ Harker section both sites overlap. As was shown in the example before, we can solve the peaks on the $z = 0.5$ by noting that the peaks fall in pairs at $(x, y, —)$ and $(2x, 2y, —)$. The position of the Pt peak is already known as $(0.25, 0.08, 0.0)$, so the other two pairs of peaks are from the Gd site. We can assign the Gd site as $(0.50, 0.10, ?)$ since we cannot assign both sites at $z = 0.0$. Also, we do not know the relative origin of the second site. A just-as-valid single-site solution for the Gd site would be at the Patterson symmetry related position (y, x, z) or $(0.10, 0.50, ?)$. In order to solve this dilemma, we must find a cross-peak between the two sites and use it to define the relative z and the origin. Since we found the Pt first and its peak is larger, we will arbitrarily decide this is our origin and that it is at $z = 0.0$. That is the end of our arbitrary choices—now we must find the corresponding Gd origin and z (actually we still have a hand choice by placing Gd at $+z$ or $-z$). Scanning through the sections, we find peaks on both sections $z = 0.133$ and $z = 0.367$. We note that 0.367 is $0.5 - 0.133$. This is the expected pattern for cross peaks in P6$_3$—for every cross peak on z we know that there is one on $1.0 - z$, $0.5 - z$, and $0.5 + z$. This can be worked out from algebraic manipulations of the symmetry operators, or we can try a few hypothetical test cases with xpatpred and then look at the patterns of the predictions. Taking the first likely cross-peak on the Patterson, we find a broad peak at $(0.25, 0.03, 0.133)$ (see Fig. 4.5). We assume this is a vector Gd $-$ Pt, so to generate the Gd coordinates from this site we add Pt + cross-peak or $(0.24, 0.08, 0.0) + (0.25, 0.03, 0.133)$ to get $(0.49, 0.11, 0.133)$. The Harker vector for this site would be $(0.49, 0.11, 0.5)$, and we see a nice Harker peak on the $z = 0.5$ section at this point, marked Gd-Gd in Fig. 4.5, that partially overlaps with the Pt vector. This site is then put into xpatpred, and the pattern of Harker vectors and cross-peaks all fall on or close to a peak in the Patterson map, so we conclude this is a valid solution. A little fiddling with the coordinates to center the Harker vectors for the Gd site

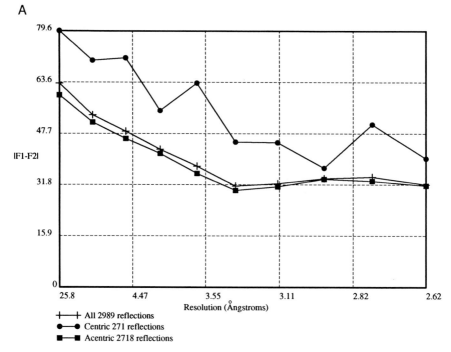

A

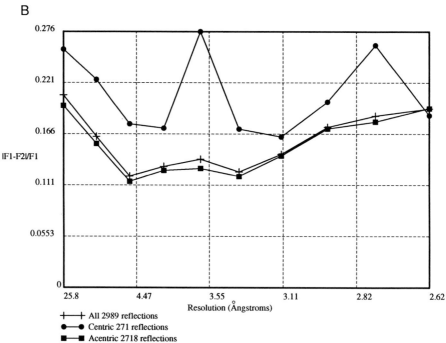

B

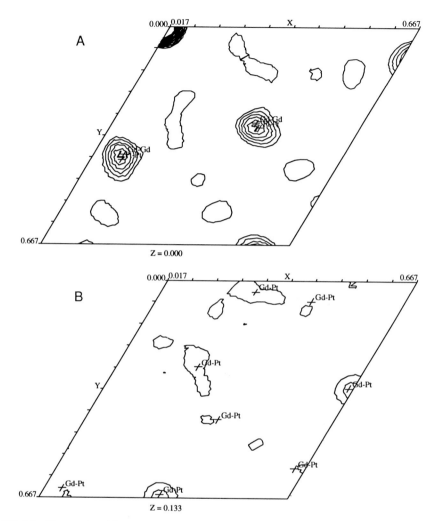

FIG. 4.5 Photoactive yellow protein double-derivative Patterson map. The crosses indicate the positions of the heavy-atom vectors. A Pt self-vector is labeled Pt-Pt, a Gd self vector is Gd-Gd, and a cross-vector between the Pt and Gd site is labeled Pt-Gd. (*Figure continues.*)

FIG. 4.4 Photoactive yellow protein double-derivative statistics. Graphs show the statistics for a double soak of K_2PtCl_4 and $GdSO_4$ merged with native data. (A) |FP − FPH| versus resolution. If the derivative is isomorphous, this graph should steadily decline with increasing resolution. The derivative seems to be isomorphous to about 3.2 Å. Also, the centric differences should be larger than the acentric differences. (B) R-factor versus resolution. This gives the degree of substitution, which is fairly low for this derivative, being only about 0.13 for the average reflection.

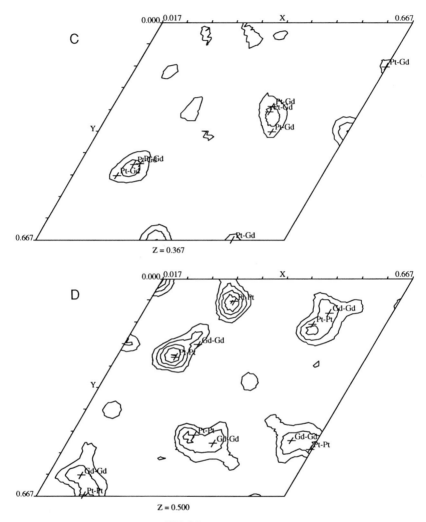

FIG. 4.5—*Continued.*

gives a better match at (0.505, 0.105, 0.133). We now have two sites: Pt at (0.24, 0.08, 0.0) and Gd at (0.50, 0.10, 0.133), which should now be confirmed. One way to confirm them is with a cross-Fourier if we have another derivative. However, in this case, since the derivative already contains the Pt site, a cross-Fourier with the Pt site will not be very informative because ghost peaks from the Pt will also give us a peak if the solution is incorrect. In this case, we went on and made a 5.0 Å protein map that looked quite good, and so we continued the solution with confidence. The structure was subsequently solved and has been published.[3]

Example 4: Complete solution of *Chromatium vinosum* Cytochrome *c'*

This protein was solved by Zhong Ren, Susan Redford, and myself using four derivatives in the final phasing out of seven potential ones. The entire screening, data collection, and phasing were mostly done in 2 months of effort using the Siemens area detector to scan the derivatives.

Crystal Growth

Several forms of crystals were grown. The first crystals out of ammonium sulfate proved to be useless as they were always multiples. Two crystal forms grew out of PEG, sometimes in the same drop: a monoclinic form with 2-dimers in the asymmetric unit and an orthorhombic form with a single dimer in the asymmetric unit.[4] It was possible to tell them apart by examining the angles between crystal faces. The orthorhombic form had two edges that met at an angle of 90°, while in the monoclinic form there were no edges exactly 90° apart. By macro-seeding, large crystals of the orthorhombic form could be grown reproducibly.

Heavy Atom Soaks

An artificial mother liquor with a higher PEG concentration was necessary for soaking or else the crystals cracked and dissolved. If the crystals were transferred to a solution at growth conditions, they dissolved, presumably because of the lowered protein concentration. The crystals were stable in a PEG concentration 2% higher than that used for growth. Once it was established that crystals were stable for long periods of time, they were soaked in 1 mM solutions of our favorite heavy-atom compounds. Each drop had two or three crystals, with the idea being that they would be mounted and scanned at different time intervals. The first crystal was usually mounted after 24 h and the rest followed an uncertain schedule depending on the progress of the other scans. The basic scanning strategy used was as follows. The

[3]McRee, D. E., Tainer, J. A., Meyer, T. E., van Beeumen, J., Cusanovich, M. A., and Getzoff, E. D. (1989). *Proc. Natl. Acad. Sci. U.S.A.* **86**, 6533–6537.

[4]McRee, D. E., Redford, S. M., Meyer, T. E., and Cusanovich, M. A. (1990). *J. Biol. Chem.* **265**, 5364–5365.

TABLE 4.3

Summary of Native Diffraction for *C. vinosum* Cytochrome *c'*

Resolution (Å)	No. possible reflections	No. collected reflections	Complete (%)	No. observations	$I\sigma(I)$
2.9	6498	5750	88.49	24205	22.6
2.3	6291	6290	99.98	28117	9.8
2.0	6214	6165	99.21	24069	4.3
1.8	6162	5931	96.25	21758	1.5
1.7	6140	5622	91.56	18547	0.8
1.6	6158	3786	61.48	9645	0.5

crystal was mounted on the area detector and aligned so that one axis was roughly parallel to the incident beam. Data were collected such that 100° of data could be scanned in about 12 h. The swing angle was set to collect 2.7-Å data, a decision that was later regretted. Higher-resolution data could have been easily collected with a larger swing angle. After about 50 frames had been collected, they were integrated and scaled to the native data that had been collected earlier (Table 4.3). If the R_{merge} of the shell infinity to 5 Å was below 0.10, the crystal was removed and another put on (the usual case); otherwise a full data set was collected. Of the full data sets, it was often found that the R_{merge} decreased with more data, but never that the R_{merge} increased with a fuller data set. Therefore, one would not make the mistake of removing a potential derivative by scaling a partial data set, but might waste some beam time on poor derivatives.

Patterson Maps

As each data set was integrated, it was merged and scaled to the native data. A Patterson map was made for each one at 100–5-Å resolution. The first Patterson maps all looked the same and were not interpretable. They had large diffuse peaks. By listing and sorting the differences, we discovered that the largest differences were due to very low-resolution reflections that had been clipped to varying degrees by the beam stop. These were eliminated in future Patterson maps by applying a filter cutoff so that any reflection with a difference greater than 100% of the average of the two observations was deleted. These Patterson maps were now interpretable. The first Patterson map solved was that of gold cyanide, and had two major sites. This entire Patterson map sectioned in *z* and the three Harker sections are shown in Fig. 4.6. The solution was slightly tricky—there are two major heavy-atom

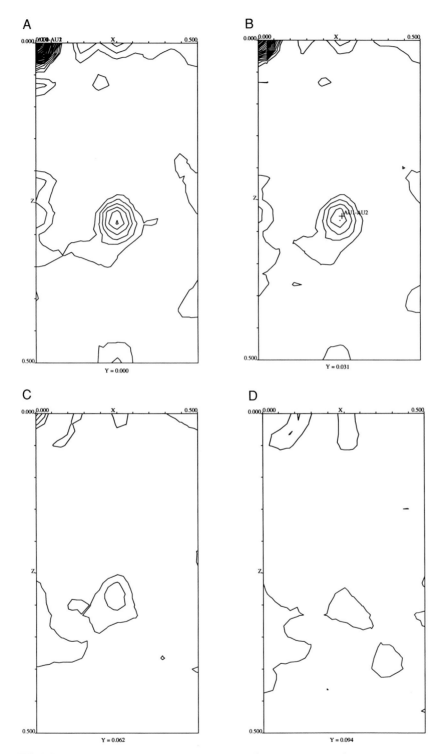

FIG. 4.6 Complete gold derivative Patterson map of *C. vinosum* Cytochrome *c'* (CVCC).

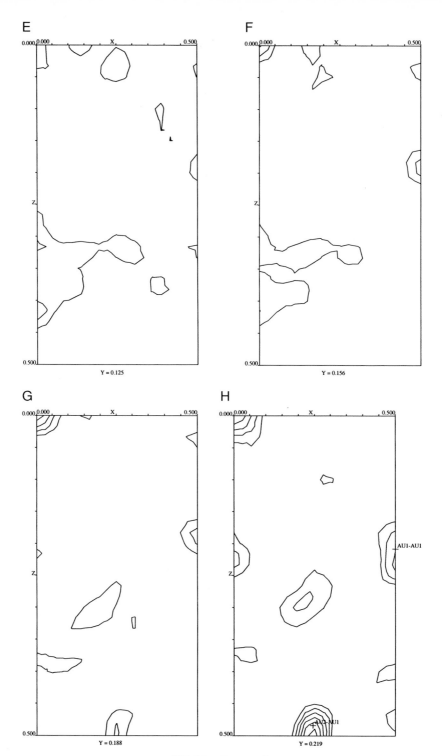

FIG. 4.6—*Continued.*

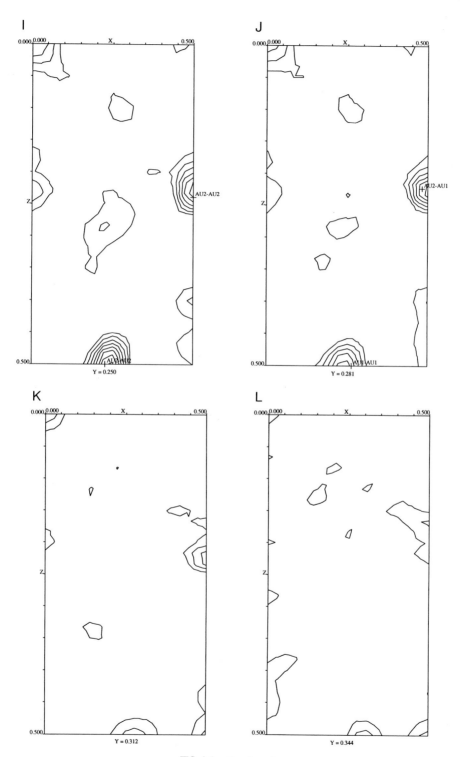

FIG. 4.6—*Continued.*

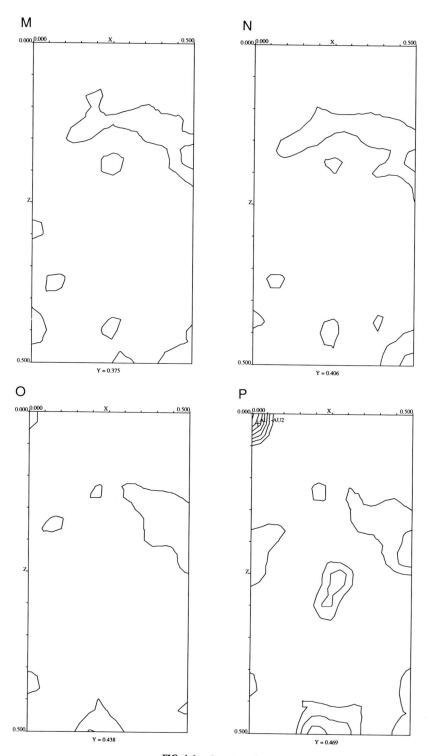

FIG. 4.6—*Continued.*

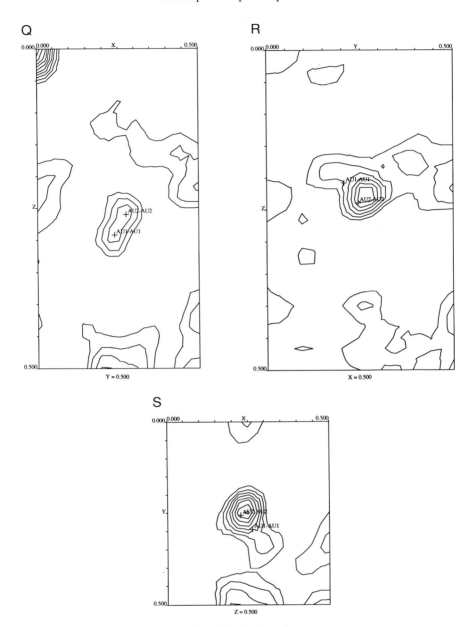

FIG. 4.6—*Continued.*

sites, but both share the same Harker peaks. (In $P2_12_12_1$ there are three Harker sections $x = 0.5$, $y = 0.5$, and $z = 0.5$, see Table 4.1.) The clue is that there are large cross-peaks on non-Harker sections, and we also knew that the protein was a dimer and, therefore, there were probably at least two sites. It is possible, if a heavy site is on the dimer axis, for there to be a single site (this was, indeed, the case for the major site in the iridium chloride, $IrCl_6$, derivative). A single site was first assumed, and the Harker vectors were solved to produce the site (0.125, 0.125, 0.125). Now if there is a second site and it shares the same Harker peaks, it could be thought of as being an origin shift relative to the first. In this space group the origin can be either at 0.0 or 0.5 in all three directions and produce the same Patterson. In addition, the sign of a site can be changed, and it will have no effect on the Harker vectors. However, if there are two sites, then doing these operations, adding 0.5, and changing the sign will have an effect on the cross-peaks between the two sites. The pattern of the cross-peaks will change with each combination of origins. The solution is entered into xpatpred and displayed on the Patterson map by writing out the predictions and reading them into xcontur as a labels file. Xpatpred predicts the Patterson vectors, given a possible solution and the space-group operators. Alternative origins can be chosen for each site. By changing the origin on the second site and displaying the resulting vectors, the correct position of the second site relative to the first could be found when the vectors best matched the pattern of cross-peaks. When the solution is satisfactory, it is saved in a solution file.

Initial Phasing

The output of xpatpred feeds directly into xheavy. The name of the data set containing the merged heavy-atom data is added, using the edit-solution window (Fig. 4.7). Now the solution can be refined against the differences. A two-site model was used that refined quickly to an R_c of 0.55. A set of SIR phases was calculated and used to make difference Fouriers of the other derivatives. At this time, everything is still done at 5-Å resolution. This is a good resolution for heavy-atom searches: higher resolutions often add more noise, and the contrast is lowered, making the search more difficult. At 5 Å there are enough reflections to over-determine the x, y, z positions of a few heavy-atom sites; too low a resolution will not have enough reflections to determine the heavy atom sites accurately or to allow them to be refined. Experience has shown that usually 5 Å is a good compromise, and we solve all of the derivatives at this resolution first and then raise the resolution later to calculate a protein map.

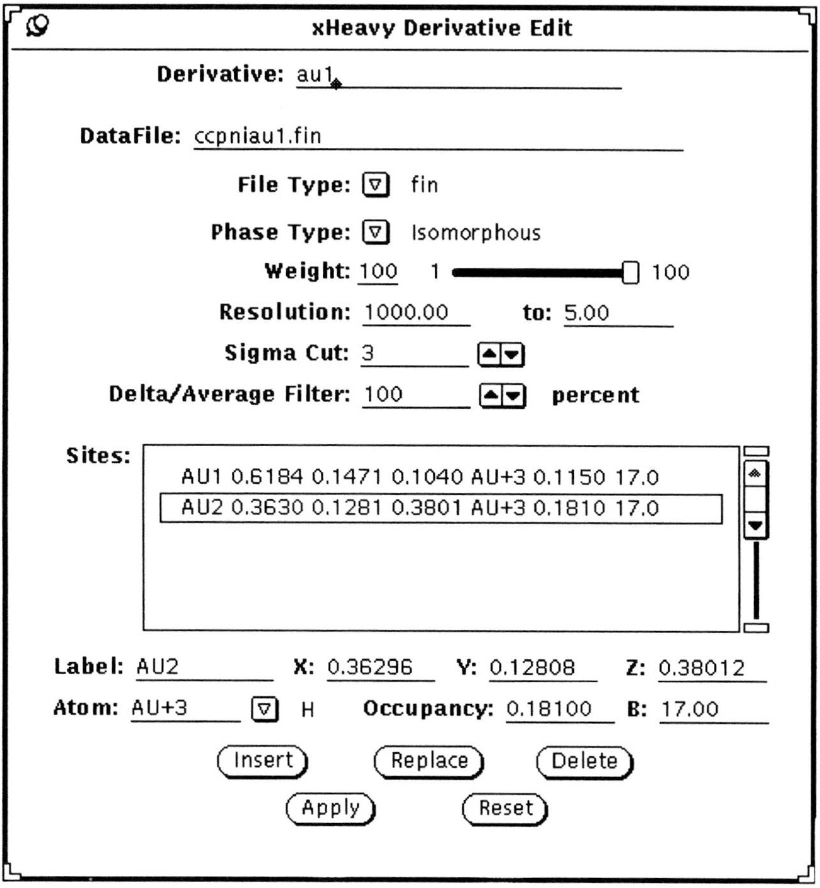

FIG. 4.7   Xheavy edit window with Au1 solution.

Search for Minor Sites

Even though the SIR phases are very noisy, they produced interpretable heavy-atom difference-Fourier maps ($|F_P - F_{PH}|$, α_{SIR}). There are many more observed differences than there are heavy atoms bound and, while noise tends to add everywhere in the Fourier, the signal builds up in the correct position. The two-site refined solution for the Au derivative was used to make a difference Fourier to look for minor sites. The highest peaks will, of course, be the original Au sites. The next-highest peaks may represent minor sites. To check this, their positions were entered into xpatpred and the predictions compared against the Patterson map. The minor peaks superimposed on peaks in the Patterson map in most positions. For minor sites, it is best to look at the cross-peaks for two reasons: (1), the Harker sections are usually noisier, and lower peaks here are less reliable; and (2), the height of a peak is the product of the two atoms that give rise to that vector. So if the major site has a scattering power of 4 and the minor sites 1, then the self-vectors of the minor peak will have height of 1 and the cross-peaks will be 4. Thus, the self-vectors may actually be below the noise level, while the cross-peaks are still visible. The minor sites accounted for some of the unexplained density in the Patterson map. However, as we went to smaller and smaller peaks on the difference Fourier, they matched smaller peaks in the Patterson, and at some point most of the cross-peaks will be missing a peak in the Patterson. At this point, it is usually worth stopping since adding these peaks will contribute little to the phasing power. In fact, a problem with adding too many minor peaks is that we may start modeling noise and not real sites, which will actually interfere with phasing.

Cross-Phasing other Derivatives

We could go on to solve all of the derivatives by examining the Patterson maps, but now that we have solved one derivative we can use the SIR phases to make difference Fouriers for all the other derivatives data sets. In order to put all the derivatives on the same origin, we will need to cross-phase anyway. The difference Fouriers are made using xmergephs. The merged and scaled derivative-data set is used as the input, and the SIR phases from the solved Au derivative is used as the input phases. We then make a map of the type Fo-Fc with xfft and display it with xcontur. It is necessary to be careful that the Fo corresponds to F_{PH} and Fc corresponds to F_P so that the peaks will be positive in the difference Fourier. The heavy-atom positions can be picked off of the difference Fourier and put in a solution file using the xheavy edit solution function. The cross-Fourier of derivative Ir1 using the SIR phases from Au1 is illustrated in Fig. 4.8. It is important, though, to check this solution against the Patterson of the new derivative to be sure that it is real (Fig. 4.9). Also, the possibility of ghost peaks exists. These are peaks that are due to using SIR phases instead of the real protein phases. Ghost peaks are found at the position of the heavy-atom sites in the derivative used to calculate the phases. So a peak found at this position in the difference Fourier might be a ghost peak, or it may really be a common site. To decide this, the Patterson of the new derivative can be checked to see if it has the correct vectors to explain the site.

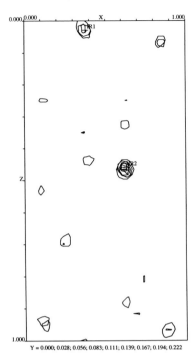

FIG. 4.8 Cross-Fourier of Ir1 with Au1 SIR phases. Crosses and labels show the positions of two iridium sites.

Refine the New Site

The new derivative can be refined at the same time as the first one. With two derivatives the phases will be improved, and if both derivatives are of good quality, we may even be able to make a recognizable protein map. However, before rushing off to make a map it is advisable to improve the solutions and to tune the derivatives further. Now that we have better phases, it is worthwhile redoing the difference Fouriers for both derivatives. This can be done with xheavy by selecting each derivative in turn and then writing out a difference Fourier phase file, running this through xfft and then examining the output the map with xcontur. Look for new peaks that are at two or more times sigma. Be careful not to pick peaks that are symmetry-related to the ones you already have in your solution.

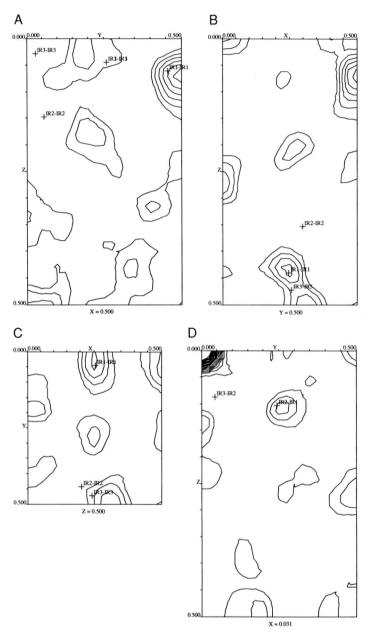

FIG. 4.9   Ir1-derivative Patterson map.

Adding More Derivatives

It is simple now to add new derivatives. Just cross-phase the differences with the latest protein phases at 5.0 Å and pick the peaks off the map, check them against the Patterson, and refine. With several derivatives, it pays to become critical and to determine if a derivative is worth including. For *C. vinosum* cytochrome *c'* (CVCC) we found seven derivatives for which we could explain the major peaks on the Patterson map. Of these, three were found to be inferior because they had low phasing power and high Rc values. Protein maps made including these derivatives were judged to be inferior to ones with them left out, so that in the final analysis only the best four out of the seven (Table 4.4) were used in the final protein map. The Pattersons of the two platinum derivatives are shown in Figs. 4.10 and 4.11.

The First Protein Map

The first protein map is always an exciting moment. The proper map to make is an Fo times figure-of-merit map at the best (centroid) phase. xheavy writes out a file that is records of h, k, l, Fo, f.o.m., α_{best}. Use xfft on this file with the Fo*f.o.m option to make the map. It is worth first making a 5-Å map and studying it to see if there are clear solvent boundaries and if a single molecule can be picked out. This will make it easier to interpret the higher-resolution maps. At 5 Å the solvent boundaries are usually obvious. There may be tight contacts that make it difficult to decide where the boundaries of a single molecule are. By symmetry considerations, it is sometimes possible to resolve these by noting that two bits of density are equivalent and, therefore, the boundary must be between them somewhere. A section of the 5-Å map of CVCC phased by the four derivatives is shown in Fig. 4.12. You can see clear solvent boundaries, and it is possible to see the dimer axis in this map and the positions of the hemes. We knew beforehand that the protein should be a four-helix bundle and a dimer. All of these features could be easily picked out. If a helix can be picked out at this stage, note its position for later reference. It will be useful to look at helices at higher resolution because they have a definite structure, are good guides of map quality, and are a good way to check the handedness.

Medium-Resolution Protein Map

After examining the 5-Å map, refine the derivatives to a medium resolution of about 3–2.7 Å. It is important to look at the phasing power of the derivatives with increasing resolution. Using a derivative to too high a resolution may actually degrade the quality of the map and not provide any use-

TABLE 4.4

Heavy-Atom Derivative Parameters for *C. vinosum* Cytochrome *c'*

Compound	Site No.	Relative occupancy	x	y	z	B-value	R-centric	Resolution (Å)
KAu(CN)$_2$	1	0.11	−0.6184	−0.1471	−0.1040	17.0	0.57	2.71[a]
	2	0.18	−0.3630	−0.1281	−0.3801	17.0		
	3	0.02	−0.8749	−0.0912	−0.3513	17.0		
	4	0.02	−0.1140	−0.0954	−0.0954	17.0		
	5	0.02	−0.6392	0.0088	−0.4134	17.0		
	6	0.01	−0.3950	−0.0983	−0.8210	17.0		
K$_3$IrCl$_6$	1	0.23	0.3771	0.0229	0.4698	17.0	0.60	5.0
	2	0.08	0.1487	0.2211	0.0734	17.0		
	3	0.08	0.6288	0.2359	0.0142	17.0		
K$_2$PtCl$_4$	1	0.22	0.6339	0.505	0.1129	17.0	0.62	2.71[a]
	2	0.13	0.6105	0.232	0.8781	17.0		
	3	0.23	0.7703	0.050	0.1043	17.0		
	4	0.17	0.0215	0.230	0.4174	17.0		
	5	0.11	0.1190	0.069	0.3753	17.0		
	6	0.03	0.1047	0.033	0.9053	17.0		
	7	0.21	0.8730	0.215	0.1349	17.0		
	8	0.15	0.5094	0.150	0.5795	17.0		
	9	0.04	0.7478	0.020	0.1164	17.0		
	10	0.05	0.7457	0.159	0.8713	17.0		
K$_2$PtCl$_2$(CN)$_2$	1	0.13	0.7399	−0.0188	0.1227	17.0	0.63	3.3
	2	0.10	0.1085	0.0321	0.8975	17.0		
	3	0.10	0.5125	0.1910	0.5674	17.0		
	4	0.05	0.4796	0.1334	0.0273	17.0		
	5	0.14	0.0289	0.2428	0.4030	17.0		
	6	0.07	0.7466	−0.0095	0.5904	17.0		
	7	0.12	0.7626	0.0432	0.1125	17.0		

[a] Only centric reflections were included past 3.3 Å resolution.

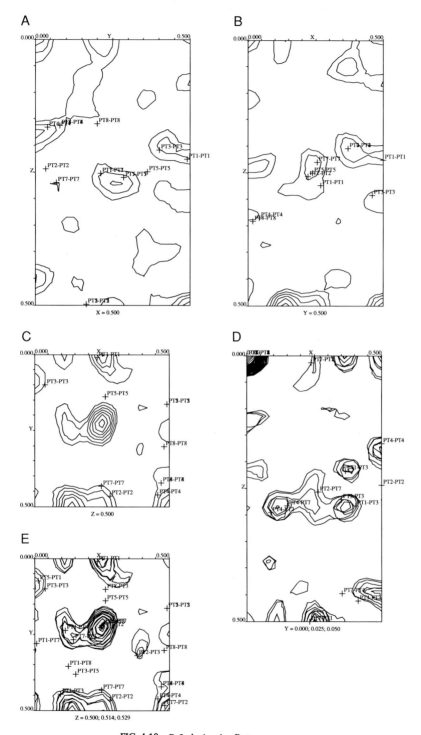

FIG. 4.10 Pt3-derivative Patterson map.

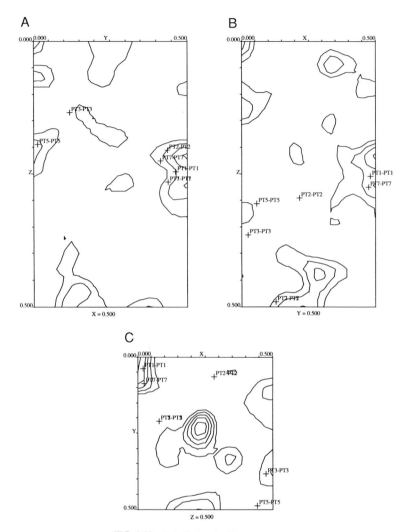

FIG. 4.11   Pt1-derivative Patterson map.

ful information. Some derivatives will be better than others. Most derivatives fall off above 3 Å. This seems to be due in part to the decreasing size of structure factors at higher resolution but, also, it can be due to non-isomorphism. In order for a heavy atom to bind, it must disturb the structure to some degree. The size of this movement relative to the resolution will determine whether the assumption of isomorphism holds. In the past it seemed

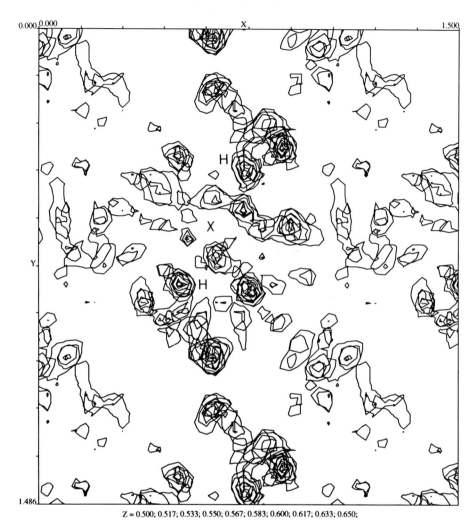

Z = 0.500; 0.517; 0.533; 0.550; 0.567; 0.583; 0.600; 0.617; 0.633; 0.650;

FIG. 4.12   Slab from a 5-Å MIR map of CVCC. The dimer axis is at the position of the "X" in the center of the figure. It is a few degrees from being coincident with z so that the two halves of the dimer are not exactly the same in this figure. The two hemes are marked with an "H" and appear as round blobs at this resolution. Running diagonally on either side are helices that appear as rough tubes of density at this resolution and are being clipped so that they are incomplete. There are clear solvent regions surrounding the dimer. The first contour level is at 1.5 times the sigma of the map.

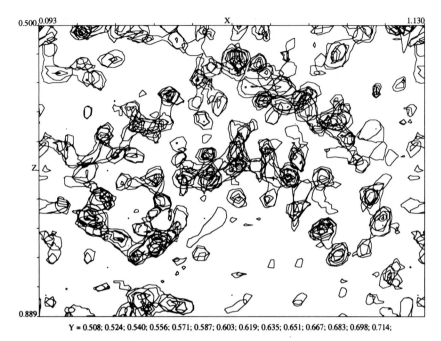

0.500 0.093   X   1.130

Y = 0.508; 0.524; 0.540; 0.556; 0.571; 0.587; 0.603; 0.619; 0.635; 0.651; 0.667; 0.683; 0.698; 0.714;

FIG. 4.13 Section of 2.7-Å CVCC map showing dimer axis. Dimer axis is vertical near center of page. Sections cut diagonally through helices can be seen.

that most heavy atoms disturbed the structure sufficiently so that above 2.8 Å, or so, the derivative was not usable. In the case of CVCC, the derivatives were used to different resolutions determined by the quality of the statistics. When the phasing power fell below 1.0, the derivative was cut off. The final MIR map is shown in Fig. 4.13. This map is also used to illustrate Figs. 3.17, 3.19, 3.26, 3.28, 3.33, 3.39, 3.40, 3.47, and 3.48.

At this point it is worthwhile using a three-dimensional graphics program such as xfit. xfit can FFT maps directly from phase files so that it is not necessary to run xfft beforehand. The turns of the helices were clear, and the side chains could be picked out. It was also evident upon inspection that the helices were left-handed. In reality α-helices are right-handed, as Linus Pauling demonstrated many years ago. When we chose our first heavy-atom solution, there was no way to choose the correct hand—both hands worked just as well. The Patterson map contains both hands due to the extra inversion center symmetry that the Patterson function adds. There was a 50/50 chance that we would get it correct, and we got it wrong. It is simple to fix, however. Just multiply all of the heavy atom coordinates by $(-1, -1, -1)$ to put the solution on the other side of the inversion center, and change the hand. Now recompute the phases and make a new map and the helix (now at the new, inverted position) will be right-handed. α-Helices make this easy; it is harder to decide with all β proteins, although tight-turns provide an important clue (see Section 3.11 and Fig. 3.42).

The other feature of the helices that we can discern is that the side-chains point toward one end of the helix. If you look at a model of an α-helix, it is apparent that the side chains point toward the N-terminal end of the helix (Fig. 3.40). This then gives us the chain direction.

Fine-tuning the map.   If your map looks perfect, then go ahead and fit it. In the case of CVCC, while the map was obviously that of a protein molecule, it was still noisy and could use improvement. The first task was to fine-tune the heavy-atom parameters to see if the map could be improved. Up to now we have not made use of the anomalous scattering. However, in the case of CVCC, this turned out to be a much smaller signal than anticipated. At first this was puzzling since we could think of no way that the differences would become smaller. Adding noise would have made the overall differences larger. The answer lay in the position of the non-crystallographic dimer axis. It goes through the unit cell at 0.25 in x and 0.25 in y and is parallel to z. It turned out that our heavy-atom positions, when reflected by the two-fold, formed a pseudo-centric array. The same is true for the Fe positions in the two hemes. Further, a test was made using a program supplied by Siemens with our four-circle area detector that tests for centric versus acentric data based on the distributions of the amplitudes. This is a common practice for small-molecule structures. The CVCC amplitudes test as centric even though the protein is clearly acentric. This is due to the presence and position of the 2-fold in the unit cell; at low resolution the protein is pseudo-centric. We tried including the anomalous data anyway but it did not improve the maps—in fact it seemed to add noise.

Solvent Flattening

Another successful technique is solvent flattening. In this method, the solvent is assumed to be flat and featureless; therefore any features in the solvent are noise and should be removed. The first trick is to find the solvent portion of the map. For CVCC we did this using the method developed by B. C. Wang as implemented by William Furey in the phases package. His programs are easy to use and come with a good manual. After editing the files to add our unit cell, we left all the other parameters at the default values. We ran `doall.sh`, and after about half an hour, the cycling was finished and we looked at the resulting map. The map appeared to be improved, with higher contrast and, of course, flatter solvent regions (Fig. 3.28). Because of the automatic solvent masking, we were aware that some side chains may have been clipped. Lysines and glutamines, which are often quite weak toward the ends anyway, are susceptible to this phenomenon. In fitting, you may want to use both the flattened and unflattened maps at the same time to check this out. It is usually easier to follow the path of the main chain in the flattened map, and the exterior side chains can be checked with the unflattened map to see if they have been clipped by the solvent mask.

"Chasing the Train"

Chasing the train is a spoonerism for tracing the chain. This is clearly the trickiest and most difficult part of solving proteins. The more experience one has, the easier this is. It is difficult for the beginner to visualize the possible paths of the main chain and to see these possibilities in the map. You must plunge in and just try it at first. Do not be afraid to throw out your first attempts at interpretation and try again. As you progress, this process will become easier.

In the case of CVCC, this was simplified by already having the structure of the *Rhodospirillum molischianum* species cytochrome c' in the Brookhaven Data Bank. Although the *R. molischianum* structure is not similar enough for a molecular replacement solution, it has the same basic fold and can be used as a guide to fitting our map. The largest peaks in our MIR map were taken to be the Fe positions in the heme. The *R. molischianum* structure is a dimer of four-helix bundles with a heme bound in the center of each bundle. Searching between pairs of highest peaks, we found a pair with the expected distance of 25 Å for the dimer pair that were not crystallographically related. We suspected from the heavy-atom work that the dimer axis was nearly parallel to z, and looking at slabs of the MIR map in z confirmed this. The helices that form the dimer interface were apparent, and we zoomed in to look at these. Using xfit, the model can then be superimposed on the map so that the Fe atoms fit onto the heme peaks. This fixes the translation and leaves one degree of freedom to be decided, a rotation about the Fe–Fe vector. This can be roughly accomplished by putting the Fe-Fe vector horizontal and in the plane of the screen and then rotating the model about x. This gives an approximate superimposition. In our structure determination, we tried rigid-body refinement with XPLOR to try to improve the fit. However, while the R-factor improved, the fit drifted away from the best superposition. It was obvious that we needed a better model. In hindsight, the failure of the rigid-body refinement is due to some large differences between the two proteins. The loops are of different size and shape, and the helices, while they follow the same general directions, do not superimpose well. The dimer angle is also different by approximately 10° in the two structures.

The superimposed model was still quite useful as it provided raw material roughly placed with which to begin fitting. An individual helix can be placed within its density by rotating about the helix axis until the side chains align. The side chains are then "mutated" to the correct ones by inserting a new residue of the proper sequence and then least-squares fitting this residue to the old model with the C_α, C, C_β, and N atoms. The side chain can then be positioned into its density by torsions about the proper bonds.

The roughly fit model was then subjected to rigid-body refinement with XPLOR. Each monomer was refined separately so that we had two groups, and then each dimer was broken into separate helices and refined. The R-factor at this stage was quite high, about 0.45 in the resolution range 20–5 Å. However, this is not unexpected for a model manually fit to an MIR map. We then started refinement at 3.0 Å, allowing the individual coordinates of each atom to refine (subject to geometrical constraints) but keeping the B-values fixed at 15.0. This is the method in the XPLOR example file positional.inp. This quickly dropped the R-factor to 0.34 from a starting R-factor of 0.51 (see Table 4.5). After additional fitting to the MIR maps, the R-factor dropped to around 0.27. At this point we tried the simulated annealing refinement of XPLOR. The first round lowered the R-factor a little, but the next round raised it. At this point, we decided we had to dig in and

TABLE 4.5
Progress of CVCC Refinement

Cycle	d_{min}	R-factor	$\Delta\phi$ from final model
0 (Start)	3.0	0.51	60.9
1	3.0	0.34	50.0
2	3.0	0.27	37.5
3 (SA refinement)	3.0	0.25	37.8
4 (SA refinement)	3.0	0.27	38.31
5	2.3	0.25	31.0
6	2.3	0.24	30.7
7	2.3	0.26	30.8
8	2.3	0.24	30.7
9	2.3	0.23	32.6
10	1.8	0.22	10.9
11	1.8	0.19	0.0

just carefully fit the MIR maps and give up hoping for an automated solution. At this point, the path followed is less clear since we sometimes backed up, and we tried fitting to combined coefficient maps, and so on. In any case, we changed the resolution to 2.3, which gave us more information to fit to, and gradually we discovered errors in the geometry and fixed them. We used a large number of omit maps and compared between the two dimers, which were fit independently. In the end, we got the R-factor down to 0.22 at 1.8-Å resolution. The maps at this point looked quite good. We suspected that some of the error might actually lie in the data set. It was collected from two crystals and merged, neither data set being complete, and we had found problems earlier with merging data from different crystals. So we completely recollected the data from a single large crystal. This dropped the R-factor to 0.19, and we identified one last problem with the model at Gly 53 that had its peptide plane flipped (see Fig. 4.14).

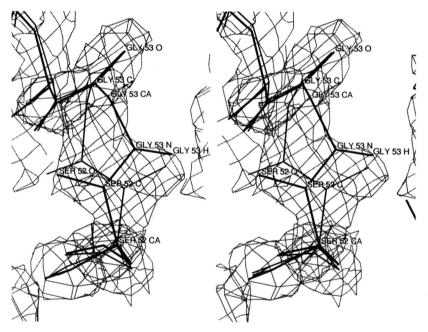

FIG. 4.14 Map showing area around Gly 53 in CVCC. The map is made with 2Fo-Fc coefficients from the phases of cycle 10 in the table 4.5. The model at this stage is shown in thick lines, and the final model is shown in thin lines. Note that the hydrogen is included on the peptide nitrogen atom; the electron density at hydrogen is so low in X-ray maps that it cannot be seen. This map clearly shows that the peptide plane needs to be flipped.

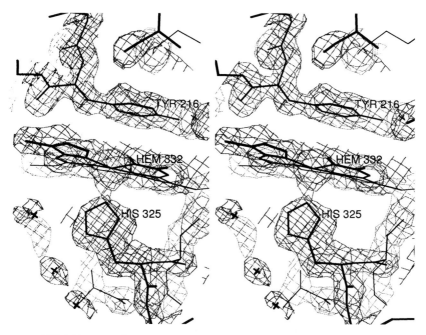

FIG. 4.15 A 1.8-Å 2Fo-Fc map showing the heme and its surrounding area.

The final structure after refinement is shown in Figs. 4.15–4.17. The R-factor statistics are shown in Tables 4.6 and 4.7. At this point, the structure is fairly accurate. However, the R-factor could be even lower as judged by other structures in the literature. At this point, we cannot find anything in the structure that needs to be fixed, but this does not mean that we have found everything.

Hindsight Is 20-20

For the purposes of this book, I thought it would be interesting to look back at the structures and compare them with the final model. In Table 4.5 the last column lists the difference in phase between different stages of the refinement and the final model. The difference was calculated for all the data between 6.0- and 1.8-Å resolution even when the model was refined at lower resolution. A separate analysis of the MIR phases reveals them to be about 45° from the final phase in the resolution range 34–3 Å. In Figs. 4.18 and 4.19 the backbone of several of the structures in the table are compared to

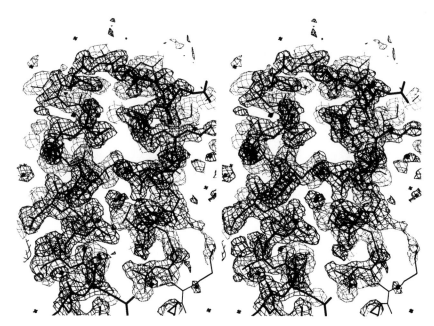

FIG. 4.16   The final fitted model of CVCC with the 1.8-Å 2Fo-Fc map.

TABLE 4.6
CVCC R-factors by Resolution for Final Model

Resolution	Range	No. reflections	Shell R-factor	Accumulated R-factor
3.74	10.00	2339	0.1718	0.1718
3.00	3.74	2766	0.1551	0.1638
2.62	3.00	2706	0.1935	0.1709
2.39	2.62	2627	0.2008	0.1758
2.22	2.39	2548	0.2140	0.1804
2.09	2.22	2404	0.2328	0.1850
1.99	2.09	2217	0.2581	0.1896
1.90	1.99	1956	0.2673	0.1930

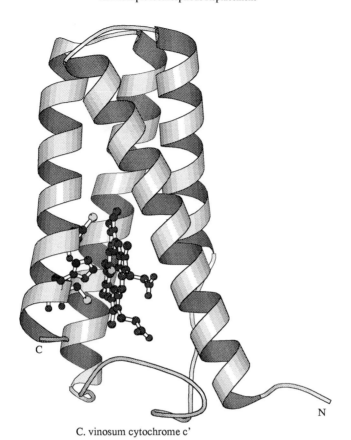

C. vinosum cytochrome c'

FIG. 4.17 Ribbon diagram of CVCC showing overall fold. This figure was produced using MOLSCRIPT. See Kraulis, P. J. (1991). *J. Appl. Crystalogr.* **24,** 946–950.

TABLE 4.7

CVCC R-factors by Amplitude for Final Model

Amplitude	Range	No. reflections	Shell R-factor	Accumulated R-factor
18.102	291.406	16933	0.2327	0.2327
291.406	564.709	2218	0.1322	0.2039
564.709	838.013	363	0.0860	0.1951
838.013	1111.316	42	0.0537	0.1934
1111.316	1384.620	6	0.0666	0.1931
1384.620	2204.530	1	0.0334	0.1930

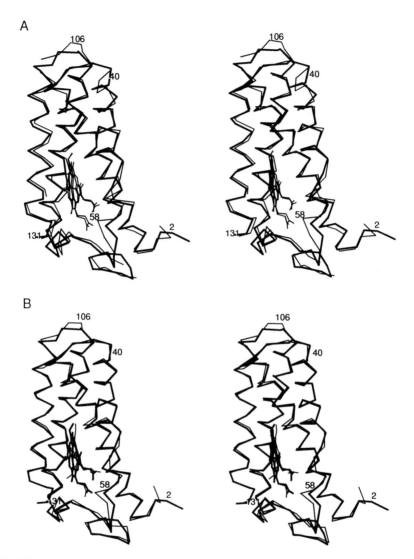

FIG. 4.18   Models of CVCC at various stages of refinement. The main chain of the final model is shown in thick lines, with the various models at different stages in the refinement shown in thin lines: (A) Cycle 0, starting model. (B) Cycle 1. (C) Cycle 2. (D) Cycle 3. (E) Cycle 4. (F) Cycle 5. (G) Cycle 10.

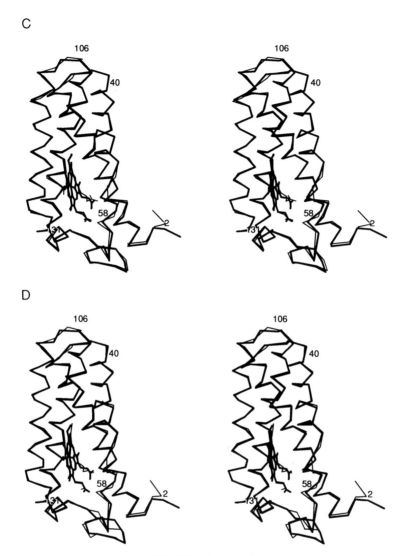

FIG. 4.18—*Continued.*

the final structure. It can be seen that we had our biggest problems with the loops and with the unusual conformation found for residues 52 through 59, which is an extended strand with a 3–10 helix in the middle. Figure 4.20 shows a map of the region around Glu 106, which was one of the last errors found, made with the phases from cycle 6 (Table 4.5). This error was fixed

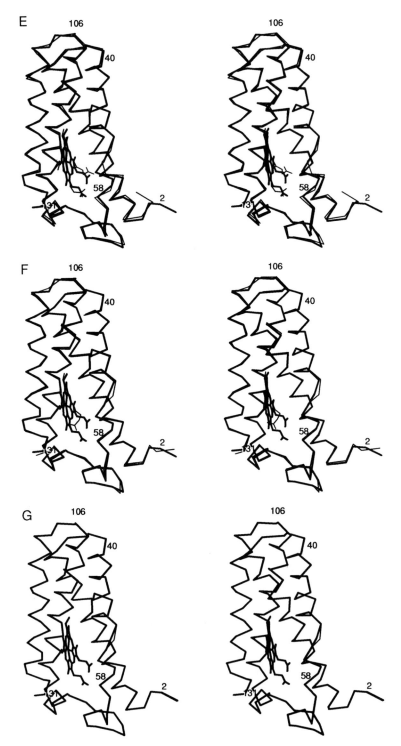

FIG. 4.18—*Continued.*

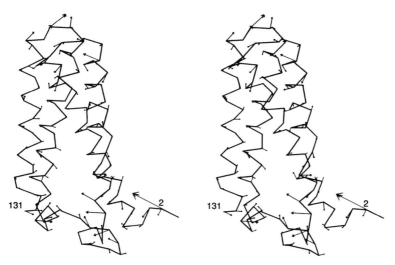

FIG. 4.19 CVCC with arrows showing model differences. An alternative way of showing the differences between the final model and the model from cycle 1. The backbone of the final model is shown with arrows drawn along the vector between the two models. The arrows are scaled by a factor of two to exaggerate the differences. Notice that most of the long arrows are in the loops between the helices.

later, but I was curious to see if it could have been fixed earlier. The map is clearly ambiguous here. It does not match well to any protein structure, and there is a break in the chain. The thin lines show the final structure in this region, and there is no density for alanine 107, although there must have been some at an earlier stage because a water has been built at into the density for the CB atom that has now completely disappeared. I had begun to consider seriously a sequencing error in the protein, and we had shortened Glu 106 to an alanine because we had nowhere to put the side chain. We had made heavy use of omit maps, so I made an omit map of the region shown in Fig. 4.20B, omitting residues 104 through 108. This map is a little better, but the break is still there, the density is too broad, and it cannot distinguish between the right and wrong structure shown. Clearly, we had a severe phase-bias problem here—all of the refinement had subtly altered the rest of the model to bring density back near the incorrect structure as well as some for the correct structure. To remove this bias, I tried "shaking" the structure by adding a small random number between −0.25 and 0.25 to all of the atoms. This resulted in an average 0.17-Å movement, which, compared to the 2.3-Å resolution of the data, was quite small and raised the R-factor from 0.23 to 0.28. However, as can be seen in Fig. 4.20C, all of the phase bias was removed. The structure became quite clear and, in fact, it appeared that the

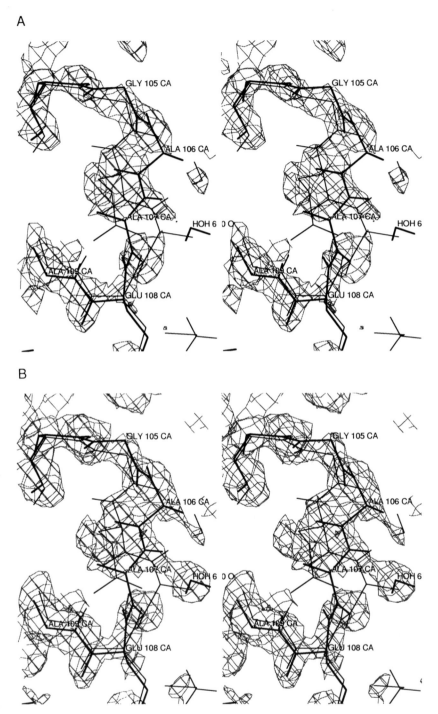

FIG. 4.20 Electron density at Glu 106 in the middle of the refinement. The model at cycle 9 is shown in thick lines, and the final model is shown in thin lines. The 2.3-Å electron density map is shown in gray lines with coefficients as follows: (A) 2Fo-Fc, (B) residues 105–108 2Fo-Fc omit map, (C) 2Fo-Fc map with residues "shaken." See text for explanation.

C

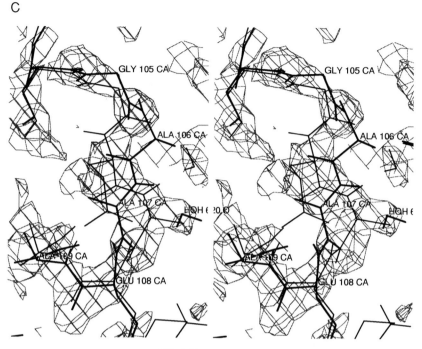

FIG. 4.20—*Continued.*

final structure could use some small adjustments to fit this density better. This method is quite simple and powerful, and we plan to use it more in the future to produce maps with reduced phase bias.

••••• 4.2 •••••
MUTANT STUDIES

A very powerful method of exploring the residues involved in function is to mutate positions and to study the effect. In order to interpret many of these mutations, it is necessary to have a structure. Doing the structure of a mutant once the "wild-type" protein's structure is known is usually much simpler since good starting phases are available from the wild-type structure to phase the mutant structure. In order to interpret mutants best there is more that can be done than simply examining a 2Fo-Fc map. The wild-type phases include the old structure but not the new one, so some caution is advised. Examination of an omit map is suggested, where the residues of interest are omitted from the phase calculation. This greatly lowers the amount of phase bias, although some can still exist (see Section 3.11 under Phase Bias). In the following example of a mutant of cytochrome c peroxidase we illustrate the procedure in solving a mutant where an Asp has been replaced by a glutamate resulting in some unusually large changes.[5] The form of cytochrome c peroxidase we used as the wild-type was actually a recombinant form with three mutations already added that do not affect its functions and is called MKT.

Example 1: MKT D235E Mutant

This mutant crystallizes in the same conditions as the wild-type and in the same form. Data were collected on our four-circle area detector and reduced with XENGEN. The crystals were not as large as those used to collect the 1.7-Å resolution native data, and so data are about 80% complete to 2.1-Å resolution. Difference maps are fairly insensitive to missing data, so the incompleteness is not a problem. The wild-type and mutant data are combined with xmerge (Fig. 4.21). This file is then "phased" by adding the phase information from the wild-type phase file with xmergephs to produce a file containing records of $h, k, l, F_{D235E}, F_{wild-type}, \alpha_{wild-type}$. The first map made was an Fo-Fc map, which in this case is the map $F_{D235E} - F_{wild-type}$. Peaks in this map that are positive represent density that was not in the wild-type structure, and negative peaks are places where density was in the wild-type but not in the mutant structure. In the simplest case, this map will reflect the atoms that were changed to make the mutant structure. Often, there are also rearrangements to accommodate the new structure. It is important to examine the difference map over a large area to see if there are any changes that are far removed from the mutant site; do not just assume that all the changes will be local. In this case the changes were far greater than expected from just the single carbon atom difference in the side chain of Glu versus Asp (Fig. 4.22). In fact, the entire helix that includes position 235 has been disturbed. Clearly, interpreting this difference map is confusing. There are large amounts of positive and negative density. One tool xfit has to help interpret the difference map is the ability to calculate the gradient of the difference map at an atom position. This then gives the direction that the atom needs to move according to the difference density at that atom. This direction is displayed as an arrow, where the length of the arrow indicates the steepness of the gradient on an arbitrary scale (Fig. 4.23). In the case of the helix, there is a net movement away from the side chain of the Asp to Glu mutation. The direction of the vectors indicates the direction of the gradient and the length indicates the relative magnitude, but the absolute magnitudes are incorrect.

[5] Goddin, D. B., and McRee, D. E. (1993). *Biochemistry* 32, 3313–3323.

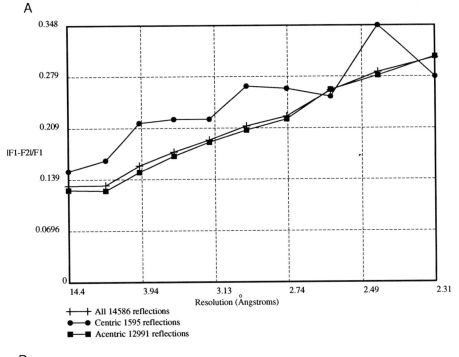

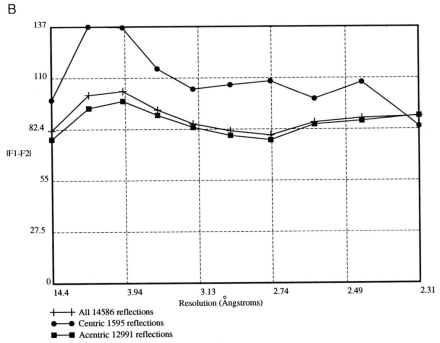

FIG. 4.21 Merging statistics of D235E.

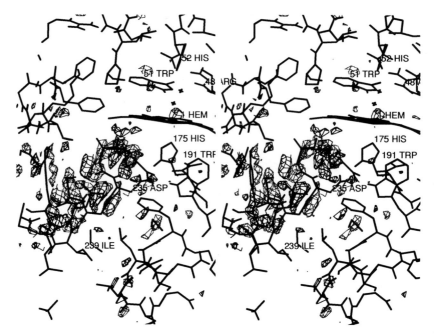

FIG. 4.22 D235E 2.1-Å difference map. Asp 235 to Glu mutant $F_{mutant} - F_{wild-type}$ electron-density map. Black contours are at $+3\sigma$, and gray contours are at -3σ. Note the movement of the entire helix.

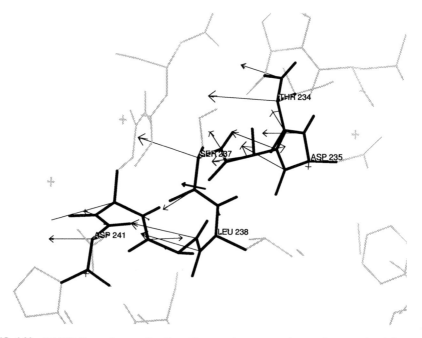

FIG. 4.23 D235E $F_{mutant}-F_{wild-type}$ Gradient Vectors. Arrows are drawn showing the difference map gradient at the main chain atoms. The arrows show the direction that the atoms move in going from the wild-type to the mutant structure. Their length is exaggerated. The thick arrow in the center shows the average vector.

In order to refit the positions of the helix, an omit map is used. Because the data are not on an absolute scale, the omit map must use the coefficients Fo, α_{omit}. Now that it is known from the difference map which atoms move significantly, they can be selected in xfit and omitted. The "omit selected atoms" option in the SFCalc window is used to calculate the phases for these residues and then to delete them from the protein phases. This map is shown in Fig. 4.24 and can be used for fitting the mutant structure. After fitting, XPLOR was used to refine the model. The final refined structure is shown in Fig. 4.25.

Example 2: SOD C6A Mutant

As another example of a mutant study, we will consider a mutant of bovine copper, zinc superoxide dismutase (SOD). We wished to make SOD more stable to long-term heating. This has proved useful in purifying the protein as most of the other proteins in the cell can be destroyed by heating at 70° C, while the SOD Cys 6 → Ala (C6A) mutant is relatively unaffected. In order to study the structural affects of this mutation we did a crystallographic study.[6]

Crystallization and Data Collection

The mutant was crystallized as was the native protein from 2-methyl-2,4-pentanediol solutions. The data were collected on our Siemens three-circle area detector to 2.1-Å resolution using three crystals. Most of the data were from a single crystal; the other two were used mainly to fill in missing data. Because the crystals are monoclinic and a quarter of a sphere of data is needed, it is difficult to get all of the data off a single mount. The unit cell was 93 × 90 × 72 Å, $\beta = 95.1°$, and we used a detector distance of 12 cm and a swing of 22.0°. The crystal was oscillated in 0.25° frames from $\omega = -50$ to 10° and then ϕ was incremented by 60° and another run of ω was collected. The data were reduced and scaled with XENGEN.

Difference Map

The data were merged and scaled with the native data using anisotropic scaling (xmerge) to produce a .fin file with h, k, l, $F_{wild-type}$, $\sigma_{wild-type}$, F_{C6A}, σ_{C6A}. The differences were fairly small, but they consistently fell off with resolution, indicating an isomorphous structure with small differences. Since the mutant is the removal of a single sulfur group, by changing residue six from

[6]McRee, D. E., Redford, S. M., Getzoff, E. D., Lepock, J. R., Hallewell, R. A., and Tainer, J. A. (1990). *J. Biol. Chem.* **265**, 14,234–14,241.

A

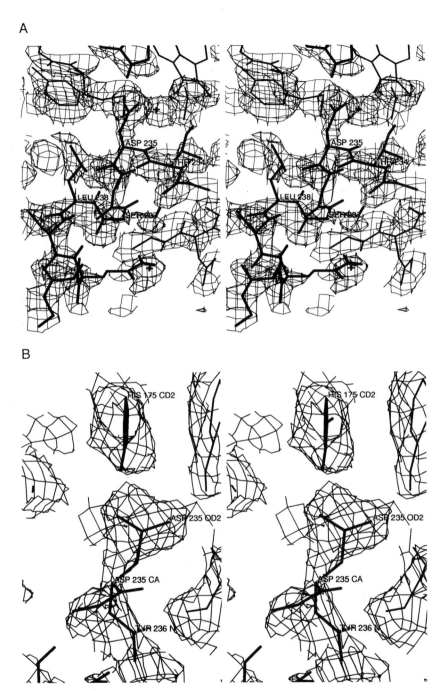

B

FIG. 4.24   D235E 2.1-Å omit map. (A) The atoms in the helix containing Asp 235 were omitted from the structure factors, and a 2Fo-Fc map was calculated. The model shown is the wild-type. Because of the large number of atoms omitted, the map is fairly noisy, but the movements of the residues can be clearly seen. (B) Close-up of residues around the mutation. The rotation of His 175 can be seen, and the new conformation of Glu 235 can be seen.

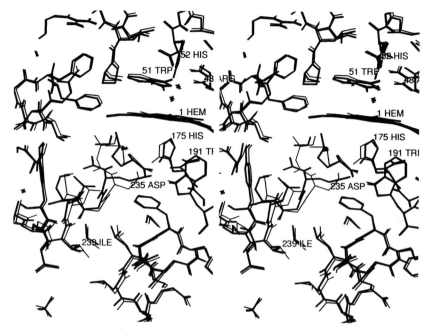

FIG. 4.25 Comparison of final structures. The wild-type model is shown in thin lines, and the D235E mutant is shown in thick lines.

a cysteine to an alanine, this was to be expected. The phases of the refined wild-type structure were merged with the data (xmergephs) and a difference map $F_{mutant} - F_{wild\text{-}type}$, $\alpha_{wild\text{-}type\text{-}calc}$ was made. The largest differences in this map were at the site of the missing sulfur atom, where a large hole was found (Fig. 4.26). The unit cell contains two dimers of SOD for a total of four monomers. There were differences around residue 6, which indicated small changes in the main chain. Another feature was a pair of negative and positive peaks on either side of Gly 145 O, which indicate a shift of this atom toward the hole left by the removed Cys 6 SG. Since it is difficult to assess the correct distance to move atoms in a difference map, an omit map was made. In order to calculate the phases for this map, residue 6 and the residues near it with peaks in the difference map were left out of the calculation. These phases were then used to calculate an omit map. To increase the signal further, the omit map was averaged using the non-crystallographic symmetry operators known from the refinement of the native protein. This map was remarkably clean and clearly showed the movements needed to fit the structure. Thus, even though the movements were subtle, they could be accurately modeled from the 4-fold averaged omit map. The model was refit using this

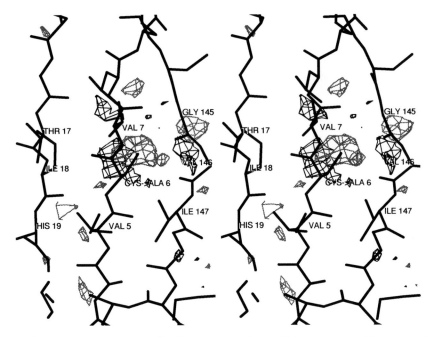

FIG. 4.26 SOD C6A mutant 2.1-Å difference map. The Coefficients are $_{C6A}$ − $F_{wild-type}$ at +3σ (black) and −3σ (gray). The model is wild-type, showing three β strands in the area of the mutation.

map and refined. The final refined structure is shown superimposed on the wild-type in Fig. 4.27.

Refinement

In a parallel study we were interested in what would happen if we just refined the structure without any refitting. We tried three different refinement methods: PROLSQ refinement, XPLOR positional only refinement, and XPLOR with molecular dynamics. None of the three programs moved the main chain around residue 6 to the correct position even though we did 2000 cycles of molecular dynamics with XPLOR at 2000° C. In hindsight, we think this is because the native model has been extensively refined and the model satisfies the geometry constraints very well. In order to move the main chain, all the atoms need to be moved *in concert* as a group. If one atom or a few atoms move in the right direction, the geometry will be distorted, and the next cycle will move the atoms back to fix the geometry. This is an example

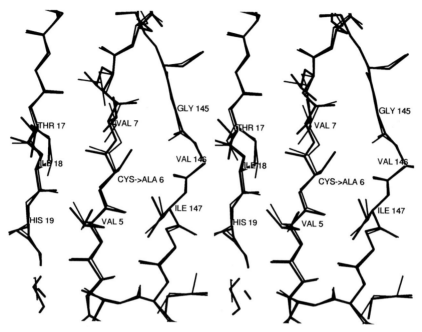

FIG. 4.27   Superposition of SOD wild-type (thick lines) and final refined C6A mutant (thin lines) structures.

of being stuck in a local minimum. On the other hand, the refit model stayed where it had been fit and refined to a lower R-value, indicating that it is in a lower minima. The lesson of this structure is not to rely on refinement programs to fix your mutant structure. It would seem that if any structure could have been done solely by refinement, this one could have.

Analysis

The mutant, while more stable to prolonged heating, has a melting temperature (T_m) that is slightly lower than that of the native structure. The increased stability to long-term heating could be explained by the lack of a the free SH group on the cysteine. A cysteine would be subject to oxidation at the elevated temperatures, could form incorrect disulfides with other SOD molecules, or could undergo a β-elimination reaction that leaves a kink in the main chain that may prevent refolding. The difference in the melting temperature was not enough to explain the loss of the energy due to the removal of the buried hydrophobic surface of the sulfur. We looked at the packing of the native versus mutant structures to see what the difference was in packing, using Mike Connelly's surfacing program MS. The mutant was actually packed better than the native; the native has a small hole accessible to 0.7-Å probe, while the mutant has no hole large enough for a probe of this size. This tighter packing may explain the better than expected stability of the mutant. The loss of the folding energy due to the sulfur-buried area is partially compensated for by the better packing of the interior.

..... 4.3 .....
SUBSTRATE-ANALOG EXAMPLE

Substrate-binding studies are an especially powerful technique in protein crystallography, revealing details about the bound state of substrates and their interactions with protein groups. However, they have several difficulties. First, the substrate must bind tightly and have a high occupancy if it is to be seen in the electron density. The substrate must be stable for a relatively long time as it takes several days to collect the data. In the case of enzymes, this means that the substrate is usually an inhibitor that cannot undergo turnover.

While the chosen substrate may bind well to the protein in solution, it may not do so in the crystal. For example, there are many cases when the substrate causes a large change in the conformation of the protein upon binding that cannot be accommodated by the crystal packing. In such cases, the crystal may crack or the substrate may not bind. Another example is if a crystal contact blocks the binding site. In these cases, it may be possible to co-crystallize the protein and substrate (i.e., grow the crystal in the presence of the substrate). These crystals may not be isomorphous with the non-substrate form, necessitating a solution by molecular replacement. An example of such a case is aconitase.

Example: Isocitrate–Aconitase

Aconitase is an enzyme containing an Fe_4S_4 cluster that catalyzes the stereospecific interconversion of citrate to isocitrate via *cis*-aconitase. The structure of the 83 kDa was solved at Scripps by Dr. A. H Robbins and Dr. C. D. Stout.[7] In order to probe the mechanism of the structure, several crystal structures of inhibitors of the enzyme have been done by Dr. H. Lauble and

---

[7] Robbins, A. H., and Stout, C. D. (1989). *Proteins* 5, 289–312; and Robbins, A. H., and Stout, C. D. (1989). *Proc. Natl. Acd. Sci. U.S.A.* 86, 3639–3643.

Dr. C. D. Stout,[8] from which an example is shown below to illustrate substrate binding studies.

Crystal Growth

The original aconitase structure was solved in an orthorhombic space group $P2_12_12$. The active form of the enzyme is oxygen-sensitive, and so all of the substrate experiments were carried out in an anaerobic hood. In the first attempts, large crystals of the orthorhombic form were soaked with substrate solutions. These crystals invariably cracked and disintegrated when the substrate was added. To overcome this, it was decided to co-crystallize the protein in the presence of substrate. The crystals were grown using the vapor-diffusion method from hanging drops using ammonium sulfate as the precipitant. Each drop was a three-part mixture of protein, inhibitor/substrate, and precipitant. In order to start crystal growth and to encourage isomorphous crystals, seeding was used. Upon mounting and taking some test shots, it was found that the crystals had a tendency to twin and that the space group was now monoclinic. In a second round of seeding, where these new monoclinic crystals were used, it was possible to get large, single crystals in the monoclinic form suitable for data collection. A variety of inhibitors were tried, and they all invariably yielded the monoclinic form. The new space group was C2. In order to facilitate comparisons, it was decided to reindex the data as B2, an alternate setting of C2 with the 2-fold parallel to c instead of b, so that the 2-fold of the monoclinic cell coincided with the 2-fold along c in the orthorhombic, $P2_12_12$ space group. The unit cell of the orthorhombic space group was $a = 173.6$, $b = 72.0$, and $c = 72.7$ Å, while the monoclinic cell was $a = 185.5$, $b = 72.0$, $c = 73.0$, and $\gamma = 77.7°$.

Data Collection and Phasing

X-ray-diffraction data were collected on the Siemens area detector. The resulting data had a d_{min} of about 2.1 Å. Data were collected on crystals grown in the presence of *cis*-aconitate, the reaction intermediate. The resulting data were 90% complete, had an average $I/\sigma(I)$ of 18.9 with last shell at a ratio of 3.8, and an R_{symm} of 6.8% on F. Since the crystals were not isomorphous with the form that had been solved previously, a molecular-replacement solution was used. The Fe–S cluster was removed from the starting model. Since its contribution to the total scattering power of this 83-kDa protein was minimal, it made little difference to the molecular-replacement

[8]Lauble, H., Kennedy, M. C., Beinert, H., and Stout, C. D. (1992). *Biochemistry* **31**, 2735–2748.

solution. However, it did provide a good check on the final solution to see if it showed up again after the solution was found. The rotation search gave a hit at 17.0σ. This is a very high hit and not surprising since the protein molecules in both crystal forms have 100% homology. The translation search also produced a large peak at 23.4σ. The rotated and translated model was then refined with XPLOR with rigid-body refinement. The starting R-factor was 0.33, which was refined to a final value of 0.22 at convergence. A 2Fo-Fc map at 3.0 Å showed clear density for the Fe_4S_4 cluster as well as new density for the substrate.

Refinement and Interpretation

The structure of the protein in the monoclinic, *cis*-aconitate co-crystallized crystals was refined to an R-factor of 0.23 at 2.1-Å resolution, at which point water molecules were added to the structure and the conformation of the side chains was adjusted as needed against 2Fo-Fc maps. Individual, isotropic B-factors for each atom were refined to give an R-factor of 0.19. A final round of simulated annealing refinement lowered the R-factor to 0.18. At this point, it was decided to interpret the density of the substrate, and a surprise was again found. Attempts to model *cis*-aconitate into the density gave poor fits. Instead, it was found that isocitrate gave the best fit (Fig. 4.28). Since aconitase interconverts citrate to isocitrate through the *cis*-aconitate intermediate, citrate was also modeled into the density without success. If the

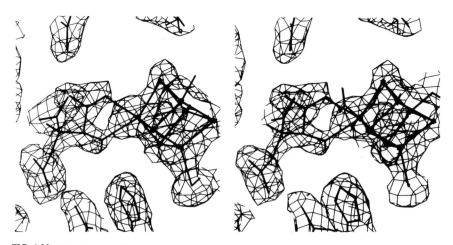

FIG. 4.28   2Fo-Fc map of aconitase with isocitrate bound. The aconitase Fe_4S_4 cluster is at the right with the bound isocitrate at the left.

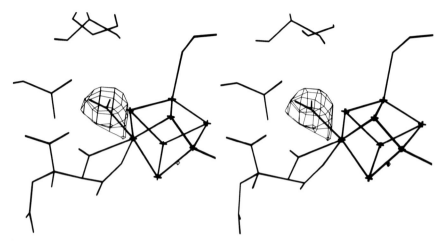

FIG. 4.29   Aconitase-isocitrate difference density. This figure is rotated about 90° downward about the horizontal from the previous figure. The difference density is interpreted as a water molecule bound in the sixth position to the Fe of the Fe_4S_4 cluster.

protein was active during crystallization then a mixture of isocitrate, *cis*-aconitate, and citrate should be found in the crystal. The enzyme was able to convert the *cis*-aconitate to isocitrate during crystallization, indicating that it was active. That only isocitrate is in the crystal was confirmed using Mossbauer spectroscopy and crystals of [57]Fe-substituted enzyme. This spectrum showed only one form in the crystal and was the same form as observed after mixing aconitase with isocitrate and rapidly freezing the sample. Thus, the crystal has somehow trapped the protein in a conformation where it binds isocitrate only and thereby shifts the reaction equilibrium. After refining the model with the isocitrate included, another Fo-Fc map was made to check for unaccounted density (Fig. 4.29). This map showed a peak consistent with a water ligand in a sixth position to the unique Fe at the corner of the Fe_4S_4 cluster. ENDOR spectroscopy indicated the presence of a water molecule bound to the Fe–S cluster in addition to the substrate, confirming the choice of a water molecule. An overall view of the aconitase–substrate complex is shown in Fig. 4.30.

In summary, there were two points that make this solution interesting from a crystallographic point of view. First, the crystals in the presence of substrate grow in a new space group. Interestingly, the crystals can be seeded with the other space group. Comparison of the cells shows a similarity in cell size and the direction of the 2-fold present in both space groups. The second point is that the crystal selects a form of the enzyme that preferentially binds

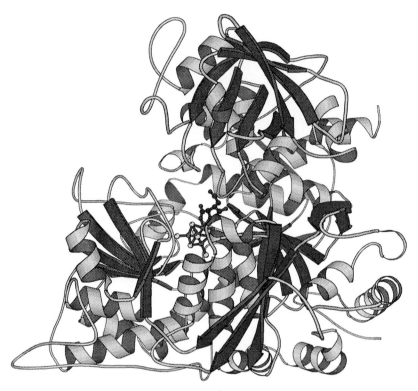

FIG. 4.30   Ribbon drawing of aconitase. The large size of aconitase is illustrated here. The arrows represent β-sheet structure, and the helical ribbons indicate α-helices. In the center of the protein, in a deep cleft, is the isocitrate bound to the Fe_4S_4 cluster.

one of three possible substrates present in solution. This could be explained if the protein undergoes a conformational change during the reaction, and the crystal traps this form over other conformations.

····· 4.4 ·····
MOLECULAR REPLACEMENT

The following steps are needed for molecular replacement:

1. Collect a native data set of the unknown crystal structure.
2. Run rotation function with probe molecule.
3. Refine rotation solution (optional).
4. Run translation function.

5. Carry out rigid-body refinement.
6. Fit sequence into difference maps.
7. Refine new structure.

Example: Yeast Cu,Zn Superoxide Dismutase

As an example of a molecular replacement solution, the crystal structure of yeast Cu,Zn superoxide dismutase (ySOD) by Drs. Hans Parge and
John Tainer is presented. When this structure was undertaken, two Cu,Zn
superoxide dismutases had been previously solved, bovine (bSOD) and human (hSOD). The structures are very similar to each other, so either one
could probably serve as a probe. Choosing the human SOD structure as the
probe molecule, the molecular replacement proceeded to a straightforward
solution. SODs are dimeric molecules, and the dimer of hSOD was used as a
probe. The first step after good-quality crystals had been obtained was to
collect a native data set. This was done using the Siemens area detector, and
the data are shown in Table 4.8. The data are complete and in the range used
by molecular replacement, 15–3.5 Å; are bright (high $I/\sigma(I)$); and have a low
R_{symm}. The data are 100% complete in this range, which is important for the
rotation function, and were collected with a high redundancy, 6.5, for the
first shell, which helps with scaling out absorption and noise.

XPLOR was used for the rotation function. The example input files
provided with XPLOR were used with only the necessary changes for space
group and file names. The rotation function yielded one hit that was clearly
above the others (Table 4.9). The top 50 hits were then put through Patter-

TABLE 4.8

Native Data Statistics for ySOD[a]

Resolution	Percentage complete	No. observations	No. unique	R_{symm} on F	$I/\sigma(I)$
4.0	99	12189	1867	3.4	75.8
3.2	100	11450	1800	5.8	41.3
2.7	100	10281	1787	10.9	18.7
2.5	100	7925	1781	21.2	6.2
2.3	100	7925	1781	21.2	6.2
2.2	72	3409	1264	24.0	3.7
Total—	95	54844	10296	8.7	27.4

[a] Yeast superoxide dismutase. Space Group R32 (hexagonal indexing). $a = b = 119.3$, $c = 75.2$, 90, 90, 120.

TABLE 4.9

Molecular Replacement Solution of ySOD[a]

	Rotation Function				PC-refinement			
Index	θ1	θ2	θ3	RF	θ1	θ2	θ3	PC
1	9.79	87.50	322.52	1.115	8.23	86.73	323.68	0.181
2	8.51	82.50	331.49	0.826	8.85	86.70	323.47	0.181
3	10.05	87.50	311.87	0.764	8.82	86.57	323.91	0.181
4	9.35	85.00	269.14	0.740	7.05	85.86	268.94	0.066
5	9.00	77.50	321.00	0.735	8.97	86.48	323.74	0.180
6	9.82	85.00	258.48	0.721	6.55	84.78	259.83	0.057
7	192.62	82.50	100.71	0.718	189.55	94.07	94.48	0.164
8	196.38	87.50	90.93	0.695	189.40	93.40	96.06	0.178
9	32.39	90.0	323.48	0.688	30.36	90.78	325.17	0.044
10	11.02	90.0	337.75	0.684	9.80	89.71	337.39	0.037

Note. Translation: TF = 0.607; [0.373, 0.305, 0.267]. Initial R = 0.48; 8.0–3.0 Å. Rigid-body refinement: R = 0.42.

[a]The top 10 rotation hits are listed showing the value of the angles and the rotation function (RF). The same 10 hits are also shown after Patterson-correlation refinement (PC). Notice that 1–3, and 5 all converge to the same value. The values of θ angles were applied to the model, and the translation function produced a clean hit at the value shown. The model was then rigid-body refined. This model was then used to fit the structure manually to 2Fo-Fc maps.

son-correlation refinement. Hits 1–3 and 5 refined to the same position with the highest correlation. The next-highest correlation was for solution 8, which is only slightly less than the highest correlation. The top hit was then used in a translation search (Table 4.9). The top hit from the translation function was then rigid-body refined from a starting R-value of 0.48 to 0.42 in the range 8–3 Å. This R-value is what would be expected considering that hSOD has many changes from ySOD. When the model was built using the dimer positioned by the molecular replacement solution and the crystallographically related molecules synthesized to check contacts, it was found that the dimer was on a crystallographic 2-fold. This causes two crystallographically related molecules to superimpose. It was clear then that the asymmetric unit of the crystal was a monomer of ySOD, not a dimer, and that the dimer axis fell on a crystallographic 2-fold. Using the dimer only had no effect on the solution in this case, it essentially amounted to a doubling in the scale factor since the extra molecule exactly overlapped another molecule. If a packing function had been run, though, the correct hit would have one of the worst packings using the dimer!

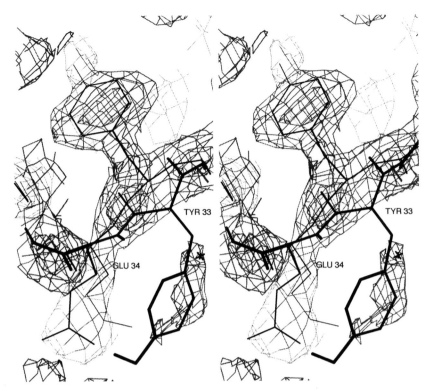

FIG. 4.31 2Fo-Fc map of yeast Cu,Zn SOD at Tyr 33. Shown in thick lines is the structure of the yeast SOD model after the first refinement following the molecular-replacement solution. In gray is the 2.5-Å resolution 2Fo-Fc map, showing very clear new density for the side chain in a completely different position than was included in the molecular replacement model. The thin gray structure shows the final model and demonstrates the excellent fit of the density to a tyrosine.

It is useful at this point to find some independent proof that the solution is correct. For this purpose the Cu and Zn were left out of the original molecular replacement probe. If the solution is correct, then there should be density at the metal sites due to the approximately correct phases. This was, indeed, the case—there was density at the Cu and Zn. An even more striking proof of the correctness of the molecular replacement phases was found at the site of Tyr 33 (Fig. 4.31). This tyrosine had replaced a glycine in the human structure, and the lack of a guiding C_β atom had resulted in its side chain being built in a rather improbable conformation. After the first round of refinement, the 2Fo-Fc map showed very clear density for a tyrosine. No information for this density was included in the molecular-replacement solution, and its presence could only be due to a correct molecular-replacement solution.

APPENDIX A
Crystallographic Equations
in Computer Code

Program 1

Read in a reflection file and filter based on $F > 2\sigma$.

```
10    read(5,*,end=100) ih,ik,il,f1,s1,f2,s2
      if ( f1 .lt. 2.0 * s1 .or. f2 .lt. 2.0 * s2)goto
      10
      write(6,20) ih,ik,il,f1,s1,f2,s2
20    format(3i4,4f8.2)
      goto 10
100   continue
      end
```

Program 2

Transform coordinates x, y, z by matrix(3,4) to new coordinates xp, yp, zp.

```
xp = x*matrix(1,1) + y*matrix(1,2) + z*matrix(1,3)
+ matrix(1,4)
yp = x*matrix(2,1) + y*matrix(2,2) + z*matrix(2,3)
+ matrix(1,4)
zp = x*matrix(3,1) + y*matrix(3,2) + z*matrix(3,3)
+ matrix(1,4)
```

Program 3

C program to strip hydrogens from pdb file.

```c
#include <stdio.h>
main()
{
        char buf[200];
        while(gets(buf)!=NULL){
                /* if not an ATOM or HETATM record write
                and continue */
                if(strncmp(buf,"ATOM",4) != 0 &&
                strncmp(buf,"HETATM",6) != 0){
                        puts(buf);
                        continue;
                }
                if(buf[12]=='H' || buf[13]=='H')
                        continue;
                else
                        puts(buf);
        }
}
```

Program 4

Fortran77 program to calculate resolution from indices—space group independent.

```fortran
        integer ih,ik,il
        write(*,*)'Enter unit cell in angstroms and
        degrees'
        read(*,*)a,b,c,alpha,beta,gamma
C   call recip once after cell entered to initialize
        call recip(a,b,c,alpha,beta,gamma)
        write(*,*)'Enter indices—enter 0,0,0 to stop'

10      read(*,*)ih,ik,il
        if( ih .eq. 0 .and. ik .eq. 0 .and. il .eq. 0)
        stop
        write(*,*)'resolution = ',resolution(ih,ik,il)
        goto 10
        end
```

```
C     ----------------------------------------
      real function resolution(ih,ik,il)
      common/recell/ raa,rbb,rcc,abg,cab,bca
      integer ih,ik,il
      real th,tk,tl

      th = ih
      tk = ik
      tl = il

sqsthl=th*th*raa+tk*tk*rbb+tl*tl*rcc+th*tk*abg+tl
     *th*cab
   $ +tk*tl*bca
      sthol=sqrt(sqsthl)
      resolution = 0.5/sthol
      end
C     ----------------------------------------
      subroutine recip(a,b,c,alpha,beta,gamma)
      common/recell/ raa,rbb,rcc,abg,cab,bca
      write(*,*)'Initializing reciprocal space
      constants'
c
c  transform to reciprocal space. stout and jensen
p.32
      salp=sin(alpha/57.29578)
      sbet=sin(beta/57.29578)
      sgam=sin(gamma/57.29578)
      cbet=cos(beta/57.29578)
      calp=cos(alpha/57.28578)
      cgam=cos(gamma/57.29578)
      root=sqrt(1.0-calp**2-cbet**2-cgam**2 +
      2.0*calp
      *cbet*cgam)
      vol=a*b*c*root
      ra=b*c*salp/vol
      rb=a*c*sbet/vol
      rc=a*b*sgam/vol
      cosra=(cbet*cgam-calp)/(sbet*sgam)
      cosrb=(calp*cgam-cbet)/(salp*sgam)
      cosrg=(calp*cbet-cgam)/(salp*sbet)
      raa= ra*ra/4.0
```

```
rbb= rb*rb/4.0
rcc = rc*rc/4.0
abg=ra*ra*cosrg/2.0
cab=rc*ra*cosrb/2.0
bca=rb*rc*cosra/2.0
return
end
```

Program 5

Subroutine to calculate structure factors by Fourier transform.

$$F(hkl) = \sum_{j=1}^{N} f_j \exp[2\pi i(hx_j + ky_j + lz_j)],$$

where f is the scattering factor for the jth atom whose coordinates are expressed as fractions of the unit cell a, b, c. In the following code, the matrix fs(3,3,nops) contains the rotation part of the symmetry operators, and ts(3,nops) contains the translation component of the symmetry operators. The reflection indices are stored in the arrays ah, ak, al; Fo is in fo; and sth contains the precalculated value of $\sin(\phi)/\lambda$ (0.5/resolution in angstroms). The coordinates are in x, y, z. Note that instead of expanding the atoms to account for crystallographic symmetry by multiplying by the symmetry operators, the indices are expanded by multiplying by the transpose of the symmetry operators instead. The form factors as a function of $\sin(\theta)/\lambda$ are stored in the table ftable(32,nforms), where each value is 0.05 greater in $\sin(\theta)/\lambda$ (these can be found in the "International Tables for Crystallography," Vol. 4). The code that follows is quite slow compared to the much faster Fast-Fourier Transform method but serves to illustrate the Fourier transform.

```
      subroutine sfcalc(natoms
$  , fs, ts, neqiv
*  ,ah,ak,al,fo,bvalue,occupancy,nrefl
*  ,ntype,x,y,z
*  ,f,ftable,nf)
c
C     *** data used in common with resolution
      common/recell/ raa,rbb,rcc,abg,cab,bca
      dimension ts(3,24),fs(3,3,24)
      dimension ih(nrefl),ik(nrefl),il(nrefl),fo(nrefl)
      dimension f(32),ftable(32,10)
      dimension ntype(natoms),x(natoms),y(natoms),
      z(natoms)real*8 acalc,bcalc
```

```
      twopi=6.283154
c
c     structure factor calculating loop == do 500 ir
c
      sumfofc = 0.0
      sumfo = 0.0
      sumfc = 0.0
      do 500 ir = 1, nrefl
        th=ih(ir)
        tk=ik(ir)
        tl=il(ir)
c
c    ########################
c
      sthol = 0.5/resolution(ih(ir),ik(ir),il(ir))
      it=sthol/.05  ;p12
c     *** precalculate the form factors for each atom
      type
c     *** by interpolating into table containing form
      factors
c     *** as a function of resolution
   23 do 53 jt=1,nf
        f(jt)= ftable(it,jt) + (ftable(it+1,jt)
        -ftable(it,jt))
      $  *( sthol-(it-1)*0.05)/.05
   53 continue
c     *** zero sums
      acalc=0.0
      bcalc=0.0
      xpart = 0
      ypart = 0
      zpart = 0
      trans = 0
          do 200 j=1,neqiv
c     *** calculate translation component of symmetry
      operator
      trans = th*ts(1,j) + tk*ts(2,j) + tl*ts(3,j)
c     *** calculate rotation by multiplying indices
c     *** by transpose (switch rows with columns) of
c     *** crystallographic symmetry operator
      xpart = fs(1,1,j)*th + fs(2,1,j)*tk
      + fs(3,1,j)*tl
```

```
      ypart = fs(1,2,j)*th + fs(2,2,j)*tk
      + fs(3,2,j)*tl
      zpart = fs(1,3,j)*th + fs(2,3,j)*tk
      + fs(3,3,j)*tl
C     *** Loop over all atoms at each symmetry operator
      do 200 i=1,natoms
C     *** calculate phase and amplitude
      phase = twopi*(xpart*x(i) + ypart*y(i)
      + zpart*z(i) + trans)
      scftmp=f(ntype(i))* exp(-bvalue(i)*sthol*sthol)
      * occupancy(i)
      acalc=acalc+scftmp*cos(phase)
      bcalc=bcalc+scftmp*sin(phase)
  200 continue
C     *** the calculated amplitude is
      fcalc=sqrt(acalc*acalc + bcalc*bcalc)
      phi = atan2(bcalc, acalc)
C     *** convert phase from radians to degrees
      phidegree = phi * 360.0/twopi
      write(6,*)th, tk, tl, fo(ir), fcalc, phidegree
  500 continue
      return
      end
```

Program 6

To calculate electron density given structure factors, the equation is

$$\rho(x,y,z) = \sum_{j=1}^{N} |F_j| \cos[2\pi(xh_j + yk_j + zl_j) - \alpha_j]$$

where $|F|$ is the amplitude and α is the phase component of the structure factor. The arrays are as above. The electron density (rsum) at a given x, y, z in fractional coordinates is

```
      rsum = 0.0
      do 200 j=1,neqiv
C     *** expand x,y,z with crystallographic symme-
      try operators
      xpart = fs(1,1,j)*x + fs(2,1,j)*y +
      fs(3,1,j)*z + ts(1,j)
      ypart = fs(1,2,j)*x + fs(2,2,j)*y +
      fs(3,2,j)*z + ts(2,j)
```

```
        zpart = fs(1,3,j)*x + fs(2,3,j)*y +
        fs(3,3,j)*z + ts(3,j)
C     *** Loop over all reflections at each symmetry
      operator
      do 200 ir=1,nrefls
C     *** sum each reflections contribution to electron
      density
      rsum = rsum + fo(ir)*
     $ cos(twopi*(ah(ir)*xpart+ak(ir)*ypart+al(ir)
      *zpart)-phi(ir))
200     continue
      write(6,*)x,y,z,rsum
```

Program 7

Read in a PDB file into atom arrays (see PDB User's Guide). This includes some lesser known but important PDB fields that are often ignored. Serno is the serial number for each atom, which increases ordinally for each atom. The atomname field is the atom name, such as CA or OD1. Alternative locations for multiple conformations can be specified in altloc such OD1A and OD1B. The residue type is aatype; e.g., ALA, CYS. There is some confusion because this field is called the residue name in the PDB user guide, but a *name* should be unique so I have changed this to type. The chain identifier, chain, is used to identify separate chains not connected to each other; e.g., A, B, C. The residue name is usually a sequence identifier, such as 1, 2, 100. The insertid field is for a single letter to identify insertions in the sequence such as 10A, 10B. The three orthogonal angstrom coordinates appear next as floating-point numbers. The occupancy and B-value are last.

```
      parameter( maxatm = 10000)
      real x(maxatm),y(maxatm),z(maxatm)
      real bvalue(maxatm),occupancy(maxatm)
      character*1 chain(maxatm)
      integer*4 resname(maxatm)
      character*4 atomname(maxatm)
      character*1 insertid(maxatm), altloc(maxname)
      character*3 aatype(maxatm)
      character*80 buf
      i = 0
10    read(10,'(a80)', end=100), buf
      if(.not.(buf(1:5) .eq. "ATOM" .or. buf(1:5) .eq.
      "HETAT"))goto 10
```

```
        i = i + 1
        read(buf,20)serno,atomname(i),altloc(i),
     $   aatype(i),chain(i),resname(i), insertid(i),
     $   x(i), y(i), z(i), bvalue(i), occupancy(i)
20      format(6x,i5,1x,a4,a1,a3,1x,a1,i4,a1,3x,3f8.3,
        2f6.2)
        i = i + 1
        goto 10
100     continue
        natoms = i
        end
```

Program 8

An awk script can be used to reorder data from one program for another. In this example we have data in the form h, k, l, FP, FPH and we want to put in the fin format. We will have to add a fake sigma field as this information is not in the original:

```
awk '{print $1, $2, $3, $4, " 0.01 ",$5, "0.01 "}'
inputfile > outputfile
```

The symbol $1 means the first item on each record. Awk works on series of records, repeating the command on each record. Items on records are separated by white space: a space or a tab. The output will again be items separated by spaces. This is what is commonly called "free format." To put out formatted output where each item is in the same column, the command can be modified:

```
awk '{printf("%4d %3d %3d %7.2f 0.01 %7.2f 0.01\n",
$1, $2, $3, $4, $5}' inputfile > outputfile
```

Note that the formatting statement is the same as that used in the C language. In general, any valid C expression can be used in awk. However, an awk program is much simpler to write and is interpreted rather than compiled.

Program 9

Awk can be used to do simple calculations. For instance, if we want to modify a phase file with records of h, k, l, FO, FC, phi so that it has records of h, k, l, $5FO - 3FC$, FC, phi to make a $5Fo$-$3Fc$ map,

```
awk '{$3 = 5*$3-3*$4; print}' inputfile > outputfile
```

Program 10

Another use of awk is to make decisions with `if` statements. In this example we want to remove all data that is below 3 σ from a file with records *h, k, l, Fo, σ(Fo)*:

```
awk '{if( $4 > 3.0 * $5) print}' inputfile >
outputfile
```

Program 11

This awk script can be used to find the minimum and maximum of a stream of coordinates *x, y, z*. Save the awk script into a file called minmax.

```
awk '
      BEGIN{ xmin = 99999; ymin = 99999; zmin = 99999;
             xmax = -9999; ymax = -9999; zmax = -9999; }
      {      if( $1 < xmin) xmin = $1;
             if( $2 < ymin) ymin = $2;
             if( $3 < zmin) zmin = $3;
             if( $1 > xmax) xmax = $1;
             if( $2 > ymax) ymax = $2;
             if( $3 > zmax) zmax = $3; }
      END { print xmin, ymin, zmin, xmax, ymax, zmax;
      }' $*
```

Now, if you needed to find the range of coordinates in a PDB file, where *x, y, z* are fields 6, 7, 8, then you can use another awk command to print *x, y, z* and pipe this into minmax:

```
awk '/ATOM/{print $6,$7,$8}' inputfile | minmax
```

The /ATOM/ field causes awk to ignore all records that do not match the pattern ATOM. In a PDB file this will reject all non-atom records that do not have any coordinate information.

Program 12

Awk can call mathematical functions such as sqrt(), as in this script to find the distance between two atoms input as *x*1, *y*1, *z*1, *x*2, *y*2, *z*2:

```
awk '{xd = $1 - $4;
      yd = $2 - $5;
      zd = $3 - $6;
      print sqrt( xd * xd + yd * yd + zd * zd) }' $*
```

Program 13

XPLOR requires that you split up your protein into separate files before you can enter the coordinates. This can be easily done with awk. In this example, the protein is split into three files containing the protein, the heme prosthetic group, and the water:

```
awk'/LYD |SER|THR|ALA|CYS|ASP|GLU|PHE|GLY|HIS|ILE|
LYS|LEU|MET|ASN|PRO|GLN|ARG|VAL|TRP|TYR/{print}'
$1 > protein.pdb
awk '/HEM/{print}' $1 > hetatom.pdb
awk '/HOH/{print}' $1 > water.pdb
```

♦ ♦ ♦ ♦ ♦ ♦ ♦ ♦ ♦ ♦ ♦ ♦ ♦

APPENDIX B:
XTALVIEW USERS' GUIDE
A Software System for
Protein Crystallography

♦ ♦ ♦ ♦ ♦ ♦ ♦ ♦ ♦ ♦ ♦ ♦ ♦ ♦

Where do I start? After you set up the environment, and assuming XtalView has been installed, type `xtalmgr &` to start the Xtal Manager program. From this program you can start all the other programs.

Setting up the Environment

In order to use XtalView, several environment variables need to be set to tell the programs where key information is stored. These should be added to your .cshrc file.

XTALVIEWHOME. This is the name of the directory where the XtalView software is installed. Several parameter files are stored here. Put this line somewhere before the path statement.

CRYSTALDATA. This is the name of a user-supplied directory that contains database files on each of the users crystals. Create two files in this directory with `touch crystals; touch projects`.

CRYSTAL. This is the name of the current crystal database file in use. If you have not created one yet, you can ignore this variable for now. This file is searched for by first looking in the current directory, then by looking in $CRYSTALDATA directory and finally by looking for $HOME/xtal_ info. By temporarily placing a copy of the file in the current directory, it is possible temporarily to override the one in $CRYSTALDATA for a special purpose.

Example:

```
setenv XTALVIEWHOME /asd/prog/XtalView
setenv CRYSTALDATA /asd/dem2/xtal_info
setenv CRYSTAL sirhp
```

Add XtalView to Your Path.   Put $XTALVIEWHOME/bin into your path in your .cshrc file.

XtalView Crystal Database

XtalView is organized around the concept of a crystal. A crystal is all data that have a common unit cell and space group. The information for this crystal is kept together in a database file. This information can then be accessed by all XtalView programs by entering the name of this file into the Crystal field or by setting the environment variable **CRYSTAL**. The file is editable and in ASCII; it consists of lines with a keyword followed by data. Xtalmgr provides a facility for editing new crystal files, or an existing one can be copied. Xtalmgr knows about all 230 space groups.

History Files

XtalView is intended to be more than just another pretty face; it is an entire system for organizing and tracking crystallographic data. The history file is key to the tracking and record keeping of data. All data files created by XtalView also have an associated history file that contains the name(s) of the input file(s), the date created, and all the values of variable parameters. The history file can be used to track all data backward and forward—where the data has been and how it was created. If an error was made at a step in the past, it can be found by searching through the history files. Intermediate data files can be deleted and later recreated by using the information in the history files. This can save a tremendous amount of disk space. A history file has the same name as its parent file with the extension ".hist" added.

File Formats

XtalView is file-based. All data are stored in files. Almost all XtalView files are ASCII with the notable exception of map files, which are in FORTRAN binary form for backward compatibility. Records are separated by end-of-lines, and data fields are separated by white space (a comma is not considered as white space). Columns are not necessary except in PDB files, the standard to which XtalView faithfully adheres for better or worse. The files are easily readable by FORTRAN or C programs. Using ASCII allows easy importing and exporting of data to and from XtalView and permits browsing of the data. It also forms the major bottleneck in running programs since every line of an ASCII file must be scanned and converted as read in or written out. Given the trade-offs, it seems a price worth paying.

As few file formats as possible are used. Standard extensions are used. Strict typing of these extensions is not enforced, although it would make the system much more robust, because it would make XtalView less backward compatible. A list of the most common file types follows, including most files used with a brief description.

.fin. The basic crystallographic data file contains: *h k l* $f1$ $\sigma(f1)$ $f2$ $\sigma(f2)$. $f1$ and $f2$ can be Bijvoet pairs, fp and fph, or any other pairable data. If one of the two observations is missing, then f should equal 0.0 and $\sigma(f)$ should equal 9999.0 to indicate the data are missing or not applicable (i.e., centric reflections do not have an anomalous signal, and this is indicated by setting the Friedel mate to 0.0). When a .fin file is read, if one of the observations is missing, it is correctly handled, depending upon how the data are to be used.

.df. A "double fin" file contains enough data to merge two data sets while preserving Bijvoet pairs: h k l f1 sigma(f1) f2 sigma(f2) f3 sigma(f3) f4 sigma(f4). Usually, the first two are native data $fp+$ and $fp-$ with $f3$ and $f4$ $fph+$ and $fph-$. Missing/unobserved data are handled as for a .fin file. For example, a centric reflection would look like this:

```
0 0 5   220.05 5.76   0.0 9999.0   157.3 9.67   0.0 9999.0
```

.phs. A "phase" file contains either h k l Fo Fc phi, or h k l Fo f.o.m. phi, where phi is in degrees and f.o.m. is the figure of merit. The Hendrickson–Lattman coefficients *A, B, C, D* are sometimes saved at the end of each record, in which case the file contains: h k l Fo f.o.m. phi A B C D.

.pdb. This is a PDB file as specified by the Brookhaven National Laboratories. Every effort has been made to remain true to the standard even though no one else seems to bother. PDB files have several arbitrary limitations like short atom and residue names, but it is the only game in town. No other format is directly supported. There may be some utilities around for converting from other formats to PDB. The definitive guide to the format is in the PDB User's Guide, which can be obtained from Protein Data Bank.[1] It also contains other valuable information on online services and new releases.

PDB coordinates are in orthogonal Cartesian coordinates. In non-orthogonal space groups there are at least two ways that the transformation from Cartesian to fractional can be done. If the model is built from scratch using XtalView, then the XtalView default convention will be used. Xtal-View programs that use Cartesian-to-fractional conversion, or vice-versa, will print out this matrix, so you can write it down. The utility matrices can also be used to find these values. If you import a model and it is in a different convention, you can override the default matrix with the line

```
ctof 11 12 13 21 22 23 31 32 33
```

To override the Cartesian-to-fractional matrix,

$$
\begin{bmatrix} x_C \\ y_C \\ z_C \end{bmatrix} = \begin{bmatrix} x_F \\ y_F \\ z_F \end{bmatrix} \begin{bmatrix} 11 & 12 & 13 \\ 21 & 22 & 23 \\ 31 & 32 & 33 \end{bmatrix}_{ctof},
$$

the inverse operation, fractional-to-Cartesian, is carried out with the inverse matrix so that only the first matrix needs to be specified.

.sol.   This is a heavy-atom "solution" file that contains a keyword format. xheavy provides an editor for creating and editing these files. They are easily read after this and contain all information necessary to keep track of several derivatives. This allows the use of individual heavy-atom files, avoiding merging them into a "superfile" with the concomitant headaches of altering or removing data.

.map.   This is an electron-density-map file. This is the FSFOUR format from Pittsburgh used in the Wang solvent-flattening procedure. It is FORTRAN binary and contains an entire unit cell. Given the very large size of electron-density maps, it is not practical to use an ASCII format. Xfrodomap will convert this to a FRODO DSN6 file (not available for all system types).

In addition, many other formats are supported here and there by various programs, such as .mu and .rfl files from XENGEN, XPLOR .fobs files, and TNT .hkl files. Xprepfin can be used to import/export .fin files to/from a variety of formats. All of your format woes can be solved by learning to use

---

[1] Protein Data Bank, Chemistry Department, Bldg 555, Brookhaven National Laboratory, Upton, NY 11973; tel. (516) 282-3629; FAX (516) 282-5751; e-mail pdb@bnlchem.bitnet or pdb@chm.chm.bnl.gov.

awk. For instance, to read in a file with h k l fp sigma(fp) (fp(-)-fp(+)) into .fin format with awk: awk '{print $1, $2, $3, $4, $5, $4+$6, $5}' import_file > imported.fin.

Plotting is done using Postscript. Postscript can be viewed on SUNs with pageview, other systems have similar commands. Services are available for preparing slides from Postscript files commercially.

Notes on XView

XtalView was written using the Xview toolkit because it can be ported at no expense to any system capable of running MIT's X11 system. This decision had absolutely nothing to do with the fact that the author's workstation has a nice program for auto-generating Xview code and runs the OpenWindows windows system, which was built on X11, using Xview. Xview provides an amazing amount of functionality. For instance, the textpane windows can be edited and stored into a file, and data can be cut and pasted into other windows without any new code being written. Xview is not specific to the any window manager, and MOTIF and Xview can live side-by-side in harmony. It will run under any X-windows manager, although there will be differences in regard to headers and footers, which are never essential in any case. Occasionally, difficulties are encountered with fonts. Not every system has the necessary fonts preinstalled. In this case, a source of the fonts will be have to be found.

X11 Features

X11 has the nifty feature of being networkable, permitting an application to run on one machine while being displayed on another. This other machine can be in the same building or at another institution as long as there is a TCP/IP connection. The window managers can be different, and the graphics (usually) can be displayed regardless. To do this, type xhost + on the machine you want to display on. Then set the environment variable DISPLAY with: setenv DISPLAY hostname:0 on the machine the program will run on. Alternatively, you can add the flag -display hostname:0 to the command line. There are many command-line options available to X applications, and this applies to all XtalView programs. For instance, the default font can be changed. A complete list of flags can be found in Volume 0 of the X11 user's guide.

Note on Textpanes

Every XtalView application has a textpane where various messages are sent. This effectively forms a log of each session. It can be saved if the user wishes by selecting the file option on the menu that appears when the menu button is pressed anywhere within the textpane.

Note on Backgrounding

If you start applications at the command line instead of using Xtalmgr, then you can run them either in the foreground or in the background. It is best to start them with the & sign at the end of the command line to put the application in the background so that the cursor is returned and you can continue using that window for input. Often, you will forget and you can then use ^Z to stop the application and then type bg to background it. If you use ^Z and then do not background, *do not use the application in any way* or your computer will hang! The only way to recover is to rlogin from another computer and kill the process. This is not a bug in XtalView but is a problem with the window manager and is beyond us users to fix. The bottom line is this: Do not let stopped windows sit around because they are ticking time bombs.

More on Input and Output Files

All files in XtalView are opened with a common routine so that I/O is consistent throughout the system. The filename can be substituted with a dash (—), in which case input is taken from stdin and output is taken from stdout. This feature allows piping between applications. However, the chain of history files will be broken since the application has no way of knowing where the stdin originated or where the stdout goes to. Pipes are faster than using an intermediate file and in some cases may be beneficial. Stdin cannot be used if the input needs to be rewound. If so, you will get an error message telling you so, and the remedy is to use an intermediate file. If you try to overwrite an output file, you will be notified with a notice box that will ask if it is OK to overwrite the file. You can accept or cancel. These notices can be confusing if you are running many jobs at once since the notice may appear after a delay in some applications where extensive computing is required, and the notice box may appear out of context. It is not always obvious which application is doing the asking.

Too Many Windows

If your computer is sluggish and appears to be continuously swapping, you are out of memory. In this case, terminate all unneeded windows whether open or closed and things should improve. It is easy with XtalView to open lots of windows. An alternative to closing windows is to obtain more memory for your machine (highly recommended). A modern graphics work-station should have an absolute minimum of 16 Mb and preferably 32 Mb. Memory is inexpensive and getting more so all the time.

Widgets Used in XtalView Windows

These are standard XView widgets, and more information can be had on them by looking in the XView manual. Because many people will be too impatient to do so, a short guide is presented here.

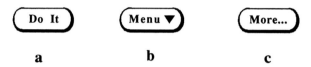

a   b   c

FIG. 1. (a) Clicking on this button initiates the action specified. The button becomes shaded and remains so for as long as the actions take to finish. (b) Menu buttons have pull-down menus associated with them. If the button is clicked on with the SELECT button, the default menu item is actuated. If the MENU button is pressed and held, a menu appears. (c) Clicking this button with SELECT pops up a window with more information/controls.

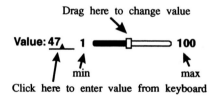

FIG. 2. The slider is used to change numerical values. They have minimum and maximum values and can be adjusted to integral values in between. Sometimes a floating-point value is desired. In these cases the value is usually ×100 of the actual value used, and this will be indicated in the legend. If you enter the value from the keyboard, which is often desirable, to enter a precise value, be sure to hit ⟨CR⟩ to enter the value.

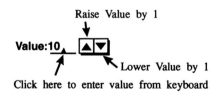

FIG. 3. Number fields are used to enter/display integer values. If you enter the value with the keyboard, be sure to hit ⟨CR⟩ to register the new value. If you click on the arrow boxes, the number is incremented/decremented by 1. If the arrow box is grayed, that means the number is at its limit and cannot be changed in that direction.

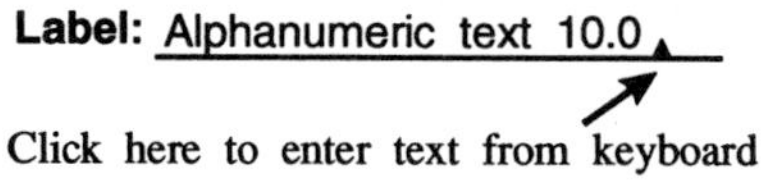

Label: Alphanumeric text 10.0

Click here to enter text from keyboard

FIG. 4.   Text fields are used to enter/display alphanumeric values.

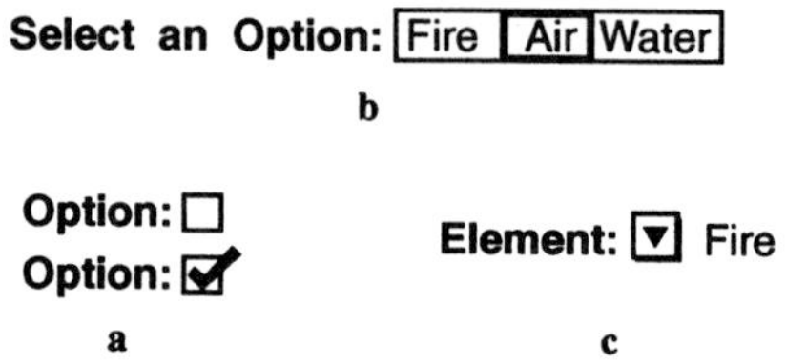

Select an Option: Fire Air Water

b

Option: ☐   **Element:** ▼ Fire
Option: ☑

a   c

FIG. 5.   (a) Check Box. The upper check box is not selected, and the lower one has the option selected or turned on. To toggle the selection, click on the square box. (b) Setting: Multiple choices can be made. In the example above, Air has been chosen as shown by its highlighted box. To select another option, click anywhere within its box. (c) Abbreviated menu button. This is similar to a setting, except that the options are accessed by clicking on the button with the downward-pointing arrow, which pops up a menu of setting buttons. This saves space when many options are needed. The currently selected option is displayed on the right, while the label for the button is on the left.

• • • • • • • • • • • • • • •

REFERENCE SECTION

Window-Based Applications

Xcontur—A Page-Based Contouring Program for Crystallographic Maps

Usage.   `xcontur file.map`

Method.   The map is contoured in sections of *x*, *y*, or *z*. The output appears in a window that is the same size as an 8-1/2 × 11 in. page. The program produces PostScript output, which can be used on any PostScript-compatible printer. The output is fairly WYSIWYG with the exception that nicer fonts are available on the laser printer. Stereo is available by shearing, and labels and bonds (lines) can be added or read from files. No interpolation is used in the contouring. If smoother sections are desired, calculate the map on a finer grid.

Crystal.   The cell information is read from the crystal file.

Cell.   The unit cell is used to scale the axes. Non-orthogonal axes are properly handled. The rows are always horizontal and the column is at α, β, or γ, depending upon the plane being sectioned.

NX, NY, NZ. This is read from the input map and cannot be changed.

Planes. This sets the sectioning direction in either x, y, or z. If you change this, then the rows and column directions are also changed so that they do not conflict. No attempt is made to detect left-handed configurations. If you stack up sections for a mini-map, you may need to stack downward or upward to get the proper handedness.

Plane. There are three fields here. The first is a numeric field that can be used to increment/decrement the section where contouring starts. The second is the value of this plane in fractions of the unit cell. If you edit this value and enter it, the nearest plane section will found. The third is a pull-down menu for choosing the three sections 0.0, 0.5, and 1.0. If the number of sections is odd, the nearest section will be given for 0.5.

Slab. This is the number of sections to contour. They will all be overlaid onto each other on the screen, although when they are plotted, they can be either overlaid or plotted on separate sheets. The slab always starts at plane and increases.

Rows. Rows go horizontally across the screen. This sets whether the direction is x, y, or z. It overrides the columns when changed if it conflicts and can be overridden by planes.

Row Min. The first value is in terms of grid points in nx, ny or nz, and the second is fractions of the unit cell. You must enter fractional values for them to take effect.

Columns. This sets the direction of the columns. It has the lowest priority and so cannot actually be used to change the column direction, but it does indicate the direction being used. Both row and plane direction changes will override the column direction if there is a conflict.

Number. The number of contour levels is set here from a minimum of 1 to a maximum of 20.

Interval. The interval between successive levels is from a minimum of 1 to the maximum value of the map.

Start. This is the level of the first contour level and may be negative. The slider ranges from the minimum of the map to the maximum. Upon loading a map, both start and the interval are set to the root-mean-square value of the map.

Draw.   The screen is redrawn when this button is pressed.

Picking in the Canvas.   If you pick with the mouse button, the fractional coordinates of that point are returned in the message window. Holding down the SELECT button causes the outline of the section to drag so that it can be positioned on the page. When using slabs greater than 0, the vertical coordinate with the maximum density value will be printed. This allows finding the x, y, z coordinate of a peak. The density value printed is an interpolated value and, as such, is almost always lower than the peak value.

View Pop-Up

Title.   The title is initially set to the map file name. It is nice to edit it and put something sensible here.

Display Type.   You can look at the contours (default) or at the raw numbers, in the map file. However, for the raw numbers only a small section can be viewed before the numbers overlap each other and become unreadable.

Row, Column Ticks.   This is the number of grid lines and/or tick marks to make counting from 0.

Row, Column Origin.   This is the position of the upper left corner of the section. It is easier to set this by dragging with the mouse in the canvas window.

Scale Ångstroms/cm.   You can manually enter a value or pull down a menu of choices. The scale is accurate for hardcopy but is correct only if you happen to have the right size monitor. Remember, though, that the canvas window is the same size as a piece of paper if you have not resized it, and so, if the graph fills the window, it will fill the paper.

Alternate Solid and Dashed Contour.   Checking this will change every other level to dashed lines for an interesting effect that can make counting levels easier.

Draw Grid of Lines.   If this is checked, a grid of lines is drawn at the frequency specified by the row, column ticks.

Auto Center.   If this is checked, the section is centered on the page automatically. It must be disabled for stereo to work properly.

Use Line Width for Depth Cue.   While the computer screen does not have enough resolution for this to be useful, it produces a nice depth cue on laser-printed output and is highly recommended for multi-section slabs. It is best to set it just before printing.

Draw Labels.   This enables/disables the drawing of labels. They are drawn on the section only if the x, y, z coordinate is on that section.

Draw Jack at Label Position.   This enables/disables the drawing of a jack at the position of the label. The size of the jack can be changed with the slider.

Draw Bonds.   Bonds can be drawn superimposed on the sections. They are drawn only if their x, y, z coordinate is on that section. The line width of the bond on the printed output can be varied from thick to thin to make it more obvious.

Draw Side by Side Stereo.   The scale is halved and two images for side-by-side stereo are drawn by shearing each successive section.

Stereo Separation.   This sets the distance from each other the two images are apart. Hint: Set the contour level very high while fiddling with the stereo parameters so that drawing is quick until you get them right, and then lower the level for the final picture.

Stereo Depth.   This sets the shear angle and modifies the apparent depth.

Apply.   This causes the picture to re-draw.

File Pop-Up

Filter.   This sets a filter for the file-list box. It defaults to *.map, and all the files in the current directory that match the filter will be displayed in the file list.

List.   This forces the file list to be updated. Use it after changing the filter or the working directory.

Directory.   This is the current working directory.

Map File.   This is the source of the map. This field can be filled from the command line as the first argument, by selecting a file name in the file list, or by typing it in.

Labels File.   The labels can be loaded from or saved to this file. It has the format of lines of x y z label, where *x, y, z* are the coordinates and label is a text string. xpatpred produces a label file that xcontur can display.

Bonds File.   Bonds can be loaded or saved. The bond file has the vu format: `lines of x1 y1 z1 x2 y2 z2 color`. Where $x1, y1, z1$ are the starting coordinates of a line and $x2, y2, z2$ are the end coordinates. The color is not used by current version of xcontur. Xfit also displays/creates vu files, and a number of utilities for producing vu files from PDB files and other sources exist. See grinchbones.

Number of Bonds.   This is for your information and is not editable.

File List.   To pick a file in the list, first click on the file field you want to load into to put the input focus there (the little triangular cursor), then select a file name with the SELECT (left) mouse button.

Print Pop-Up

Print to.   You can select a file or a printer. Very large pictures may need to be sent to a file and then printed with the -s option because they are too large to spool directly.

Printer.   This is a valid name of a printer connected to your computer. Leave it blank for the default printer.

File.   This is the name of the file. The directory in the Files pop-up is added unless the name starts with ". /" or "/ ."

Print Sections Separately.   If this option is checked, then each section is printed on a separate page. This saves having to print them one by one by selecting the entire slab and this option and then printing.

Labels Pop-Up

This can be used to edit the labels list. The functions are: insert/replace/delete. Use the Files pop-up to load a labels file.

xdf—Disk Full Meter

Usage. `xdf partition`. xdf can be used to monitor disk usage. After it starts, it updates the disk usage of the partition it was started with. The usage is shown on the gauge, and the partition is displayed on the footer.

Example. `xdf /usr`. This shows the amount of /usr that is used. Since xdf updates every 3 min (to avoid over-using computer resources), a sudden jump in disk usage will not show up immediately.

Xedh—Electron-Density Histogram Plotter

Usage. `xedh [file.map]`. This is a normalized electron-density histogram plotter. It can be used to compare histograms between phase sets and between different crystals. The histogram for a correctly phased protein with a given percentage solvent and resolution range is characteristic. Any deviations are due to phase errors. By comparing your phase set with those of other well-refined proteins in the same resolution range, you can guess the amount of phase error. For example, when comparing right- and left-handed maps that include anamolous scattering information, the correct hand will have the sharper histogram.

If a solvent mask in B. C. Wang format exists, it can be used to separate out the protein and solvent components of the histogram.

The histograms can be saved and loaded in the Files window.

These methods are still being developed. In the future, example histograms may be available and more information on determining phase errors will be available.

xfft—Calculate an Electron-Density Map by the Fast-Fourier Transform Method or a Patterson Function Map

Usage. `xfft phasefile mapfile`

Crystal. A crystal file is needed with the cell and symm records present. Note that symm records should have centering expanded explicitly (i.e., C2 should have eight symm operators, not four, with the last four $1/2 + x$, $1/2 + y$, z of the first four).

Unit Cell. This is obtained from the crystal file. However, it can be overwritten by manually entering a value. But there is no way to overwrite the symmetry operators.

Defaults. A defaults file, hardwired as **xfft.defaults,** can be saved or loaded. Also, defaults can be reset to startup values. This is handy when you recalculate maps often with the same parameters. Note that you can save a defaults file, copy it to a new name, and then copy it back at a later date. In this way several standard methods can be saved.

Direction of Planes. The output can be in x, y, or z planes. This is only necessary for possible use of the output map by non-XtalView programs because all XtalView programs can read all three sort orders.

Map Type. There are several choices of map coefficients available.

Input File. This file can have several different formats. For Fourier maps either the phs format (h, k, l, Fo, Fc, phi) or XPLOR format can be used. For Pattersons, these can be used plus fin and df formats.

Type. This specifies the type of input file. For Pattersons with df formats there are three choices for defining the difference to be used in computing the Patterson coefficients. Which you use depends upon what is in each of the four columns in the df file. Normally, columns 1 and 2 are native data and their sigmas and columns 3 and 4 are isomorphous data (a column is a pair of f and $\sigma(f)$). To make an isomorphous difference Patterson map, use $(3 + 4) - (1 + 2)$, which means the average of 1 and 2 minus the average of 3 and 4 or FPH $-$ FP. Note that when calculating the average, missing observations are not included.

Resolution Filter. Set these values to your desired limits. Order is unimportant since computers are perfectly capable of figuring out which number is the min and which is the max (the other one).

Outlier Filter. This gets rid of outliers when making Patterson maps. Specifically, if the absolute value of the difference is greater than $x\%$ of the average of the two observations, the reflection is rejected. This is important when making Patterson maps so that a few bad apples don't spoil the whole bunch. A value of 100% is a conservative value to use. Note that on numeric fields you must enter return to register the edit if you edit the data with the keyboard.

Read Phase File. This button reads in the data. It is usually not necessary to use it, though, since the calculate button (see below) can figure out if the data need to be read in. However, sometimes you may want to find out the grid before calculating the map, or you may have changed the input data file and want to force a rereading of the input data. In any case, after reading the data, some useful information is printed in the message window for your edification. Also, if you change any of the data above this button after calculating a map, you must use this button to have that change take effect.

Grid Number in X, Y, Z. This is the number of points in x, y and z to be used in calculating the FFT. The FFT method used requires these to be multiples of 2, 3, 5, and 7 and nx to be even. The program restricts you to these values if you use the little arrows by the grid values to increase or decrease them. Usually, you do not need to set nx, ny, and nz directly, as explained in the following.

OR Percentage of Minimum Resolution. This sets the grid to the next largest value that will give a grid spacing equal to the minimum d-spacing or resolution times the percentage. For example, if the input data had a minimum d-spacing of 2.0 and 50% is chosen, the grid will be set as close to 1.0 Å as possible. If the cell edge was, say, 48.3 Å, then 48.3 is tried, but since this is not an integer it is rounded up to 49. An FFT requires that the grid be at least 50% of the minimum d-spacing (actually it has to be twice the maximum index input, but it works out to the same thing as the reader can verify). The map will be smoother if 33% (the default) is used. More than this usually has little benefit in smoothing the map and requires much more storage since the size of the map goes up as the cube of the grid size. Set this field to zero to prevent overriding the nx, ny, nz fields.

OR Approximate Spacing in Ångstroms. If both nx, ny, nz and percentage of minimum resolution are 0, then this field is used to set nx, ny, nz by finding the nx, ny, nz that gives the closest grid-spacing in ångstroms that is also a valid FFT grid. For example, if you have a 77.77-Å cell edge and ask for a 1.0-Ångstrom grid, then 78 is tried, but since this is not a multiple of 2, 3, 5, or 7, 80 is used.

Output Map File. The XtalView convention is to use the extension .map. This file is in FSFOUR binary FORTRAN format.

Calculate. This is the "Apply" button. When it is selected, the FFT is calculated and the map is output. If the FFT is large, then it may take several minutes to return. If you change various values, the program tries to ascertain if the data need to be reread. It usually decides to reread even when it is strictly unnecessary.

Advanced. If you want coefficients of a type that is not available, you can always precalculate them, put them in the Fo column, and then make an Fo map. For example, to make a 3Fo-2Fc map using **awk** and **xfft**, use

```
awk '{print $1,$2,$3,3*$4-2*$5,$6}' input.phs
   | xfft - output.map
```

In XtalView a dash (—) means use stdin for input files or stdout for output files. At other times a binary file with the phases is used. For example, to use a Furey format phasit output file (fort.31) using **dumpphasit**, which writes its output on stdout, use

```
dumpphasit fort.31 | xfft - output.map
```

Unfortunately, if you use stdin or stdout for I/O, then the history file will know only that stdin was used but has no way of picking up the original file used and so the thread of data files will be broken at this point.

Xfit—Display/Manipulate Models and Fit to Electron Density

Usage. xfit [-bw] [-std] [.pdb .vu .phs .map .vp .script .bones]

All of the files on the command line are loaded according to their extension. If you have files with different extensions, they can be loaded using the files option. Xfit accepts PDB file, phase files, map files, and vu (vector or line) files. The -bw flag forces the program to be black and white even on a color display. The -std flag forces the standard color mode on SUNs with GX graphics. Some of the more notable features are

- Model-fitting to electron density.
- Built-in FFT to calculate density from structure factors.
- Built-in FFT to calculate structure factors to make omit maps and update phases.
- Difference-density gradient calculations.
- Least-Squares fitting of models or portions thereof.
- On-the-fly omit maps.
- Fast contouring of up to 20 maps.
- Up to 50 models. Residues can be edited and model geometry refined.
- Simultaneous refinement and fitting—geometry anneals as the model is fit.
- 100 arbitrary background objects (ridgelines, ribbons, surfaces, etc).
- X11-based—runs on ordinary (translate inexpensive) desktop workstations—8-bit color or black and white.
- Anything on the screen can be printed with color or black and white PostScript.

- A script language allows building complicated figures or saving viewpoints with multiple objects, and also saving of different defaults.

Xfit is a large and complex program with 11 windows and many possible operations.

Environment Variables

CRYSTAL and XTALVIEWHOME are required. Optionally, XFITDICT can be used to specify a PDB file for residue prototypes (any high-resolution file with all 20 amino acids plus any prosthetic groups, waters, etc., that you need). PWD is used to override the current directory. XFITPENTDIR gives the name of the directory containing pentamer programs and data. It is usually $XTALVIEWHOME/pdbvec. Users can copy pdbvec to another location and change the structures used in the pentamer database by changing this environment variable.

Colors

Xfit figures out the screen depth and uses color if the depth is 8-bits or greater. The color database is loaded from ./colors.dat if available, then from $XTALVIEWHOME/xfit/colors.dat. Normally, the latter is used, but for special purposes the file can be overridden with .colors.dat. However, none of the built-in default colors will make sense if ./colors.dat is completely different. This option is normally used to change the shades of the colors to those more favorable for different display devices. For example, I use a separate colors.dat for making slides. A little white has been added to the red, and the blue has been made closer to cyan, in order to show up better on film.

SUNs with GX cards use color double-buffering to make the display three times faster. The drawback of this is that only eight colors are available and there is no depth cueing. You can switch back to the full-color mode with Canvas.standard canvas. It is better, though, to start in full-color mode with the -std command line option to avoid possible colormap flashing.

Mouse

The mouse is used for rotations and translation, picking, and model manipulation. There should be three buttons on your mouse; from left to right they are SELECT, ADJUST, MENU. (The order of the buttons can be changed by lines in the ~/.Xdefaults file. See an Xview or Open Look manual for more information.) Picking is done by clicking with SELECT. Rotations are done by dragging with SELECT. Z rotations are done by dragging left and right in the upper portion of the screen (it has a different color, except on black and white and SUNs with XGL). ADJUST changes purpose, depending upon the mode (in the lower right corner of the screen). At startup, the program is in CENTER mode and dragging ADJUST centers the screen. If SELECT and ADJUST are held simultaneously and moved vertically, the image is zoomed. MENU brings up a MENU that can be used to change the mode. The modes TRANSLATE, ROTATE, and TORSION are operative if a portion of the model has been selected for fitting. MOVE and TRANSLATE work by dragging ADJUST in the same way as for SELECT. For torsion, a vector around which to torsion is first picked (with SELECT). First, hit the tail of the vector (bond), then the head, and then select TORSION in the menu. All the atoms bonded to the head and currently selected are included in the torsion group. Dragging ADJUST left and right torsions the group about the selected vector. Stacked torsions are not supported, although once you get used to the click-atom, click-atom, select-torsion routine, changing torsions is quickly done. (Note that the pop-up menu can be pinned, which makes it much easier to use.) Finally, several mouse options are available in the control window. There are three mouse-event modes. **Fast** assumes the picture can keep up with every mouse event and draws each one. **Jerky** skips ahead to the latest event, ignoring earlier ones so that the picture "jerks." **Skip contours** does just that if the picture gets behind, allowing smooth rotation to the desired view, and then when the event queue is empty the contours are re-drawn. Which mode to use depends on the computer being used, so some experimentation is in order. The size of a mouse event can be set in the control window also.

Some Definitions

Xfit uses an atom stack for many operations. All operations use a prefix notation. First, the operands are selected and put on the stack, and then the command is given. Atoms are pushed onto the stack in two ways: by picking on the screen and by picking from the atom list in the Model window.

Internally, xfit uses the following hierarchy to describe models: a molecule is all the atoms given in a single PDB file; a *chain* is all the atoms with the chain-ID column set to the same value in the PDB file; a *residue* is all the contiguous atoms with the same residue-ID in the PDB file; and an *atom* is a single line in the PDB file that begins with either ATOM or HETATM. Additionally, the user can specify a *group* in the model window, which is an arbitrary grouping of residues. What is drawn on the screen are *lines* derived from *bonds* and *points*. Any atoms within a specified radius in the *same* residue are bonded. The exception is peptide bonds that are made if the nitrogen–carbon distance is reasonable. If two adjacent residues contain only single CA atoms, they are bonded if they are less than 4.5 Å apart. What is left over are points (such as water) that are drawn as three short orthogonal lines or a jack. Other polymers, such as DNA and carbohydrate, are not connected (well, this is a *protein* crystallography program . . .). Atoms have a name such as CA or CB, an occupancy, a B-value, and *x, y, z* coordinates. Residues contain an array of atoms and have a name such as 1 or 152 and a type such as ALA or HEM or H2O. Chains are specified by a single character. Molecules are referenced by the original filename.

PDB files have a record CONECT, which can be used to specify a bond between any two atoms. These residues are read and the specified bonds put in a special list so that these bonds are always made regardless of geometry. This is useful for connecting prosthetic groups to the protein, such as thio–ether linkages or metal–ligand bonds. If you make a bond in xfit, use the bond command and then save the model. The CONECT information is written at the end of the file and will be read the next time the model is loaded.

Contours

Contours are iso-values in an electron-density map. Five levels are provided in xfit. The map is scaled such that one σ is equal to 50. The startup contour levels are 1,2,3,4,5 σ and can be adjusted later. The contouring algorithm is fast but not always pretty (for instance, it makes no attempts to resolve saddle points). To pretty the contours, use a finer map. Three times the maximum *h*, *k*, and *l* are usually about right (the maximum index can be found from (cell edge)/d_{min}). If you have one long axis, then this one can be made two times and still look fine. A large number of contours can cause rotations to slow. One useful option is to select the **skip contours** option in the control window. This will cause drawing contours to be skipped if the program cannot draw the contours between mouse events and then re-draw them when rotations stop. Since the model is usually what is being used as a guide to the eye for rotations, dragging the map along is not always necessary and can speed up rotations considerably.

Ridgelines

If you have access to GRINCH, use mkskel to produce an ASCII ridgeline file or use the ASCII save filename option in interp. Later versions of Xtal-View have a program xskel to make ridgelines with. This file can then be converted to a form usable by xfit with grinchbones (see the grinchbones entry in this manual). This is read in as a vu file. Later versions of xfit support ridgelines directly in a manner closer to that of the GRINCH interp program. If you have a ridgelines button, then you can read and save ASCII format GRINCH files. You can also interactively change the contour level and box size. To pick an edge, set the pick mode to edges. To color an edge, select the type you want to color the edge, select the color edges pick mode, and then pick the edge you want to color. An entire side chain can be colored at once by setting the pick mode to edges, picking an edge to place it on the stack, then selecting color side chain. The last color side chain can be undone with the undo color button. The regular model building features of xfit can be used to build model over the ridgelines.

Background Objects

Complex background objects can be built by layering individual vu objects in the view window. Each vu object can be turned on and off using the show window. For example, you can make a vu object with side chains and another with main chain. The side chains can then be blinked on and off by clicking on the appropriate line in the vu list. Individual residues of interest can be colored individually. Multi-colored objects are composed by building up from individual pieces. Objects with a higher number overdraw lower-numbered objects. Models overdraw vu objects, so the model must be turned off to see coincident vu objects. Hint: Use a lot of individual objects instead of a few complicated objects.

How to Color

First, pop up the color window. Second, select a color by clicking on it. Third, select the correct object (if necessary). Finally, Apply the color by hitting the appropriate button and the object is set to the current color. There are two places where the option color C* only is given. This means that only the carbon atoms are colored and the others are colored by atom type in the usual way (i.e., nitrogen is blue, oxygen is red, hydrogen is white, iron is brown, sulfur is yellow). In addition, in the view window the option color-by-B-value is given. In this case, the coloring is according to B-value in the same way as described for normalized B-values in the section on colorbyB.

In the color window you will see four shades of each color. This lets you see the effect of depth-cueing on the color. On 8-bit machines these are the depth-cueing colors. A line is drawn in four different intensities depending upon its average z-value. On 24-bit machines the line is depth-cued by interpolating from the front to back. Colors.dat contains four intensities for each color. For instance, for red there are four entries: red, red1, red2, red3. To change the depth-cueing, you change the four colors where red is the foreground and red4 is the color of the farthest point. This is probably best done by a computer program from a file containing just one entry for each color. Each color display works best with a different amount of depth cueing.

PostScript coloring is always done in terms of RGB triples, and the color for the foreground is taken from colors.dat. Depth-cueing is done by multiplying the RGB values by a scale factor from 0.25 to 1. The value 0.25 can be changed in the control window. I have found that different color PostScript devices work best with different values for the minimum. For instance, the Dicomed slide maker needs a value of about 0.1, and the QMS-100 color printer works better with about 0.3.

Least-Squares Fitting

This can be used to superimpose two models or portions of the models. It can be used as a comparison tool. It can also be used to speed up fitting. The idea is to rough in the model using edited residues with only CA or CA and CB residues (put these in the dictionary with a type such as CBA). Then standard protein pieces such as α-helices, β-sheets, or tight-turns are read in as extra models (up to 49 of these plus your working model) and then least-squares-fit to the markers. Overlaps are deleted, and then these are read out and concatenated to form the whole molecule. The initial geometry will be better, giving a faster refinement. More complicated fittings that use unconnected segments can be done with the program pdbfit instead and then loaded into xfit.

FFT

This is a reworked routine from Lynn Ten Eyck that provides a fast way to calculate electron density and structure factors. The electron-density calculation is performed with the FFT pop-up, and structure factors are calculated on the sfcalc pop-up. The FFT pop-up can be used to calculate electron density directly from structure factors using coefficients such as Fo, Fc, Fo-Fc, 2Fo-Fc. It is usually faster than reading a precalculated map file and lets you redefine the grid and the resolution range, at any time. The same phase file can be used to make both an Fo-Fc and a 2Fo-Fc map. It also saves a lot of disk space. Say good-by to map files! The whole unit cell is available, so there is no need to worry about map boundaries. Make sure the correct crystal is specified beforehand!

Structure factors can also be calculated by FFT. In this case, the only information needed to make maps are the Fos and the coordinates of the model. Make up a phase file with *h, k, l,* Fo 0.0 0.0 and read this in. Make sure the correct crystal is selected before reading. Then use the sfcalc window and select **calculate all and scale.** You can now make omit maps and update phases.

There are two ways to calculate structure factors: FFT and summation. FFT is fast and introduces an error typically about 1%. In general, this is the way to go. The only time the summation is faster is when you are doing a single atom. The FFT is done in two parts: first, a model of the electron density is built from the atom coordinates and their structure factors, and then this map is inverted by FFT. The FFT is very fast, and for the entire model the rho building step takes the longest. However, for a single residue the rho building is very quick and the entire FFT time is less then the corresponding summation time. The only instances in which summation is used are when extreme accuracy is needed (given the error in our coordinates I cannot see when this would be needed) and if there is not enough memory to build the entire unit cell's rho. The FFT needs $8 * nx * ny * nz$ bytes of memory, no matter how many atoms are being calculated. (Note: In centered space groups it is common not to calculate the centered symmetry operations to save computer time. In this case, Fc will be lower than absolute scale. For instance, in C2 the Fc will be one-half of its correct value.)

Omit Maps on the Fly

To make an omit map select the atoms to be omitted (see below: How to Select Model) and then select the **omit current atoms** option in the structure factor (sfcalc) window. Structure factors are calculated for the currently selected atoms and subtracted from the model structure factors. Then use FFT to make either an Fo map or a 2Fo-Fc map. This assumes that you have read-in some phases and that Fc has been previously calculated on an absolute scale (Fo is scaled to Fc). This can be done using **calculate all and scale.** Note: partial models (such as missing residues or waters) cannot be used to make omit maps with Fo-Fc or 2Fo-Fc maps because the relative scale of Fo to Fc is incorrect. Instead, use Fo maps. This is also true of models that still have a large amount of error, where Fc is still inaccurate (i.e., R-factors above about 0.25).

Updating Phases on the Fly

As the model is moved it changes the phases, since these are calculated from the atom positions. You can either recalculate the phases from all of the model using **sfcalc.calculate_all_and_scale** or you can update just the porition being fit, which is much faster. In this case, after you move the currently selected atoms, you can select **sfcalc.update_current** and then re-FFT the map. **Update all** subtracts the old position from the phates and then adds the new position. As the residue is moved, you can calculate the corresponding map and compare it. For example, if there are two likely positions for a side chain, you can move it to both positions and see which best matches its corresponding map. When you are finished fitting, select **update current** just before applying the fit to match the map.

Difference Map Gradient

At the bottom of the SFCalc pop-up is a section used for calculating the gradient of the difference density at an atom. This is represented as an arrow in the direction of the gradient scaled to its steepness or magnitude. It shows the direction an atom will move in the first cycle of a crystallographic refinement. The command implements the equations in Stout and Jensen, 2nd ed., 349. It is useful for such things as analyzing mutants to look for the direction of movements and to look for concerted movements of groups of atoms. The command works on the currently selected model (see below). It uses a slow summation method, so it can take quite a long time for large number of atoms. You must first have loaded some phases. The command works on the map currently selected at the top of the window. Set the **Deriv vu number** to a free vu object. The **Sign of Delta** is usually Fo-Fc, but to reverse the direction of the arrows, select **Fc-Fo.** The **Scale Arrows x 100** slider sets an arbitrary scale for the length of the arrows. To start the calculation, select the **Derivatives of Current Atoms** button.

How to select Model

Model is selected for fitting or structure-factor calculation by first putting the appropriate atom(s) on the stack and then selecting one of the fit options. To put an atom on the stack, pick it on the screen or select it in the model window (see below). If you select **atom,** the atom on the top of the stack is selected. If **atoms** is selected, all the atoms on the stack are selected. **Residue** selects the entire residue that contains the atom on the top of the stack. **Residues** selects all residues that contain an atom on the entire stack. **Group** selects all the residues in the same group as the residue containing the atom on the top of the stack. Groups are set in the model window. **Molecule** selects the entire model. The selected atoms are referred to as the "current" group. Bonding is irrelevant. Use the **clear stack** button before setting up atoms or residues. If you are dissatisfied with your selection, hit **Cancel. Reset** puts you back to your starting positions and **Apply** commits your fit. However, the last 20 fits can be undone with **Swap.** When model is selected for fitting, its old positions are saved in one of 20 slots. Select the slot containing the model to be undone and **swap** will toggle between the saved and fit model. Judicious users will save their model to disk frequently with the **Save** button.

Show Window

Objects can be hidden by selecting them on the list in the show window. Useful information about each object is also given to identify it. An object that is difficult to identify can often be found by toggling it on and off. Hide all objects that are not needed to speed up the rotation. Object blinking is set up here (see Blinking below). Also, contacts can be cleared here.

Canvas Window

Objects are drawn in the canvas window. On SUN systems with a GX or GS graphics accelerator the canvas can use the card or use the standard drawing mode. The GX card cannot show as many colors because most of its color map is consumed in a double buffering scheme so that it cannot depth cue. Another reason for using the standard canvas is that it is closer to what is plotted on hard copy. A ruler is drawn at the bottom with the scale in ångstroms. Objects can be dragged over to the ruler for measurement. In the upper left is a gnomon that shows the laboratory axes. In the lower right a small advertisement appears. The MENU button pulls up a menu that can be pinned and left up permanently if desired. The use of the pull-up menu eliminates the constant back-and-forth motion between the screen and a menu off to the side. The menu items are the following:

Center at Pick.   Centers at the last atom or edge picked.

Center Mode.   Puts mouse in center mode, where ADJUST (middle mouse button) moves the center of screen.

Translate Mode.   Current atoms can be translated with ADJUST. Note the difference between this and center—they are easily confused. A common mistake is to forget to go back to center mode to move the screen center.

Rotate Mode.   Current atoms can be rotated with ADJUST.

Torsion Mode.   Two ends of a current bond are first picked. The atoms attached to the second end will become a group that can be torsioned around the selected bond. The atoms must all be in the current group.

Contour. Contours at the center of the screen using the values in the contour window.

Atom. The last picked atom becomes current, or selected.

Atoms. All of the atoms in the stack become current.

Residue. The residue to which the top atom belongs becomes current.

Group. The group to which the top atom belongs becomes current. Groups are set up in the model window.

Molecule. The entire molecule of which the top of the stack is a member becomes current.

Rotate X Clockwise 90 Degrees.

Rotate X Counter-Clockwise 90 Degrees.

Rotate Y Clockwise 90 Degrees.

Rotate Y Counter-Clockwise 90 Degrees. The two rotate Y commands are especially useful for centering an object in density. First center in one view then rotate Y 90 to fix the third direction and then rotate Y − 90 back to the starting view.

Rotate Z Clockwise 90 Degrees.

Rotate Z Counter-Clockwise 90 Degrees.

Labels Window

Labels can be controlled and edited in the labels window. Just to be confusing, the control window contains the control for labeling picked atoms or not. If an arbitrary label is desired, first pick an atom near where you want the label to go. Then pick the atom. The new label will show up at the bottom of the list in the label window. Select the label in the label list. You can then edit the fields for the label string and its offsets in x and y and select replace. The offsets in x and y are pixels from the x, y, z coordinates of the label that are independent of the rotation matrix and the scale. Labels can be deleted by selecting them on the list and clicking on delete. All labels can be removed by selecting the appropriate command on the main window.

To edit a label, select it on the list, edit the field you wish to change, and then pick replace. New labels can be entered here. The current center can be popped and used for an arbitrary label. You can also set the label style here. The point size of the labels when plotted can be selected. All labels will be of the same size. The pixel offset on x and y is useful for moving labels into clear space so they do not overlap other objects. When labels are plotted, a drop shadow is plotted around them to make them stand out if they do overlap other objects.

A special type of label is a *contact* distance. These change to reflect the current distance as the model moves and are not editable. (See Controls for more information.)

Controls Window

Several miscellaneous controls are located here. **Mouse damping** controls how far the model rotates with a mouse movement. The **drawing mode** fast draws every mouse event. If the computer is taking too long to draw each event, then the model may keep moving after the mouse has stopped. The jerky mode will prevent this by skipping ahead to the end of the event queue. For an unknown reason, this mode does not work well on some SGI machines. The skip-contours mode will skip drawing the contours if the event queue gets too long. This allows the model to be rotated into position, and then the contours will redraw when the event queue empties. The **pick** mode defaults to only picking the active model. This allows picking of superimposed models by locking out the others. When in the **pick all** mode, the closest atom is picked. If two atoms are in the same position, then the last one in the display list is chosen. The pick mode is also controlled by a toggle in the ridgelines window, which switches the pick mode from model to ridgelines. In order to pick symmetry-related edges, the mode must be all.

The **symmetry radius** controls the size of the box that is searched for symmetry-related atoms. Making this very large will slow things considerably because the volume goes up as a power of three. The **rocking range** and **rocking rate** are controlled in this window. Xfit uses a timer that is supposed to be machine-independent but, in fact, turns out not to be, so the rocking rate is different on different machines. The **depth cueing** slider controls how thin the back lines are relative to the front lines when plotting in the depth-cued mode. It also controls the percentage saturation of the back color to the front color when plotting in the color mode.

Contacts are also controlled in this pop-up. A contact is a dashed line drawn between two atoms with a distance at the mid-point that shows the distance in ångstroms. You can turn on/off drawing contacts whenever the distance function is used. You can also enable contacts to be drawn for picked atoms. A maximum radius for this search can be set. The mode can be off, all atoms within the specified radius, or only between atoms in a different residue (inter-residue). As the model is fit, these distances will update.

Blinking

In the show window are controls that allow you to set up two blink groups and then alternately blink them on and off. First select all the groups you want in one set and make sure the others are not selected. Choose **set group 1.** Now deselect group 1 and select group 2 and choose **set group 2.** Objects not included in either group will not appear. Objects in both groups will stay on all the time. You can turn blinking on and off, and you will not need to reset the groups unless you want to change them. The speed of the blinking is set in the control window.

Refine Window

The refine window is split into two parts. The upper part is for real-space refinement and the lower part is geometry refinement (early versions of Xfit lacked geometry refinement). In real-space refinement you select the map against which to refine the active model. You can select **translational, rotational,** or **torsion search.** The torsion must be already selected or nothing will happen. In good cases the model will be moved in the mode specified until the sum or electron density at each atom is maximized. This works well for torsion searches with only one axis to search (i.e., Phe works well but Arg does not). Translate and Rotate often try to move into density that is connected but already belongs to another atom (i.e., the neighboring residue).

Geometry refinement (minimization) can be done in two ways. The entire model can be refined by simply selecting **refine active model.** Be sure to have the model saved as this cannot be undone. The ideal bond distance and angles are taken from the residues in the dictionary file (see below) by using them as guides. This is done to allow different ideal lengths for different refinement packages. Also, this allows one to have all hydrogens, no hydrogens, or partial hydrogens. The residues in the dictionary should be refined by the refinement program you are using. This will then maximize the agreement between Xfit and the refinement program so as to correct geometry. Planes are specified in the geometry file with the PDB extension record PLANE, which specifies the residue type and the atom types in the plane. An example is

```
PLANE ASP 4 CB CG OD1 OD2
```

Since planes are supposed to be flat, it is not necessary to specify the ideal value. Connectivity is assessed by a variety of critera. First, if in the original PDB file A TER or record was found, then this marks a C-terminus and no attempt will be made to form a peptide bond. Second, if the chain is letter changed while scanning the file, this also marks a C-terminus. Third, if the residues have recognizable names such as an amino acid (e.g., ALA, Ala, ala, or A), then a search will be made for the proper atom types for a peptide bond and plane (CA, C, O, +CA, +N, [+H]), and if the nitrogen and carbon atoms are within 5.0 Å (probably too generous, but I was thinking of really rough models), a peptide bond, angles, and plane are formed. The algorithm for drawing lines representing bonds is different and less stringent in order to speed up rebuilding of lines for display purposes. If incorrect bonds are drawn on the screen, they will *not* show up in the refinement.

The second refinement method is to refine a portion of the model. Click on (or select an atom in the model window), and put two residues on the stack. Now use each of the **pop** buttons to get these two residues in the start at and end at fields (you cannot type these values in). Then select **set up refinement residues,** and the selected residues will be highlighted on the screen. When you push **refine,** a round of refinement will be done. If you select the **refine while fitting** option, then the selected portion will be refined in the background while you fit. You can then select a residue for fitting within the segment being refined, and as you move it around, the rest of the segment will follow along and the geometry will anneal as you fit.

Dictionary

The dictionary file is simply a PDB file with an example of each residue. If the environment variable XFITDICT is set, then this file is loaded upon start-up as the dictionary. The dictionary can also be loaded and appended to in the File window. This is not used as a geometry file. New residues are found by finding the first residue in the dictionary that has the same name. When a residue is replaced, a residue of the same new type is found in the dictionary and least-squares-fit to the old one. The standard dictionary is in $XTALVIEWHOME/xfit/dict.pdb. The dictionary can be appended when loaded or it can be replaced. The append mode allows you to add a prosthetic group without having to mess up the standard dictionary. The first residue in the dictionary is used if there are multiple copies. All of the unique residues in the dictionary appear in the model window from a pull-down menu button to the right of the residue-type field. Selecting one of these names puts this value in the residue-type field and can then be accessed with the insert commands.

Script Commands

Xfit can accept commands from a script file. The script commands can substitute for many of the interactive operations and, in addition, some operations can only be done from the script. One use of the script is to return to a given view. The save view command writes a script file that, when run, returns Xfit to a given viewpoint. Load view is just another method of running this script. The other common use is to build complex figures with many parts. The script also gives more control over linewidths, allowing different widths to be set for each object. There are also some animation commands, such as `roty`, that rotate the view around *y*. The script window can be accessed from `view.edit_script`. Scripts can be run and edited by loading into the script window. The script can be altered and then re-run to see the effect of the edits. Every time the script is run, it is first saved and the old script moved to `file.script%`; that is, the file name with a "%" added. Scripts can also be loaded from the command line if they have the extension ".`script`" or ".`vp`" (for viewpoint), at which time they are run. For instance, the command `xfit my.pdb save.vp my.map` will first load the PDB file, then switch to the new viewpoint in `save.vp`, and finally load and contour the map. The size and location of the canvas can also be saved. A line starting with "#" is interpreted as a comment.

```
crystal value
loadpdb modelnumber file
loadpdbcenter modelnumber file
loadmap mapnumber file
      load fsfour format maps
loadmapphases mapnumber file
      load map phases (follow with fftapply)
plottofilecolor file
plottofileb&w file
plottofileb&wdepthcued file
plottoprintercolor printer
plottoprinterb&w printer
plottoprinterb&wdepthcued printer
# object building commands:
color value
      value is either a the xfit number or a name
      i.e. yellow is 56 or ''yellow''
      this set the current color buildvu uses the
      current color to paint its lines, therefore
      set the color first
```

```
colormodel modelnumber
      model modelnumber is recolored to current color
colormodelatom modelnumber
      as above except that atoms are colored by atom
      type
      except C* atoms which are set to current color
vulinewidth value
      sets the line width in plotted output for all
      subsequent
      buildvu commands
buildvu key modelnumber vunumber first_res last_res
chainid atomfilter
      use * to indicate any
      atomfilter should be separated by commas (i.e.
      ''CA,CB,C,N,O'')
      key = one of:
      side = protein sidechains
      main = protein mainchain
      other = non-protein
      all = side,main and other
      link = link CA's
      peptide = peptide planes
      hbond = H-bonds
examples: to build the backbone of model 1 and show 2
key residues:
   color red
   Buildvu link 1 1 **** links CA's of model 1
   color green
   Buildvu side 1 2 18 18 ** side chain of residue 18
   of model 1
   Buildvu side 1 3 218 218 ** side chain of residue
   218 in green
# viewpoint transform commands:
zoom value
frontclip value
backclip value
translate x y z
rotation vxx vxy vxz vyx vyy vyz vzx vzy vzz
   the way to get this is view.save to a file and then
   read in the file to the script editor
transform
   the point of view commands don't take affect until
```

   transform is given so that all transformations are
   applied at once
rotx value
roty value
rotz value
   these take effect immediately—used to make
''movies''
stereo on
stereo off
canvasrect left top width height
   This set the position and size of the canvas window. The first two set the position of the upper
   left corner of the window and the next two set the
   width and height. The units are pixels.
show and hide objects:
showmap mapnumber
hidemap mapnumber
showmodel modelnumber
hidemodel modelnumber
showvu vunumber
hidevu vunumber
bond modelnumber1 resid1 atomid1 chainid1 modelnumber2
resid2 atomid2 chainid2
   draws a line between two atoms. chainid can be *
atom labeling commands:
labelpoints value
   sets the point size of label in plotted output
labelstyle value
   sets the label style according to:
      0 = GLU 100 CA
      1 = GLU 100
      2 = GLU
      3 = 100 GLU CA
      4 = 100 GLU
      5 = 100
      6 = CA
      7 = none
atomlabel modelnumber resid atomid chainid
   forms a label of the type set by labelstyle (default 0)
map contouring commands:
maptocontour mapnumber

   sets the map to contour for all subsequent commands
   a map load will overwrite
contourlevels value
   levels are turned on and off by adding numbers as
follows
      1= level 1
      2= level 2
      4= level 3
      8= level 4
      16=level 5
   thus to turn on levels 1 and 3 only is 1 + 4 = 5.
contour1level value
contour2level value
contour3level value
contour4level value
contour5level value
   contour level 5 at value
contourradius value
contourcolor value
   set level value of the current map to the current
color
contourmap
   contours current map at current center
FFT commands:
resmin value
resmax value
coefficents value
   Note misspelling. Value is one of the following:
     Fo
     Fc
     2FoFc
     Fo-Fc
     Fo*Fo
     Fo*fom
     3Fo-2Fc
fftnumber mapnumber
fftnx value
fftny value
fftnz value
fftapply

Example 1

```
# This script was used to make two electron density
# maps with and without
# figure of merit weighting to illustrate the
# difference
crystal cvccp
# make the model thicker than the map to aid in
# distinguishing
maplinewidth 0.7
modellinewidth 3.0
loadpdb 1 cvccp.topdb.pdb
#the viewpoint lines are from the save viewpoint
#option
translation 24.965 34.205 68.649
rotation 0.6916 -0.3000 0.6570 0.6543 0.6455 -0.3941
-0.3059 0.7024 0.6427
zoom 31.10
frontclip 4.60
backclip -3.30
stereo on
transform
#load the map phase and transform
loadmapphases 1 cvccp.mir.phs
resmin 3.0
coefficents Fo
fftapply
# set the contour options and contour the map
contourlevels 1
contour1level 50
contourradius 10
contourmap 1
plottofileb&w mir.fo.ps
# change the coefficients of the FFT and compute a
# second map
coefficents Fo*fom
resmin 3.0
fftapply
contourlevels 1
contour1level 50
contourradius 10
contourmap 1
plottofileb&w mir.fofom.ps
```

Example 2

```
#this example builds up vu objects. A backbone of the
#A chain is built
#and then particular side chains are displayed and
#labeled. If a mistake
#is made it is much easier to edit the script than to
#start over in interactive mode.
#Also the script can be saved to re-make the figure at
#a later time.
stereo on
loadpdb 1 cvccp.pdb
labelstyle 1
hidemodel 1
color 56
vulinewidth 2.0
buildvu side 1 1 ** A*
buildvu main 1 2 ** A*
buildvu other 1 3 ** A*
vulinewidth 0.9
buildvu side 1 4 ** B*
buildvu main 1 5 ** B*
buildvu other 1 6 ** B*
atomlabel 1 16 CG A
atomlabel 1 16 CG B
atomlabel 1 18 CB A
atomlabel 1 18 CG B
atomlabel 1 132 C1A A
atomlabel 1 132 C1C B
atomlabel 1 21 CD1 B
atomlabel 1 21 CZ3 A
translation 30.810 34.796 60.196
rotation 0.8504 -0.5256 -0.0256 0.5256 0.8507 -0.0075
0.0257 -0.0071 0.9996
zoom 19.8
frontclip 2.60
backclip -8.10
transform
```

Model window

The currently active model is listed in the model window. When a residue is selected, the atoms in that residue show up in the atom list. When an atom is picked in this window, it is put onto the atom stack. It can then be used for stack-based operations such as center or distance. This is the quickest way to center on a particular residue. When a residue is selected, it is highlighted and becomes the focus residue and its information is listed in the bottom of the window in the **type** and **name** fields. Several residues can be highlighted, but only the last one is the focus residue, the one that is in the lower fields. Highlighted residues can be grouped together in a single group. An entire group can be moved as a unit after being selected (see above). A group number above 0 must be chosen since 0 means ungroup. Before selecting a group, it may be wise to clear all selections first to prevent residues that have scrolled off the list from being accidentally added. Residues may deleted with the **Delete** button. They can be undeleted at any time by a second application of delete. The only way to clear the list of deleted residues is to write the model out and read it back in again. The **insert** button has a pull-down menu with a number of features. You can insert a new residue before and after the focus residue. First select the insertion point. Then edit the name and type of the residue. All of the types in the dictionary will be listed in the pull-down menu at the end of the type field. Then select **insert.before** or **insert.after.** The new residue will appear in the center of the screen. **Insert.Replace** is similar except the new residue is least-squares fit to the focus residue and the old residue is deleted. Thus, a replace can be undone by undeleting the old residue and deleting the new residue. At the end of the insert menu is a **Pentamer** pull-right menu. If it is grayed out, then the pentamer information and programs have not been installed (see above). The pentamer option allows least-squares fitting a pentamer with one that has the closest geometry from a list of well-refined structures. The middle-three residues can then be replaced with the pentamer residues. The only atoms that need be in the target pentamer are CA atoms. The quickest way to build a protein model is to place single CA atoms (type MRK in dict.pdb) in the density. Keep the CAs in order as much as possible as their order matters for the pentamer fitting. Pick the first residue of five contiguous residues in the residue list, and then select **insert.pentamer.best_fit_pentamer.** A pentamer should appear as a separate model least-squares fit to the target residues. If it is acceptable, then select **insert.pentamer.replace_middle_three** to replace the new residues in your model. The side chains of the new residues will be whatever was in the structure database. You then replace these side chains by picking the residue to be changed in the residue list, select the correct type with the little pull-down button by the type field, and select **insert.replace.** The side chain is then torsioned into its density.

View Window

This is used for controlling the viewpoint and for building alternative representations of model. At the top of the window are the viewpoint matrices at the time when the **View** button on the main window was selected. It does not update as the model is turned as this would seriously impede performance. Instead, you must select the **View** button again to get the current matrices. The views can be saved and loaded from a file. The viewpoint is saved as script commands and these can be copied into a script. Also, the load command actually just runs the script command on the file so that you can add other script commands to the viewpoint file if you wish. You can move to a specific x, y, z by entering these into the translation field and selecting transform. It is less likely that the rotation matrices can be entered successfully and produce a useful result because errors will distort things severely. Note that the script commands for the rotation are the same as that used by the program MOLSCRIPT by P. Kraulis. The translations are of the opposite sign.

   Plan view splits the view into the standard viewpoint in the bottom half and a from-the-top view in the top half. This is useful for docking operations as you can see and control all three axes and translations at once. This feature is not available on all display devices/modes.

   Stereo turns **split stereo** on and off. The stereo can be made either cross-eyed or wall-eyed with the angle slider. The separation can also be controlled. The separation can better be changed by scaling the canvas window width. You can also select left- and right-eye only views. These are used for making stereo slides and are only useful for photographic purposes. If hardware stero is supported, then there will be a fourth selection available, **hardware**. On SGIs this is done using Crystal Eyes or StereoVision glasses. As they do not work well with windows, Xfit rearranges the windows to work best with these glasses and to assure that the stereo off button is available. When **stereo off** is selected (make sure the cursor is not in the upper half of the screen), the windows return to the original position. You can then do such things as edit the model and load files and then return to stereo viewing again. Picking is less reliable in hardware stereo due to perception problems and is best done in less crowded views.

   Make and **Edit Script** are used for script commands (see above). Make script is supposed to generate a script that when run will return you to where you are. It will load models and maps and restore the viewpoint. Unfortunately, because the original order in which these operations are done makes a difference, the resulting script file may need to be edited in order to get back what you have. It provides a great starting point, though, for complicated figures. The biggest problem is that loading a model changes the screen center. All model loads should be done first before viewpoint commands. The edit script window allows you to edit the script and re-run it. To save the script use the MENU mouse button while in the editing window and select **File.Save_New**.

Below this are the controls for **build vu.** This is a very versatile and useful command. The user selects which model he wants to build the vu object from and the residues between which to use. The default is to use all of the model. The user then selects one or more options. An atomfilter can be set. The syntax for this is a list of names separated by spaces, such as "C N CA O," which is the same as the main chain option (actually main chain also requires a peptide). An asterisk can be used as a wild card and means that the rest of the name does not matter—that is, C* matches C, CA, CB, CG1, ...CG* matches CG1, CG2, ... In Xfit the model is never changed for viewing purposes. Instead, a vu object is built and the original model is hidden if desired. If you want to have seven side chains all in different colors, then seven vu objects are built. There can be up to 200 vu objects. Long lists of vu objects can be saved in a single file and read in as a single vu object later. By this means the 200 object limit can be overcome. The script command buildvu corresponds to these controls. However, if you use a script, then you can edit your mistakes without starting over. You can edit the file, run it with "xfit file.script," find any errors, and then fix them in the edit script window and rerun the script. This is then saved with the textapane menu available by pressing the MENU mouse button while in the script editing window. The other advantage of the script commands for building vu objects are that you can change the line width between objects. While you cannot see this on the screen, when you plot it the different widths show up. This is very useful for making black-and-white publication figures where two objects need to be clearly distinguished.

Plotting

The current canvas can be plotted to a PostScript file or printer using the plot command in the Files window. First enter the name of the printer or file, then select the plot type, and finally select one of the two options in the pull-down menu selected, using the MENU mouse button on the Plot button. Additional information can be plotted as a second page using the **control.stats on/off** option. This will add a second page with the viewpoint and the names of the objects used and the map levels. You can select to plot in color mode, black and white, and black and white with depth cueing. In the depth-cued mode the width of the line indicates the distance from front to back, with objects in back being thinner. Color is very useful for making slides. These can be sent to a commercial slide service. The advantages are that the exposure is always right and the resolution is usually much higher than on the screen. The disadvantage is that the screen colors and the slide colors may differ substantially. Blue shows up especially dark on slides. Use cyan instead. The PostScript file can be edited if desired because it is a list of commands in PostScript language. The file has been set up to allow easy editing. Each object is bracketed by "BEGIN type # name" and END. You search for each object by looking for the string BEGIN. The most common field to edit is the line width. You can also edit the set gray level. Black is 0 and white is 1 and 50% gray is 0.5. The labels are at the end of the file, and you may wish to edit these. It is common to want to move labels around a bit to prevent them from overwriting other objects. Objects in the file are drawn starting from the front of the file and going to the end. Thus, a later object overwrites an earlier one. You can change the order in the file if you wish. Labels have a shadow drawn around them, so they will be readable even if they fall on an object of the same color. The label point size is set in the Labels window.

Memory

Xfit uses dynamic memory allocation for virtually everything. The only limits are the size of memory available in your machine (including swap space) and the largest number representable by an integer on your machine. If you run out of memory or cannot allocate errors, close all unused windows. Every open window takes up memory. Next reread over objects that are not being used, instead create new ones and hide the old ones. Re-FFTing a map with a coarser grid can free up a lot of memory. If performance becomes sluggish, prompting swapping (often you can hear your swap disk chuckling merrily away), then more RAM may be the answer. The cheapest way to speed up your workstation may be to invest in more memory. I get good performance with 32 Mb. Less than 16 Mb is not recommended. If you are calculating structure factors by FFT, then at least 32 Mb is needed. Finally, most work stations are set up with small swap spaces, which becomes a liability later. If you have a choice, set up at *least* a 100-Mb swap space.

More on Color

It is advisable to modify only the color shades in ./colors.dat. For instance, red may be modified to add some white. It is possible to customize the colors completely if you have a minimal knowledge of the C language and rewrite colors.h along with ./colors.dat and then re-compile. The colors occur in groups of four that are used for depth cueing on 8-bit color systems. To change to degree-of-depth cueing from front to back, it is possible to change ./colors.dat to give more or less saturation.

Importing Other File Formats

Vu files.   Any arbitrary line information can loaded with a vu file. The format is lines of

```
x1 y1 y2 x2 y2 z2 color
```

x1, y1, z1, and x2, y2, z2 are the end points of a line and color is either an integer to a corresponding entry in colors.dat (see the footer message in the color window) or an ASCII name such as "red," "yellow green," or "white." Hint: break long lines into short segments so that they will not be completely clipped when the viewport is small. A "#" as the first character in a line is interpreted as a comment.

MS surface file (.surf). Surf files can be converted to .vu files with awk:

```
awk '{ print $5,$6,$7,$5,$6,$7,'' white''}' file.srf
    > file.vu
```

This command prints the fifth, sixth, and seventh fields of the surface file twice and adds a color. Since the beginning and end of the vector in the vu file are the same, a dot will be drawn the size of a single pixel at the position of the surface point, making a "dot surface." You may find this command in a shell script in

```
$XTALVIEWHOME/util/srftovu.
```

Other three-dimensional Arrays. Arbitrary three-dimensional arrays can be loaded into Xfit by reformatting into a map file. These can then be contoured as if they were electron-density maps. A corresponding crystal entry will be needed. To write this array, you will need to use the information in the programmer's guide. One limitation is that the map is assumed to start at 0, 0, 0 and extend for one cell edge in each direction. This may mean making an array bigger than is needed and padding it with 0's.

Xfrodomap—Reformat XtalView Maps for FRODO

Usage. `xfrodomap [file.map file.dsn6]`
This command does not work properly on non-SUN machines. This is used to reformat a map from XtalView format to FRODO DSN6 format. It also orthogonalizes the map, interpolates it to a grid, and extracts a portion of the cell. If non-crystallographic symmetry is known and in the crystal file, the map can be non-crystallographically averaged.

Enter the Bounds in and Grid in angstroms. If you have fractional bounds, use `cvtxyz -f x y z` to return Cartesian coordinates. The map is interpolated to the new grid, but if the input map is not calculated on a fine enough grid, Xfrodomap cannot make up for it. Calculate the input map on an equivalent of finer grid.

Non-crystallographic symmetry. If the non-crystallographic symmetry box is checked, then records with the keywords ncrsymm1, ncrsymm2

... ncrsymm*n* are searched for. The format is nine numbers describing a 3 × 3 matrix r11 r12 r13 ... r32 r33 followed by three floats describing a translation (angstroms). The identity transformation is required, for example

```
ncrsymm1 1.0 0.0 0.0 0.0 1.0 0.0 0.0 0.0 1.0 0.0 0.0 0.0
```

Note.   The output of PROLSQ non-crystallographic symmetry search is in the correct form. There has not been time to check other programs. The translation is applied after the rotation.

Xheavy—Refine Heavy-Atom Derivatives and Calculate Phases

Usage.   `xheavy [file.sol file.phs]`

Xheavy is used to calculate Multiple Isomorphous Replacement phases and to refine heavy atom solutions. The MIR phasing is done in an iterative manner where SIR phases for each derivative are first calculated, these are combined, new errors are estimated based on the combined SIR phases, and finally new phases using the new E estimates are output. The refinement method used is a recursive correlation search. Each heavy atom is moved in turn on a finer and finer grid until no improvement in the correlation between the observed and calculated difference is found, and then the relative occupancies are refined. The B-values are not currently refined. Residual maps and difference maps can be calculated to locate more heavy-atom sites.

There are two windows for Xheavy, the main one and an edit pop-up. On the main window is the **Crystal** field, which is the key to the file containing this crystal types parameters. **Unit Cell** is used as a verification that you have the information for the crystal loaded. Do not edit this here but edit the crystal file instead with **xtalmgr. Directory** is where the program looks for files. The **Derivative File** is a solution file where the heavy-atom parameters are saved. You can **Load** and **Save** the solutions files. The save button is a menu button that gives you the option of saving a single derivative (the one that is highlighted in list) or all of them, or outputting all of them in a PHASIT (a Bill Furey program from his PHASES package) format file. Several derivatives can be put into a single solution by loading their individual solution files and then saving all of them in a new file. **Output Phases** is the name of a file that will contain the output phases. Next is the list of **Derivatives,** where a single derivative can be selected by clicking on and highlighting it. Below are three buttons for manipulating the list. Use **New** to start a list from scratch. **Edit** is used to enter/replace parameters and sites for the selected derivative. Selecting it pops up the edit window (see below) and saves the current data in a buffer. **Delete** deletes the currently selected derivative.

With **Method** you can choose, in the order they are meant to be used, **Correlation Refinement, Calculate Protein Phases,** or **Map coefficients.** Correlation refinement is unique to this program but has proved very robust and successful. It is independent of scale factor so that it is not necessary to start with a close approximation of the occupancy as in least-squares methods. Since it is a search method, it finds the global minimum (within the search radius) and does not get stuck in a local minimum. (For the true global minimum use xhercules, which searches the entire unit cell using the same correlation function). Protein phases are calculated in a several-step procedure. First, the single isomorphous replacement phases are computed for each derivative. These are then gathered together to get a rough estimate of the protein phase. Fc is then rescaled using the protein phase to account for the extra scattering power of the heavy atom (i.e., the scale should be greater than 1.0—for a well-occupied derivative it is from 1.05 to 1.10). Better error coefficients are then calculated, these are used to rephase each derivative, and then the final protein phases are gathered from these. The method is similiar to lack-of-closure error refinement except that the only parameters changed are the errors and the scales. The map coefficients should be calculated for a selected derivative after the phases are calculated. The map coefficients option creates a phase file for the selected derivative with the coeficients: h, k, l, $|FP-FPH|$, $(FPH-FP) - FHcalc$, α protein. Use xfft with the Fourier type Fo to make a heavy-atom difference map or Fo-Fc to make a residual map. These are used to look for new sites. Finally, the **Apply** button is used to start one of the methods and **Abort** is used to stop it in the middle if something goes wrong.

   Solution Editor.   Solutions can be read form xpatpred or entered with the editor. The derivative picked on the list of derivatives is edited when the **Edit** button is pushed. First the **derivative** title is entered. A fin or df file containing the difference information must be specified in the **DataFile.** Tell the program which one it is with **File Type.** Additionally, you can specify which of the columns in the df file contains the phasing pair. One of three **phase types** is specified next, **isomorphous, isomorphous anamolous** (must use a df file), and **native anamolous.** As of this writing the **native anomalous** option does not work. You can downweight this derivative relative to other derivatives with the **Weight** slider. There are two **resolution limits** between which data is used. **Sigma Cut** is used to remove small weakly observed reflections. If either of a pair of a measurements is below this value, the pair is not considered. Then comes a filter for removing ridiculous differences by comparing the value $(|F - F2|/(F1 + F2)/2) * 100$. If this is greater than the value in the filter, then the reflection is rejected. A value of 100 is about right for most derivtives. Derivatives with large differences can use a larger value of about 130. Below this is the list of heavy-atom **Sites.** They are edited with the controls that follow. Start the list by first entering the values you want and then select **Insert.** Insert will insert at the end of the list if nothing is selected, otherwise it inserts after the selection (however, never select a line in the middle of entering values because it will overwrite what you have just done!). To edit, first highlight the atom in the list you want to change, edit the values, and then select **Replace.** To delete an atom, select it and then use **Delete.** Next to the **Atom** type is a little pull-down menu button (with the triangle) that lists all of the available atom types. It is better to use the pull-down menu and select one than to type it in and risk typing errors. The choice of atom type is non-critical in the isomorphous and native anomalous case because it mainly just changes the scale factor. It does matter when using the isomorphous anomalous phasing calculations because it sets the ratio of f'' to f'. If your atom does not exist, choose the closest one. See the following for more information about adding new scattering factors. When you are done editing, you can either push **Apply** to save the edits or close the window. (Although this may be window-manager dependent, apply is safer.) If you mess things up, you can recover to the last saved position with **Reset.**

Verbose output is given in the message window. Do not forget that this can be saved using the **MENU** button in this window to pull up the **FILE** submenu.

Notes.   This program is under active development. As mentioned, the native anomalous phasing has yet to be implemented; the function is there to remind me of my shortcomings and to tease users. The program has been used to solve several structures. In particular, the correlation refinement works very well. To date, no one has trusted it as the only source of phasing, however. So far, everyone has used the PHASIT output option to confirm the phases with that program. Comparison of maps shows very little difference between the phases calculated, but the calculation is so important that no one was willing to rely on the newer program. Also, PHASIT prints more statistics, which I promise to add in some future version. Right now, one has to rely on Rcentric and figure-of-merit as guideposts. Adding new scatterers is cumbersome because the program has to be recompiled. I should add all of the elements (which I will do if someone has an online listing). The file sfcalc.h in $XTALVIEWHOME/xheavy is where they are saved. There is a listing of atom names followed by the coefficients as found in Vol. IV of the "International Tables for Crystallography" in Table 2.2B on p. 99. When you see the density of this table, I think you will understand my reluctance to type all of the data in! Follow these steps to add a scattering type:

1. Increase the value of MAXFTABLE by 1 in its #define line.
2. Add the key for this scattering type in the atype list. It must match the type you are going to use to identify it in the edit window.

3. Add each of the coefficients in Table 2.2B to the end of the declaration list. These operate in a manner analogous to a FORTRAN data statement. There are nine arrays, *a*1, *a*2, *a*3, *a*4, *b*1, *b*2, *b*3, *b*4, *co* which correspond to the table entries of the same name.

4. Add the anomalous scattering values to the arrays frano and fiano. Find these in Table 2.3.1, p. 149 of the same volume. The entry $\Delta f'$ for the wavelength of X-rays you use (probably CuKα) is put at the end of the frano list and $\Delta f''$ at the end of the fiano list.

5. Remove the old object files (rm *.o) and type `make` (on non-SUN systems it is usually necessary to source `../Setenv` before compiling).

6. Confirm the installation by first rechecking all of the values you entered a second time and by starting the new xheavy and confirming that the list in the edit pop-up next to the atom type has your new type.

Xmerge—Merge and Scale Two Data Files

Usage. `xmerge data1.fin data2.fin`

data2.fin is merged and scaled to data1.fin. The data in file 2 is scaled to match that in file1. The data in data1.fin are referred to as f1 and those in data2.fin are f2.

Crystal info needed. Cell.

Method. The data is scaled into *n* bins based on $\sin(\theta)/\lambda^2$. Two scaling methods are available: single and anisotropic. In single scaling a single parameter is used to scale the data in each bin such that

$$\text{sum}(f1) = \text{sum}(f2) * \text{scale}.$$

If Bijvoet pairs exist, they are averaged for scaling purposes.

In anisotropic scaling, first single parameter scaling is done and then a six-parameter scaling such that the scaling parameter is a function of the three crystal indices. This helps minimize errors due to differential absorption, etc. The formula used to for the scale factor *s* at each *h, k, l* is

$$s = h * h * a11 + k * k * a22 + l * l * a33$$
$$+ h * k * a12 + h * l * a13 + k * l * a23,$$

where *a* is found by a least-squares fitting procedure. The fitting procedure may fail if too few data are used. In this case, you have two choices: decrease the number of bins to put more data in each bin or use single-parameter scaling.

Fin File 1. The "standard" data set, as the second fin file is scaled to this one. Normally use your best native data here.

Fin File 2.   This is the fin file containing the data to be scaled to fin file 1, normally derivative or mutant data.

Output File.   This is the merged and scaled data. The name should indicate which data were used to make this file. (See the example below.)

Output Type.   You can output your data in either fin or df format. fin format merges $f1$ and $f2$ (Bijvoets $= f+$ and $f-$) in the input fin file so that the output fin file also has only two fields. All fields are preserved in a df file so that Bijvoet information is preserved for both input files. Since df files can always be turned into a fin file with xdftofin, this is the more useful of the two options.

Output.   Reflections in common means that reflection must be in both file 1 and file 2. **All reflections** are all of those in file 1 and those in common in file 2. If a reflection is in file 2 but not in file 1 it is not used or output.

Number of Bins.   The data are divided into bins based on $\sin(\theta)/\lambda^2$. This slider sets the number of bins with 10 as the default. Note that dividing by the square does not put equal numbers of reflections into the bins as would $\sin(\theta)/\lambda^3$ but provides a better distribution in reciprocal space. If there are too few reflections in a bin, the number of bins should be decreased to put more reflections in each bin. For anisotropic scaling, about 500 reflections per bin are good, with 50 being a good minimum. Too many bins will decrease the signal by scaling away any differences, and too few may not adequately scale the two data sets. Note: On sliders, if you edit the number with the keyboard, be sure to hit $\langle CR \rangle$ to register the new value.

Sigma Cut.   The data are excluded from the scaling equations if their amplitudes are less than sigma cut times the sigma of the amplitude. Usually this makes little difference because these data are weak anyway.

Scaling Type.   This sets the scaling method used, as explained previously. Anisotropic scaling is actually single scaling followed by anisotropic scaling. If there are too few reflections for anisotropic scaling, you will get errors. In this case, decrease the number of bins or use single scaling.

Graph.   Two results are graphed in the graphics window: delta, the $|f1 - s * f2|$ (dashed line) and the R-factor $|f1 - f2 * s|/f1$ (solid line) for each shell. For a heavy-atom derivative, the delta should start high and decrease with resolution. If it starts to increase, this may mean that the derivative is non-isomorphous past this resolution. The R-factor should change smoothly with each bin. If not, then use fewer bins.

History File.   The history file contains, in addition to the normal information, a list of the scale, r-factor, absolute delta for each resolution bin.

Examples.   To scale a native data set to a derivative data, set

```
xmerge ccpnat4.fin ccppt1.fin ccpnat4pt1.df
```

You can use this file as input to xfft to make a Patterson map, which can be contoured with xcontur.

Advanced.   In some instances you may wish to scale data in a single fin file to itself, for example, if you wish to scale Bijvoet pairs. In this case you must first split the file and then remerge and scale. For example, to scale unscaled.fin to scaled.fin (remember a fin file is records of $h, k, l, f+, s(f+)$, $f-, s(f-)$), use the following commands:

```
awk '{print $1,$2,$3,$4,$5,$4,$5}' unscaled.fin
     > unscaled_fp.fin
awk '{print $1,$2,$3,$6,$7,$6,$7}' unscaled.fin
     > unscaled_fm.fin
xmerge unscaled_fp.fin unscaled_fm.fin scaled.fin
```

xmergephs—Merge Fin and Phase File. Phase Derivatives and Mutants

Usage.   `xmergephs fin_file phs_file`
The data in fin_file are phased with the phases in phs_file, and the resultant structure factors are written out in another phase file. The phase file can be a .phs file, an XPLOR phase file, or a TNT *hkl* file. Options are available for using isomorphous or Bijvoet differences.

Quit.   This is the normal quit button.

Directory.   This is the directory.

Fin File to Phase.   This is the fin file to be phased ($h\ k\ l\ f1\ \sigma(f1)\ f2$ $\sigma(f2)$).

Phase File.   This is the file with phases to be used for phasing the fin file. It can be a .phs file, an XPLOR format file with the entry FCALC=, or a TNT *hkl* format file. The phase is normally expected to be in degrees, although if you use xfft in the next step, radians are satisfactory.

Output Phase File. The output is written to this file. The $f1$ and $f2$ fields in the fin file are written out $h\ k\ l\ f1\ f2$ phi unless otherwise specified (see below).

Swap f1 and f2 (Isomorphous Fourier). If $f1$ is the native FP in your fin file and $f2$ is the derivative FPH, then $f1$ and $f2$ should be swapped so that the output is $h\ k\ l$ FPH FP phi, and then you should make an Fo-Fc Fourier map, which will show the positions of the heavy atoms if protein phases are used (double-difference Fourier).

You may also want to swap mutant data that have been merged with native data to make a $F_{mutant} - F_{native}$ difference Fourier.

Add 90 to Phase (Bijvoet Fourier). If the fin file contains Bijvoet differences ($h\ k\ l\ f+\ \sigma(f+)\ f-\ \sigma(f-)$) and the input phases were in degrees, then a Bijvoet difference Fourier can be made by adding 90 to the phase and swapping $f1$ and $f2$. Centric reflections will be handled correctly if they have 0.0 in the $f-$ position. If you have used XtalView the whole way and started with a XENGEN mu file, then they will be correct. If $f+$ is equal to $f-$, then the reflections will be output, but since their difference is zero, they will not contribute to the Fourier.

Merge. Click here when you are ready to go. The program can take quite a long time if the two input files have a different sort order because the program searches for matches in indices between the two files. Note that if you have equivalent indices that are not numerically the same (e.g. $h\ k\ l$ and $-h\ -k\ l$), the merge will fail!

Bugs. The aforementioned problem that occurs if the indices are equivalent but not identical causes the program not to find a match. This can be fixed by using a routine like stdref and may be incorporated in future versions (especially if there are complaints).

xpatpred—Predict Patterson Peaks from a List of Sites

xpatpred allows you to enter and edit a set of sites and output these in a form displayable by xcontur. The sites can be written out in a solution file and refined using xheavy. Different origin choices can be selected, allowing a rapid way to try all possible origins and to choose the correct one visually.

Usage. `xpatpred solutionfile`

Crystal. The symmetry operators need to be present in the crystal data file in the symm record.

Prediction File. This file is a list of labels that can be read into xcontur with the **load labels** option under files. The labels are in fractional coordinates in u, v, w. The label indicates which sites it is derived from in its name. (i.e., pt1-pt1 is a self-vector of the site pt1, and pt1-pt3 is a cross-vector between pt1 and pt3). The labels are transformed so that they are between 0 and 1. They are generated by looping through all symmetry operators in a pairwise fashion. It is not guaranteed that all possible symmetry-related labels will be generated, so if a peak is not labeled, a symmetry-related one should be checked.

Solution File. The solution file can be loaded or saved. The solution file is a unique format to XtalView that is reminiscent of the pdb file in that each record starts with a key word. The solution files can be shared with xheavy in order to refine the heavy-atom sites. When a solution is written out, any origin choice is added to it, and when it is read back in, the origin is reset to 0, 0, 0. Coordinates are fractional in solution files.

Site List. The sites are listed in a scrolling list. Clicking on a site selects it and places its values in the fields below, where it can be edited. **Insert** takes the information in the individual fields and places it in the scrolling list at the end. **Replace** replaces the currently highlighted line with the data in the fields. Be careful to select a line to be replaced first and then edit it (if you forget, you can insert a new line and then delete the line it was supposed to replace.) **Delete** removes the highlighted line—the data are lost forever. (Solutions can often be saved, however.) Each line contains the atom label, x, y, z, origin choice, atom-type, occupancy, and B-value. There are eight origin choices of adding 0.5 or not to x, y, z. B-value and occupancy are not directly used in this program.

Label, x, y, z, B-Value, Occupancy, Atom-Type. These are text fields that can be used to change or enter data. To replace data in the scrolling list, select the line first, edit it, and then replace it.

Origin. This is a pull-down list of eight choices. First select the line to be changed, then select the origin, and then replace the line. To see the effect in xcontur, write out the prediction and reread it in xcontur. Usually this is used to find the relative origin of a second choice while holding the first constant. Self-vectors will be unaffected by an origin choice, but the cross-vectors are dependent upon the relative origin choice. With two sites, only four of the choices will make a difference, the other two being hand choices and not detectable with a Patterson map. With three or more sites, all eight choices may make a different pattern of cross-peaks.

Not all the origin choices are valid in every space group. The possible origins are listed in the "International Tables for Crystallography." The origin choices given are valid for orthorhombic space groups. However, if an incorrect origin is chosen, either it will make no difference (e.g., adding 0.5 to y in monoclinic makes no difference) or it will cause the self-vector not to match and so will be detectable. Other origin choices will have to be entered by hand, unless they are added to a future version.

Xprepfin—Import/Export Fin Files from/to Other Formats and Reformat Fin Files

Usage. `xprepfin in out`

Quit. Use this as usual.

Directory. This is the directory.

Input File. The input file can be either a fin or XENGEN mulist.

Input Format. Set this to the correct type.

Use Data. There six options as to how to switch the input to make the output:
As is: No change to input.
Avg f1 and f2: If $f1$ and $f2$ exist, then they are averaged; if one is missing, then the output is set to the other one.
F1 then F2: If $F1$ exists, then use it first otherwise take $F2$.
F2 then F1: If $F2$ exists, then use it first otherwise take $F1$.
F1: Always use $F1$ if it exists.
F2: Always use $F2$ if it exists.

Switch $F1$ and $F2$. Select **SWITCH** to do this. This operation is done after Use Data.

Reduce. The data indices are reduced to produce a unique value, the reflections are sorted on h, k, l; and any duplicates are merged. Unfortunately, the algorithm used does not always put the data in the quadrant you desire.

Data Are. This controls how some operations behave. If the data are Bijvoet pairs, then when the indices are switched it may be necessary to reverse the Bijvoet pair. Also, when a centric reflection is output, Bijvoet pairs are averaged and placed in the first field, and the second field set to zero since centrics do not have an anomalous scattering signal. Obviously, this should not be done for isomorphous-pair data.

Output Format.   There are three choices. Pick the one you want.

Output File.   The name of the output file goes in here.

Apply.   When you are ready click here.

Uses.   This is the main point for entering data into XtalView. It is really set up for XENGEN mulist files (the output of makemu $-f$). The UCSD multiwire already uses .fin files.

Bugs.   There should be more formats supported, but since I do not know about them yet, they have not been added—yet.

Xresflt—Resolution Filter

Usage. `xresflt [file filtered file]`
Any file starting with records beginning with *h k l* can be filtered. The proper crystal must be entered. The data between the two limits are output.

Xrspace—Reciprocal Space Viewer for Examining Completeness of Data, Differences between Data Sets and Intensity Patterns

Usage. `xrspace [file]`

Environment.   Xrspace reads the environment variables CRYSTAL and CRYSTALDATA. If CRYSTAL is present, then this is used to set the crystal type, or else it can be manually entered after the program starts. CRYSTALDATA is the default directory containing the information file of CRYSTAL. When looking for the information file of the crystal type, it first checks the current directory and then $CRYSTALDATA/crystal. See the XtalView documentation (if any) for more details. It works best on a color display. It does work on a black and white, but needs improvement.

Database Requirements.   Xrspace looks for the keywords cell and symm in the database/info file. If cell is present, it is read and placed in the Unit Cell field. It can be manually overwritten if desired. The symmetry operators are read from the line symm, which should be entered in the format used in the "International Tables," Vol. I. Xrspace converts these symmetry operators to Patterson symmetry by adding a center of symmetry.

Example:

cell 48.5 76.3 92.1 90.0 104.0 90.0

symm $x, y, z; -x, y + 1/2, z;$

Operation. The application presents a window that is divided into three main areas: the control panel, a message window, and a canvas where the data is displayed as specified by the control panel settings. The control panel is meant to be read from top to bottom. Messages are printed in the textpane at the bottom. Note that the textpane has the ability to save output by means of its menu. The data are displayed on a color display coded by a heat scale to represent intensity—as the data increases in intensity, it goes from black to red to yellow to white. The width of a spot is proportional to F and its area is proportional to I.

Quit. Click to quit. You get one chance to change your mind. This button is not OPEN LOOK approved, but I like it anyway.

Notebook. This is a hopeful-looking button that currently does nothing. Suggestions would be appreciated for what it should do. Actually, it should be a defaults button, but I have not got around to fixing this yet.

Crystal. Check that the correct crystal is entered here. A corresponding database file should be in the directory CRYSTALDATA—typically ~/xtal_info. If it needs to be changed, click on this field to move the cursor here and type in the new crystal and hit ⟨CR⟩. If at any time you want to force reading the crystal file (for instance you just edited it and fixed something), hit ⟨CR⟩.

Unit Cell. The proper unit cell should be displayed here. You can manually type in any cell and override the one found in the data base file.

Directory. This is the current directory.

File. This is the current input file from the command line or manually entered. Use this also for cut-and-paste work.

Read. Clicking this button causes the file to be input.

Type Setting. Click on the appropriate file type: .fin = fin file (*ih*, *ik*, *il*, $f1$, $s1$, $f2$, $s2$), .mu = XENGEN mulist file from makemu command, .df =

"double fin file" (ih, ik, il, $f1+$, $s1+$, $f1-$, $s1-$, $f2+$, $s2+$, $f2-$, $s2-$), and .urf = urefls XENGEN file from integrate.

Color. This is no longer operative. Because of color limitations, it proved impractical to use dynamic color maps, and we are still learning how to make this button work with a static color map. Do not give up hope yet.

Display Dataset. This switches between $f1$ and $f2$ or $|f1 - f2|$ (Δ). It is useful for looking at merged heavy-atom datasets or for anomalous differences.

Planes of. This sets the sectioning direction. You can use the three principal directions or the major diagonals.

Level. This slider sets the plane level (i.e., if the plane direction is h and the level is 0 then the $0kl$ plane is displayed).

Slab. This sets the number of planes that will be displayed. Normally they are drawn on top of each other, but see Offset of Vertical Axis.

Spot Size. This scales the size of all the spots up and down. Set this so that spots do not overlap.

Resolution. This controls a ring which is drawn at the specified resolution. It does not control the maximum resolution of the screen; that is set by the limits of the input data.

Symm Setting. This controls whether symmetry is used to generate symmetry-related spots or whether only the unique data are displayed.

Offset of Vertical Axis. This lets you offset planes from each other so that they are not drawn on top of each other. Try $x = 2$ and $y = 3$ and make the spot size smaller for an orthographic display of several planes (slab > 1).

Draw. Redraw the screen. After first reading the data, this must be clicked. Most of the other buttons and sliders redraw automatically. Pay attention to the message at the bottom of the window (left footer). If it does not say "Ready for new input," just wait patiently; the screen is redrawing.

General. Clicking on a spot will display its indices and intensities in the message window. These are the unreduced indices. Do not get too far ahead of the program with mouse input. Look at the footer to see if you are going too fast.

Uses.

Looking for symmetry.

Checking systematic absences. Reduce your data in an analogous space group that does not have the absences (i.e., use P222 for $P2_12_12_1$).

Looking at heavy atom differences. Are they uniform throughout space or are there just a few large differences around the beam stop? Use the delta function after running xmerge.

Comparing two merged native-data sets. Where are the differences?

Planning data collection strategies. After a run (or part of a run) has been integrated, you can look to see where the data are. Turn off symm, and you can see the edge of the Ewald sphere when looking down the axis around which you are collecting data. Use the diagonals to see if a diagonal looks better. The axis down which the data have the narrowest range is probably the one closest to which you are collecting data. Experiment: Integrate frames 1 thru 50 and run `xrspace *.urf`; then integrate frames 50 100 and run a second `xrspace *.urf`. This shows you the direction in which data are collected. If you want the h axis and you are going away from it, go the other way!

Finally, use snapshot to do a screen dump. And then do `ras2ps -C snapshot.rs > snapshot.ps; lpr -Pcolor -s snapshot.ps` to get a beautiful color print of your data on the QMS100 (at least at Scripps MB). Careful, the color printer is expensive.

General. Used over the network, xrspace normally runs on the machine you want to display it on. However, there may be times when you want to run it on one machine and display it on the machine you are actually looking at. First type `xhost +` somewhere on the machine you are using to display on. Then type `setenv DISPLAY display_hostname:0` on the remote machine and then run xrspace. Or you can use `xrspace -display hostname:0 [file]`. Xrspace accepts all the standard X windows command-line arguments. There are many and they can be used to control such things as fonts and where the window is displayed. See Volume 3 of the X Window System books.[2]

Bugs. Drag-n-drop does not work.

It needs color, although it can work well on black and white.

When a file is loaded, the display should redraw—the draw button must be used.

[2] O'Reilly, T., Quercia, V., and Lamb, L. (1988). "The Definitive Guides to the X Window System. Vol. 3: X Window System User's Guide." O'Reilly Associates Inc., Sebastopol, CA.

Check the input queue for new instructions and quit drawing if the user has changed one of the controls. (Private note to author: Hint: use XPeekEvent()).

Please let the author know of any suspected bugs. Also, if you have a useful feature, let him know, and the next version may include it!

Xtalmgr—Top-level Application for XtalView

Usage. `xtalmgr`

Xtalmgr is used to start the other XtalView applications. It allows creating, editing, and deleting of crystals. Projects can be organized using the project feature, and all data related to the same project can be grouped into a project. File filters are used to help the user weed through the usual forest of files. Autonaming can be used to make up a sensible name for an output file from one or two input files. A history of commands is kept to simplify repetitive actions and to keep track of what you have done.

Environment Variables. Xtalmgr needs the environment variable XTALVIEWHOME and CRYSTALDATA. It sets the variables PWD and CRYSTAL as it operates.

Main Window

Quit. Quits after asking if that is what you really want.

Project. The name/title/label of the current project is listed here. This is followed by an **Edit** button that pops up the project edit window. A pull-down setting button is at the end of the line and lets you set the current project from the list of all defined projects.

Crystal. This is the current crystal. Use the **Edit** button to edit the crystal database file. Use the pull-down setting button to set the crystal from the list of defined crystals.

Directory. This is the current directory. Entering ⟨CR⟩ will change to that directory and update the file listings.

Applications. This is a pull-down setting button that contains all the window-based applications available. Setting one of these updates the command line and the file filters. If a file name can be added on the command line, the appropriate filter will be set, and all the files matching it will be listed. If no file argument is needed, then the filter is left blank and no files are listed.

Utilities. These are non-window-based programs that can be used as filters and to provide some functions not available in the applications. All the options should be specified on the command line because no other method is provided for entering data.

Command. The current command line is displayed here. XtalView applications are set up so that the command arguments are almost always optional filenames for the input(s) and output. It follows all the usual UNIX conventions. This can be automatically generated by the sequence: (1) Select an application. (2) Select input and output arguments. (3) Add Args. (4) Run Command. Sometimes editing this line is the quickest way to enter a command. The command can also be set by picking a line in the history window.

List Files. Pushing this button causes a listing of all the files in the current directory using the file filters.

Auto Name Output. A name for the output file will be made up if this button is pushed based on the input arguments. If one input argument is specified, the name is made by substituting the default extension for the output with the input. If two have been specified, then the name is made concatenating the two input files without extension and adding the output extension.

Add Args. This adds the arguments to the command line by concatenating to the end and adding a semicolon.

Run Command. This causes the command line to be sent to the operating system and executed. Errors will be output in the original window used to start Xtalmgr.

History . . . . This pops up the history window, which is a scrolling list of commands previously entered. If one is selected, it will be placed on the command line for editing/rerunning.

Input Argument 1. This is the current value of the first input argument. It can be set by selecting a file name in the list below it.

Input Argument 2. This is the current value of the second input argument. It can be set by selecting a file name in the list below it. If more than two input arguments are desired, they can be added here by entering from the keyboard or directly on the command line.

Output argument.   This is the current value of the output argument. It can be set by selecting a file name in the list below it.

Filter.   This is the filter that will be used in listing the files in the scrolling list below. To list all files, set this to "*". The filters are automatically set when an application is chosen.

Project Editor Pop-Up.

Project.   This is an arbitrary name useful to you in identifying a project.

Home.   This is the name of the directory where this project will live.

Crystal.   This is the default crystal for this project. It can be changed later if more than one crystal is being used in a single project.

Replace.   This button replaces the current project with the information in the edit window.

Create.   This creates a new project and adds it to the list.

Delete.   The current project is deleted from the list. There is no undo.
Note. The project list is recorded to disk after all operations so that if Xtalmgr should stop suddenly or if the system crashes, the list will be current upon restarting Xtalmgr.

Crystal Editor Pop-Up

Crystal.   This is the current name of the crystal. This is used as a file name and so it cannot contain a space, *must not* begin with a dash, and should not contain special characters such as $><, :;]$ [{ } ( ) & * $|?~$. On some operating systems it is possible that the name is limited to maximum length.

Title.   This should be a title for this crystal that makes sense to you, such as "e. coli. miraclease E.C. 1.31.29.2 P212121 form," or something equally imaginative and informative. This is used for your information only.

Unit Cell.   This consists of six floats describing the unit cell in ångstroms and degrees. The order is *a, b, c,* α, β, γ. No assumptions are made, enter all six values. A note on accuracy: Whatever accuracy you enter here will be preserved; although later applications will report the cell to hundredths accuracy, they will actually use the accuracy entered here.

Space Group.   The name of the space group is entered here. If ⟨CR⟩ is entered on this line, a search through the data base of all 230 spacegroups is made. The name should then be in lower case, and subscripts should be entered as simple numbers; i.e., $P2_12_12_1$ would be entered as p212121. If the search is successful, then the symmetry operators will be updated.

Space Group #.   This is used to search for a space group based on the entry number in the "International Tables for Crystallography." All 230 spacegroups are in a table. If you enter a value with the keyboard, be sure to press ⟨CR⟩ to enter it.

Find Space Group by Number.   Press this button when the Space Group # file is set to find the symmetry operators and name for this number space group.

Symmetries.   The symmetry operators are listed here. It is recommended that you use search by number or name to enter this line for the sake of accuracy and formatting. The format is each symmop in lower case as listed in the "International Tables for Crystallography" separated by semicolons and ending with a period (i.e., *x, y, z; −x, y, −z.*). The unitary operator *x, y, z* is always included.

Other Fields.   These are fields that are used for special purposes. You can enter information for your own use or if you can link the library routines for XtalView into your own program to access this information. Each line is a keyword followed by data. See the "XtalView Programmer's Manual" for further information.

Keyword.   This is the keyword for this line.

Data.   These are the data for this line. There cannot be a new line character in the line or it will be split into two lines when written out.

Replace Field.   This replaces the selected line in the scrolling list with the data in the fields **keyword** and **data**.

Create Field. A new line is appended to the scrolling list.

Delete Field. You will never be bothered by the selected line again. There probably is little reason to use this as old fields never hurt. (The first field encountered with the correct keyword is used to extract the data in question; if a field with the same keyword occurs later, it will be ignored.)

Scrolling List. This is the list of all of the records that do not match the special ones discussed previously (title, unit cell, space group, symmetries). Selecting one with the mouse puts its information into the **keyword** and **data** fields.

Update This Crystal. All the information in the window is written into the crystal file named in Crystal. If the crystal file does not currently exist, it is created.

Files. Three files are used to store information for xtalmgr. The data on applications are stored in $XTALVIEWHOME/xtalmgr/applications. This is not meant to be user edited. New applications are installed by entering in this file in a special format. The project data are stored in the file $CRYSTALDATA/projects. They can be edited using the project editor window, but in times of crises or for reordering the lines, it can be edited. The first lines give a description of the file format. Crystal information is kept in the file $CRYSTALDATA/crystals. It is a simple list of crystal file names. All the file names should actually exist in $CRYSTALDATA. The only way to delete a crystal from the list is with an editor. You can delete the crystal file is you wish—however, there is no need to do so, and it may turn out to be useful later.

Non-Window-Based Applications

ColorbyB—Prepare a vu File from a pdb File Colored by B-value

Usage. `colorbyB pdbfile xfit.vu [-normalize | -n]`
The input pdb file is converted to colored lines using the same routines as in Xfit except that the lines are colored by B-value. If the -normalize flag is present, then the B-values are normalized before coloring. This is highly recommended for comparison purposes as it scales out any effect of the overall B.
Unnormalized B's are colored in intervals of five on B-value from cool to hot with the following color scheme:

0–5	cyan
5–10	blue
10–15	orchid
15–20	pink
20–25	red
25–30	gold
30–50	yellow
50–	white

Normalized B's are color-based on the sigma deviation from the mean using the following color scheme:

< −2.0	cyan
-2.0– −1.0	blue
-1.0– −0.5	orchid
-0.5– 0.0	pink
0.0– 0.5	red
0.5– 1.0	gold
1.0– 2.0	yellow
> 2.0	white

The B-value coloring can be very useful for comparing mutations with the native structure. This same coloring can be done in Xfit under View.

Cvtpdb—Convert a pdb File from Cartesian to Fractional Coordinates and vice versa and Coordinate Transformations

Usage. `cvtpdb crystal [-f| -c | -t | -r | -e] < file.pdb > newfile.pdb`

−f. Convert from fractional to Cartesian.

−c. Convert from Cartesian to fractional.

−t vx vy vz. Translate by the amount specified in *vx vy vz*.

−r r11 r12 r13 r21 r22 r23 r31 r32 r33. Rotate by matrix:

$$\begin{pmatrix} r11 & r12 & r13 \\ r21 & r22 & r23 \\ r31 & r32 & r33 \end{pmatrix} \begin{pmatrix} x \\ y \\ z \end{pmatrix}$$

−e θ1 θ2 θ3 = rotate by the Eularian angles θ1, θ2, θ3.

Note the non-standard use of crystal on the command line instead of reading from the environment; this forces you to think since it is critical for correct operation. Converting to fractional is especially useful when applying crystallographic transformations. After transformation, convert back to Cartesian. You can pipe the output of one cvtpdb into another to chain operations together. For example, to generate the second symmetry related molecule in $P2_1$ (-2, $y + 1/2$, $-z$) and translate it to the unit cell at $+1, 0, -1$, the command is

```
cvtpdb crystal -f < molecule1.pdb |\      !to fractional
cvtpdb crystal -r -1 0 0  0 1 0  0 0 -1 |\ !rotate by
    -x,y,-z
cvtpdb crystal -t 0 .5 0 |\      !translate by x,
    y+1/2, z)
cvtpdb crystal -t 1 0 1 |\      !translate by +1, 0 -1
cvtpdb crystal -c > molecule2.pdb    !convert back to
    Cartesian
```

Such a complicated command is best done in a shell file and is very handy for generating packing diagrams. Do not enter the stuff after the "!" It is for my comments.

Cvtvu—Convert a vu File from Cartesian to Fractional Coordinates and vice versa

Usage. cvtvu crystal [-f |-c] < file.vu > newfile.vu

− f. Convert from fractional to Cartesian
− c. Convert from Cartesian to fractional

Cvtxyz—Convert x, y, z from Cartesian to Fractional Coordinates and vice versa

Usage. cvtxyz [-f | -c] [x y z] |[< file.xyz > newfile.xyz]

− f. Convert from fractional to Cartesian
− c. Convert from Cartesian to fractional
CRYSTAL must be set in environment to the correct crystal.

If x, y, z are given, then these are converted and the output is reported to stdout, otherwise stdin is converted to stdout. The input is expected to be three floats on separate lines. An example of cvt*xyz* would be to use it to convert from fractional bounds read from xcontur to Xfrodomap, which expects the bounds in Cartesian coordinates.

Deh—Remove hydrogens from a pdb file

Usage. `deh < pdb_with_hydrogens`
`> pdb_without_hydrogens`
The output of XPLOR contains hydrogen atoms at some positions, and often it is desirable to get rid of them. A hydrogen is decided on if either column 12 or 13 is "H." This also will remove HG + 2. Does anyone know of a more reliable way to find hydrogens? (Note: Xfit handles hydrogens from XPLOR, and if you leave them on, you will not have to rerun generate.inp.)

Dumpphasit—Reformat PHASIT Format Output Files to XtalView

Usage. `dumpphasit phasit file > xtalview.phs`
PHASIT is Bill Furey's (University of Pittsburgh) phasing program from his PHASES package.

Deriv—Calculate the Derivative of a Difference Map at Each Atom and Output a vu File with Arrows Showing the Direction and Relative Magnitude

Usage. `deriv file.pdb refl.phs out.vu [resmin]`
`[resmax]`
The input is a PDB file (.pdb) and a phase file (.phs), and the output is a vu file suitable for xfit. If one resolution limit is given, then it is considered the minimum resolution. If two are given, data between them are used. If none is given, all data are used.

Purpose. This is the same function as is provided within Xfit. It calculates the derivative of the difference electron density of the two amplitudes given in the phase file. If the file contains F_{obs} and F_{calc}, this is $F_{obs} - F_{calc}$; another useful construct is $F_{mutant} - F_{native}$. An arrow is drawn at this position in the direction of the gradient of the difference map, with the length indicating the steepness normalized for the atom type (do not take these literally; they are only relative indicators). This is the direction a refinement program would move the atom in the first cycle if no stereochemistry is taken into account. Also, this is not very useful at resolutions far from atomic resolution (say below 2.5 Å). An offline version of this program is provided for those who want to use it for some other program (converting the vu file is their responsibility; remember it is $x1$ $y1$ $z1$ $x2$ $y2$ $z2$ color). If you want to do an entire molecule, this would take quite a long time and would stop Xfit while it was being done. Since Xfit does not provide a mechanism for saving the vectors it calculates, if you will be using them over several sessions, they should be precalculated with deriv.

Grinchbones—Convert a GRINCH ASCII Format File to a vu File for Displaying as a Background Object in Xfit

Usage. `grinchbones < grinch.ASCII > xfit.vu`
The input is a University of North Carolina, Chapel Hill Graphics Lab GRINCH ASCII format file (made with the command `ASCII save file-name` in interp or using the ASCII option in mkskel), and the output is an xfit vu file. The output uses the standard GRINCH color scheme for interpreted edges (i.e., main chain is green, side chain is violet, etc.), except that unknown edges are colored in bins of density from blue to red (the same as the default color scheme for the first map in Xfit). This can be used as a background guide for model building in Xfit. Contours can be used for close-up views, and detailed fitting and ridgelines can be used for larger views, allowing a larger portion of map to be viewed since ridgelines require fewer lines than contours.

The coloring scheme is close to that of GRINCH's except for unknown edges: M (main), green; B (bridge), brown; S (side), purple; O (carbonyl), red; F, cyan; G, yellow; R (residue), blue; U (unknown), 0–14 blue, 15–28 purple, >28 orchid.

Matrices—Get the XtalView Cartesian/Fractional Conversion Matrices for Use in Some Other Program

Usage. `matrices crystal`
This routine returns the Cartesian-to-fractional, and vice versa, matrices used by XtalView. This is provided because there is more than one way this can be done for non-orthogonal space groups (i.e., $a * bc$ versus $abc*$), and when importing/exporting coordinates, it may be useful to have these (as when you send your brand-new structure off to the databank). The output is self-explanatory. The matrices are meant to be the same as the conventions used by XPLOR and FRODO, but I give no guarantee.

Mu2fin—Convert a XENGEN mulist File to a XTALVIEW .fin File

Usage. `mu2fin < mulist > file.fin`
This converts a XENGEN mulist file to a .fin file. It reads the Bijvoet pair data from the mu file and ignores the averaged F that also appears in the mulist. Make the mulist with `makemu -f`. *The —f flag is a must.* This pro-

gram is actually not needed since Xprepfin does all this and prepares a history file, so it is recommended over mu2fin.

Resflt—Resolution Filter

Usage. `resflt limit1 [limit2] < file > filtered_file`
Limit1 and limit2 are in angstroms. This program can filter any ASCII file with *h, k, l* as the first three fields (.fin, .phs, .mu, etc.) based on *d*-spacing. If one limit is given, this is taken as the minimum and the other limit is set to 1000 Å. If both limits are given, then *d*-spacings between them are output. The line is copied to the output exactly as input; the only requirement is that *h, k,* and *l* appear first and are separated by a space or tab.

Stfact—Calculate Structure Factors from a PDB File

Usage. `stfact   pdbfile   refllist   output.phs [reslimit] [reslimit]`

Environment.   The value of CRYSTAL must be set to the correct crystal for the unit cell and space group (see Xtalmgr).
If one limit is given, it is considered the minimum resolution; if two are given, then the data between them are used; and if none is given, all input reflections are used. The input is a PDB file (.pdb) and a list of reflections in phase file format (.phs). The value of F_{obs} in the input is passed to the output, and F_{calc} ϕ can have any value, including zero (but not blank). The output is a new phase file, will be on an absolute scale (i.e. F_{obs} is scaled to F_{calc}), and is suitable for use in Xfit for making maps, including omit maps.

Purpose.   This is a completely general structure factor calculator. This is the same structure factor calculator as used in Xfit and is atom-based. Since this can be quite slow compared to an FFT structure factor, calculation in this program is provided for by doing the calculation offline (perhaps as the last step in a command file of a refinement job). If you are using XPLOR, it is faster to calculate the structure factors using XPLOR's FFT. If you use it as a command file, be sure to setenv CRYSTAL (e.g., `setenv CRYSTAL miraclase`) before running stfact.

Urf2xfit—Convert a XENGEN urefls File to a vu File for Display in xfit

Usage. `urf2xfit [urefls] [xfit.vu]`
If no files are given, then the program uses stdin and stdout. If one name is given, then that file is converted to stdout.

Environment. CRYSTAL is the crystal file. It is used to get the unit cell for scaling the axes. The present version does not handle non-orthogonal space groups correctly but always displays *h, k* and *l* as orthogonal axes. Still, this may be useful.

Purpose. By converting a urefls file to a vu file it can be displayed in three dimensions with Xfit. The three axes *h, k,* and *l* are displayed as well as a jack at the position of each reflection. Data that are missing become apparent as well as the curved nature of Ewald sphere, which is often neglected in data-collection strategies. Each run can be turned into a vu file and displayed in a different color (using the recoloring option in Xfit) to see how multiple runs overlap. The output is colored in four bins of intensity from minimum to maximum: blue, cyan, yellow, white. The color scheme needs expanding, and coloring by phi value would be useful.

Xplortophs—Reformat XPLOR Phase Files

Usage. `xplortophs [xplor.fcalc.file] [xtalview.phs]`
This program scans an XPLOR phase file and reformats it for Xtal-View. The XPLOR file is generally produced with the XPLOR script `fcalc.inp`.

•••••••••••••••
XTALVIEW PROGRAMMER'S GUIDE
Adding a Program to XtalView

Adding a program is done by editing the list of programs kept in $XTALVIEWHOME/xtalmgr/applications. To add a program, this file is edited and the new program added to the list. When Xtalmgr starts, it reads this list and displays these choices in the two pull-down menus: Applications and Utilities. Applications are expected to have an XView format icon associated with them and are expected to be window-based programs. Utilities are non-window based and run in a text window. A utility can be a simple filter such as a reformatter for importing/exporting data to non-XtalView programs. The new program must be in the user's path. On most systems this means $XTALVIEWHOME/bin is the logical place to put the program or is a symbolic link to the executable program. The line for the program contains other lines. The format for a window-based application in the file is

```
program icon-name arg1 arg2 arg3
```

Arg1–3 are put into the file filter lines in the three scrolling window lists in Xtalmgr and act as file filters. Commas are used to separate items for multiple arguments. For example:

```
xprepfin images/xprepfin.icon *.fin,*.df,*.mu — *.fin
```

The dash indicates that there are no arguments for the second list. The first field will contain all files in the current directory that match "*.fin *.df *.mu." Non-window applications are similar except that the image name is replaced with a short description of the format. Spaces in this description are replaced with commas, which are replaced by spaces when Xtalmgr loads the file.

Using the XtalView Database in Other Programs

The data in the crystal file can be accessed in other programs by linking in subroutines for this purpose in the lib directory. The program can be in either C or FORTRAN. To link, add the following to the end of the compile or load line: $XTALVIEWHOME/lib/crystlib.a. In order to pass messages from the lib routines, a routine emess is expected. This should consist of the following C subroutine:

```
#include <stdio.h>
int emess(message)
char*message;
{
   fprintf(stderr,''%s\n,''message);
}
```

Put this in a file and compile with cc-c emess.c. Then add emess.o to the compile line:

```
f77 -o program program.f emess.o $XTALVIEWHOME/lib/
crystlib.a
```

Fortran Subroutines (lib/f77subs.c)

```
xf_getcell(crystal, a, b, c, alpha, beta, gamma)
character *80 crystal
real*4 a, b, c, alpha, beta, gamma
```

This returns the unit-cell information for the crystal file Crystal. That is, it opens the file ./crystal if it exists. If not, it tries $CRYSTALDATA/crystal and reads the information on the line starting with cell.

```
xf_get_ctof(crystal,ctof,ftoc)
character *80 crystal
real*4 ctof(9), ftoc(9)
```

This returns the matrices for transforming coordinates from Cartesian, orthogonal to fractional (ctof), or vice versa (ftoc). The matrices are used:

```
xf = xc*ctof(1) + yc*ctof(2) + zc*ctof(3)
yf = xc*ctof(4) + yc*ctof(5) + zc*ctof(6)
zf = xc*ctof(7) + yc*ctof(8) + zc*ctof(9)
```

```
xf_getnequiv(crystal,nequiv)
character *80 crystal
integer*4 nequiv
```

This gets the number of symmetry operators (or equivalent positions) from the crystal file. It is meant to set up for the next routine xf_getsymmop(), which gets the character representation of the symmetry operator. The University of Pittsburgh routine translate can be used to read this line and is included in code written by B. C. Wang and Bill Furey. Other similiar codes may be found; otherwise, use xf_scansymmops(). Or you may want to use it just to list the symmetry operators for verification by the user.

```
xf_getsymmop(crystal,nop,symchar,ns)
character *80 crystal, symchar
integer*4 nop, ns
```

The ASCII representation of the symmetry operator found at the nops position is returned. ns is the length of the string symchar, which would be 80 in this case. This routine is called in turn for each symmetry operator (see xf_getnequiv). For example, if the crystal file, myprotein, contains the line

```
symm x,y,z; -x,y,-z.
```

and the call xf_getsymmop(myprotein, 2, symmop, 80) is made, symmop will contain " − x, y, − z" afterward.

```
xf_scansymmops(crystal, nops, symmops)
character *80 crystal
integer*4 nops
real*4 symmops(2304)
```

To access the iop symmetry operator at position irow, icol, use the formula

```
symmops(iop+192*((irow-1)+4*(icol-1)))
```

Thus, to find the equivalent x of the second symmetry operator,

```
x2 =    x * symmops(2+192*((1-1)+4*(1-1)))
     + y * symmops(2+192*((2-1)+4*(1-1)))
     + z * symmops(2+192*((3-1)+4*(1-1)))
     +     symmops(2+192*((4-1)+4*(1-1)))
```

or you can use xf_transform, which returns the transformed x, y, z of the symmetry operator number iop:

```
xf_transform(x, y, z, symmops, iop)
real*4 x, y, z
real*4 symmops(2304)
integer*4 iop


function integer xf_iscentric(h, k, l, symmops, nops)
integer *4 h, k, l
real*4 symmops(2304)
```

This function returns 0 if a reflection is acentric given the symmops that are from xf_scansymmops. If the reflection is centric, then it returns one of the allowed phases/15. The other allowed phase is $+180°$. Thus, if the reflection is centric and the allowed phases are 0 and 180, the function returns 180/15, or 12. Since xf_iscentric is an integer function but starts with the letter x, be sure to tell the compiler it is an integer with integer*4 xf_iscentric in the variables declaration part of the program.

Finally, to get some arbitrary data from the crystal file you can use

```
function integer xf_getdata(crystal,key,data,ndata)
character *80 crystal
character *20 key
character *1024 data
integer *4 ndata
data ndata/1024/
```

crystal is as in the preceding, key is the keyword of the data line you are searching for, and data is a string to contain the returned data. ndata tells the routine how long the string data can be before overflowing. This function returns 0 if unsuccessful and the length of the data if successful.

C Routines in crystlib.a

Many of the library routines expect a routine emess() to be available for sending messages. It should be of the form

```
int emess(mess)
char * mess;
{
    fprintf(stderr,''%s\n'',mess);
}
```

Alternatively, it can send the message to a window, as with xview `textsw_insert()` for textpanes.

```
    int get_cell(crystal, A, B, C, Alpha, Beta, Gamma,
textwin)
    float *A, *B, *C, *Alpha, *Beta, *Gamma,
    char * crystal;
    Xv_opaque textwin;
```

This puts cell from Crystal in A, B, C, α, β, γ. It returns 0 if unsuccessful, otherwise 1. Error messages go to the textpane `textwin`. If there is no textpane, use a null value for textwin. Non-xview programs should use `get_celln`.

```
int get_celln(crystal, A, B, C, Alpha, Beta, Gamma)
float *A, *B, *C, *Alpha, *Beta, *Gamma,
char * crystal;
```

```
char * get_data(crystal, key)
char * crystal;
char * key;
#include ''lib/Xguicryst.h''
```

`get_data` returns a pointer to the start of the data in the crystal file that matches key. It returns NULL if the a line starting with key cannot be found. The pointer is valid until the next call of `get_data`, which is also called by other XtalView routines, so you should copy the data if they are to be used later.

```
if(get_data(cystal,''cell'')) strcpy(save,
get_data(crystal,''symm''));
```

```
int scan_symmops2(crystal, symmops, opstring)
char * crystal;
symmops[3][4][MAXSYMMOPS];
char opstring[MAXSYMMOPS][MAXSTRING];
#include ''lib/Xguicryst.h''
```

Scan_symmops2 gets the symmetry operators for crystal and returns the number of symmetry operators. If unsuccessful, it sends messages to stderr and returns 0. symmops contains the matrices, and opstring contains the ASCII string equivalents. Even if you do not need opstring, you must allocate the string or the subroutine will crash for lack of space. Example of using the matrices:

```
x2 = x*symmops[0][0][2] + y*symmops[0][1][2] +
z*symmops[0][2][2] + symmops[0][3][2];
y2 = x*symmops[1][0][2] + y*symmops[1][1][2] +
z*symmops[1][2][2] + symmops[1][3][2];
z2 = x*symmops[2][0][2] + y*symmops[2][1][2] +
z*symmops[2][2][2] + symmops[2][3][2];
int addcenter2(n,symmops)
int n;
float symmops[3][4][MAXSYMMOPS];
```

addcenter2 changes the symmetry operators to the equivalent Patterson space group by adding a center of symmetry and negating all translations. The new number of symmetry operators is returned, and the starting number of symmops should be in n.

```
int get_ctof(ctof,crystal)
float ctof[3][3];
char *crystal;
```

This gets the Cartesian to fractional transformation matrix for the crystal. Usually, this is derived from the unit cell by the same algorithm as used by FRODO and XPLOR, in which case a 9 is returned. This matrix can be overridden if there is a line "ctof" in the crystal file, in which case the return value is 1. If 0 is returned, the operation failed. The most likely value for this is the wrong number of numbers on the ctof line. There should be 9. Usage of the matrix is

```
xf = x*ctof[0][0] + y*ctof[0][1] + z*ctof[0][2]
yf = x*ctof[1][0] + y*ctof[1][1] + z*ctof[1][2]
zf = x*ctof[2][0] + y*ctof[2][1] + z*ctof[2][2]
or use transform(ctof, &x, &y, &z).
```

To get the opposite matrix for fractional to cartesian, invert this matrix with uinv:

```
uinv(ctof,ftoc)
float ctof[3][3];
float ftoc[3][3];
```

```
int iscentric(h, symmops, nops)
int h[3], nops;
float symmops[3][4][MAXSYMMOPS];
#include ''lib/Xguicryst.h''
```

This routine returns 0 if a refection is acentric given the symmetry operators and the Miller indices in the vector h. If the reflection is centric, then it returns one of the allowed phases/15. The other allowed phase is $+180°$. Thus, if the reflection is centric and the allowed phases are 0 and 180, the function returns 180/15 or 12. A sister function that uses individual integers for the indices is

```
int iscentrichkl(h,k,l, symmops, nops)
int h, k, l, nops;
float symmops[3][4][MAXSYMMOPS];

int ReadFin(fp, refl)
FILE *fp;
FIN **refl;
#include ''lib/Xguicryst.h''
```

This routine read in a fin format file fp into an array of FIN structures. FIN is defined in Xguicryst.h. The number of records read is returned. The routine dynamically allocates all the space needed so that the address of a pointer of type FIN is sent—that is, in your code you should have

```
FIN *refl;
int nrefl=0;
nrefl = ReadFin(fp, &refl);

int ReadMu(fp, refl)
FILE *fp;
FIN **refl;
#include ''lib/Xguicryst.h''
```

This routine read in a XENGEN .mu (mulist) format file fp into an array of FIN structures. Similiar to ReadFin above.

```
int ReadDF(fp, refl, control)
FILE *fp;
FIN **refl;
int control;
#include ''lib/Xguicryst.h''
```

Similiar to ReadFin (above) except that it read in a df (double fin) format file. Since there are four columns (a column is the pair F, $\sigma(F)$) in a df file and only two in a fin file, control specifies how this information is read:

   If control = 0—Average 1 with 2 and 3 with 4 (zeros are ignored in the averaging)
      1—Return 1 and 2 only (usually the native anomalous)
      2—Return 3 and 4 only (usually the derivative anomalous)

```
int ReadPhs(fp, refl, control)
FILE *fp;
REFL **refl;
int *hmin, *hmax, *kmin, *kmax, *lmin, *lmax;
float *scalefc;
#include ''lib/Xguicryst.h''
```

   This routine reads in a phase file and dynamically allocates space (see ReadFin above). A refl structure is defined in Xguicryst.h. ReadPhs returns the number of reflections read. It also returns the minimum and maximum of the indices and the scale for Fc to Fo. If the scale is to be applied,

```
   nrefl = ReadPhs(fp, &refl, &hmin, &hmax, &kmin,
&kmax, &lmin, &lmax, &scalefc);
   for(i=0; i< nrefl; i++) refl.fc *= scalefc;
   OR
   for(i=0; i< nrefl; i++) refl.fo /= scalefo;
```

   The latter is usually more useful since it puts the data on an absolute scale.

```
float *fft3d(refl,nrefl,mapheader)
REFL *refl;
int nrefl;
MAPHEADER *mapheader;
#include ''lib/Xguicryst.h''
#include ''lib/fft.h''
```

   This routine performs a Fast Fourier Transform of data in refl with information in mapheader (defined in Xguicryst.h). The following map-header data must be set:

```
nx, ny, nz
a, b, c, alpha, beta, gamma (from get_cell)
resmax, resmin
nsym, symops (from scan_symmops2)
maptype (as defined in fft.h)
```

nx, ny, nz must be factors of 2, 3, 5, 7 and *nx* must be even. A pointer to a one-dimensional array of size *nx* ∗ *ny* ∗ *nz* is returned that contains the entire unit cell. When you are through with the array, it can be deallocated with `free()`. The array is sorted *nx* fast, *ny* medium, *nz* slow. The point *ix, iy, iz* is indexed with

```
rho = map[nx*( ny*((iz+64*nz)%nz) + (iy+64*ny)%ny )
+ (ix+64*nx)%nx]
```

The modulo arithmetic guarantees that *ix, iy, iz* wrap around the unit cell properly. The program must link in fftlib.o to access the FFT routines. The routines are very fast. On most workstations the combination of ReadPhs, fft3d is faster than reading a precomputed map from disk.

```
int fsread(rho, nx, ny, nz, rhoscale, rmsrho, fp, err)
int rho[]
int *nx, *ny, *nz;
float *rhoscale;
float *rmsrho;
FILE *fp;
char *err;
```

This reads a map file into the array rho. Map files are in the same format as those calculated by Bill Furey's FSFOUR program. Rho must be preallocated to *nx* ∗ *ny* ∗ *nz*. This information can be had from a call to `fssize`. The map will be scaled so that the root mean square of rho is 50.0. Rhoscale can be used to retrieve the original scale of the map. The string err will contain any errors. All three possible sort orders are autodetected and are resorted to *x* fast, *y* medium, *z* slow when read in. Example:

```
fp = fopen(''mapfile,''''r'');
nmax = fssize(fp);
map = (int *)malloc(nmax*sizeof(int));
rewind(fp);
nmap = fsread(map, &nx, &ny, &nz, &rhoscale, &rmsrho,
fp, err);
if(nmap <= 0 || nmap != nmax){
   emess(''Error in map file: ''); emess(err);
}
fclose(fp);
```

The array map is indexed as described in fft3d (see above).

```
int fssize(fp)
FILE *fp;
```

Returns the total dimensions *nx* * *ny* * *nz* of the map stored in the map file fp.

```
int fswrite(rho, fp, mh, direction, centric)
float map[];/* sorted x fast, y medium, z slow as-
cending */
FILE *fp;
MAPHEADER *mh; /* see Xguicryst.h */
int direction;
int centric;
```

This routine writes out a map file on XtalView/FSFOUR format. Note that the map in this case is an array of floats. This is for historical reasons. The map format is written out as integers, and floats are rounded to the nearest integer, so the map should be scaled so that this will not result in all 0's in the output (i.e., the root mean square of the map should be above at least 10.0—XtalView uses 50.0). Direction controls the output sort order. Zero is planes of *y*, 1 is planes of *x*, and 2 is planes of *z*. Centric should be set to 0. The following information is used in the mapheader structure:

a, b, c, alpha, beta, gamma
nx, ny, nz
scale, nsym

The symmetry operator positions in the file are filled with zeroes.

```
float rescalc(ih, ik, il, a, b, c, alpha, beta, gamma,
init)
int ih, ik, il; /* the Miller indices */
float a, b, c, alpha, beta, gamma; /* the unit cell */
int init;
```

This routine returns the resolution in angstroms given the Miller indices and the unit cell. The first time it is used init must be 1. Afterward init is not 1 unless a new unit cell is used, in which case init should be set to 1 one time.

INDEX